Selected Methods of Trace Metal Analysis:

Biological and Environmental Samples

JON C. VAN LOON
Departments of Geology and Chemistry
and
The Institute for Environmental Studies
University of Toronto

A WILEY-INTERSCIENCE PUBLICATION

JOHN WILEY & SONS

New York / Chichester / Brisbane / Toronto / Singapore

Library of Congress Cataloging in Publication Data:

Van Loon, J. C. (Jon Clement), 1937–
  Selected methods of trace metal analysis.

  (Chemical analysis, ISSN 0069-2883 ; v. 80)
  Includes bibliographies and index.
  1. Trace elements—Analysis. I. Title. II. Series.

QD139.T7V35 1985   543   85-3279
ISBN 0-471-89634-9

Printed in the United States of America

10 9 8 7 6 5 4 3 2 1

To the fragrant memory of my mother (Ruth) and to my father (Jack)
and to Evelyn, Cliff, Hazel, and Corine

PREFACE

This book is meant for the practicing analyst. Theory has been kept to a minimum and a descriptive approach has been used. In most cases procedures are complete so that no other source is necessary to finish an analysis. To avoid misinterpretation, procedures have been left as much as possible in the words of the authors. The authors' description of equipment and reagents used is also given. Again this allows the analyst to have the full available information, thus maximizing the chances for success.

Emphasis in the procedural selection is on those methods that had evidence of a thorough testing. In particular, procedures were sought that had been evaluated using standard reference samples or through interlaboratory comparison studies.

A large percentage of all trace element analyses are done by atomic absorption spectrometry or atomic emission (mainly with ICP source) spectrometry. Most laboratories involved in trace element analysis of biological and environmental samples are equipped with atomic absorption and/or ICP emission spectrometers. This book includes only procedures employing these two techniques.

It is difficult to structure a book such as this unambiguously. The chapters are organized according to sample type, so that overlap and repetition have been minimized. Sample types covered are botanical and zoological, foods, water, air, clinical, and soils, and related samples. When a procedure covers more than one sample type, it has been placed in the most obvious chapter.

A chapter on chemical sample preparation has been included because chemical sample preparation is often one of the major sources of error in an analysis. There is a bewildering array of alternative procedures for the different sample types appearing in the literature. Thus a thorough discussion of the various sample preparation approaches is necessary.

The determination of the *chemical forms* of the elements in biological and environmental samples is the main key to future advances in these subject areas. However, relatively few good procedures have yet appeared for this purpose. The best of these have been collected and placed in a separate chapter.

Finally, I am painfully aware of the subjective nature of organizing and presenting a book of this nature. Undoubtably, good procedures and promising approaches may have been omitted. If so, I apologize. My only defense is that the subject matter is based on 20 years of experience in this field.

I would like to express my deep appreciation to the copyright holders of the material quoted herein for giving permission for its use.

JON C. VAN LOON

Toronto, Ontario
June 1985

CONTENTS

Selected Methods of Trace Metal Analysis

CHAPTER

1

INTRODUCTION

Public demand for protection from the adverse effects of pollutants is increasing. The news media frequently run horror stories about possible exposure of people to very harmful substances. In many of these accounts the key issue is the level of the pollutants. In others the controversy encompasses the new discovery of a dangerous chemical in air, the public water supply, or food. In this public arena the analytical chemist is often a principal combatant. As analytical instrumentation becomes more highly developed, detection limits push steadily lower. Hence pollutants thought to have rather local distribution are turning up in very small amounts almost everywhere. The urgent problem is then whether these levels are harmful to humans and other organisms in an ecosystem. Thus the following questions must be answered:

1. What are the effects on plants and animals of long-term low-level exposure to harmful substance?
2. What are the pathways of these substances in the environment?

The analytical chemist should be the cornerstone of a multidisciplinary team who must answer these questions.

It is only in relatively recent times that man has greatly disturbed the geochemical environment. Primitive man began using metal about 6000–5000 B.C. However, only the native metals of gold, silver, and copper were employed at this time. Around 4500 B.C. metallurgical extraction of metals from ores was initiated. This greatly increased human exposure to rock and ore dusts, metal vapors, and other toxic vapors (e.g., SO_2). Such exposures, however, were principally local in nature and affected relatively few people.

The recognition of the toxic properties of geochemical materials was not a particularly triumphant chapter in the history of medicine. It was more by trial and error that this relationship developed. In fact, it was not until 1831 that a link was drawn between the severe disabilities (called miners' disease) shown by lead miners and the lead metal ores being mined. But it was the Industrial Revolution and particularly the large demand for metals generated by World Wars I and II that focused great attention on geochemical toxic hazards. The recent upsurge in the use of rarer elements in metallurgical chemical industries poses a very serious health hazard. Very little is known about the environmental and clinical properties of these elements.

Of the toxic elements, arsenic, lead, mercury, and cadmium have been studied the most closely. Arsenic, mostly as arsenious oxide, has during the course of history been frequently administered as a poison and has often been implicated in the death of important historical figures such as Napoleon Bonaparte and Mozart. Many a mercury compound, such as mercurochrome, has been widely used as a drug and yet it was discovered in several parts of the world, particularly in Minamata, Japan, that ingestion (from shellfish) of mercury compounds exhausted by industry resulted in the poisoning of large numbers of people. This metal causes severe neurological impairment that yields symptoms popularized by Lewis Carroll in his comic character the Mad Hatter of *Alice in Wonderland*. (Hatters, using their bare hands, rubbed mercuric chloride into felt in preparation for the production of hats.) Of the toxic elements, it is lead, however, that has received the widest dispersal worldwide, mostly as an antiknock additive to gasoline. This fact is dramatically illustrated in the historical lead and mercury data for polar glacial ice shown in Table 1.1. Lead poisoning causes symptoms similar in some respects to those of mercury poisoning.

A symposium, "Some Recent Developments in the Bio and Environmental Chemistry of Some Trace Elements," was held in Corpus Christi, Texas, in November 1979. Zingaro (1) in his account of the conference states that "the most obvious message to emerge from this meeting was the need for accurate and reliable analytical (chemistry) techniques." Here then is the challenge for the analytical chemist into the foreseeable future. What follows is an attempt to collect the best presently available procedures for trace metal analysis.

There are very few official methods for trace analysis in the environmental field. When a determination is needed as a result of a critical problem, it is often necessary to develop a suitable method. Because this is a time-consuming process, many crises pass without adequate trace metal analytical data. It is important that this situation be changed, so it is essential that ample research be devoted to trace metal analytical procedure development.

Table 1.1 Pb and Hg in Melted Glacial Ice

	Metal (ng/L)	
Year	Pb	Hg
1720	10	<10
1820	40	<10
1900	100	<10
1965	500	10

A trace metal analysis procedure should, ideally, have the following qualities:

have suitable detection limits,
be relatively fast,
be relatively low in cost,
be applicable to a wide range of samples,
be relatively specific,
be applicable in most analytical laboratories (not need special equipment),
be simultaneous multielement,
have relatively long linear working range, and
be accurate (tested on SRMs).

Trace metal analysis of environmental and biological samples is a very broad topic. It is therefore important to outline clearly what is to be included in this coverage.

1.1 THE ELEMENTS

1.1.1 Toxicity and Essentiality

Industrial processes have resulted in the introduction of metals into the environment. From the processing of ores to the manufacture of finished products containing metals, metal loss results in contamination of air, water, and soils. Subsequently these metals can be ingested by humans and other animals.

Ingested metals may invoke no response, act as a stimulant, or result in toxic response. Even essential metals in high enough dosage can be toxic. Metals can react with protein and DNA and RNA, affecting the metabolic processes, and with other substances, resulting in physiologic changes. Metals may cause enzyme inhibition and produce changes in the rates of catalytic decomposition of metabolites.

The action of a metal in an animal system depends on its ability to absorbed and excreated, its valence, and its electrochemical behavior. Toxicity due to metals results in changes, some of which may be irreversible. Acute toxicity, caused by a large dose of metal, produces sudden symptoms, often with irreversible damage. Cumulative effects of metals or chronic poisoning occurs as a result of long-term exposure to lower levels of metal. In this case the periodic of no symptoms is followed by low-level symptoms that then can gradually progress to symptoms indicating severe problems. Chronic poisoning can often be reversed by removal of the metal exposure.

Figure 1.1 is a periodic table of the trace metals of present biological and

Li	Be											B*				F*	
												Al*	Si*				
			Ti	V*	Cr*	Mn*	Fe*	Co*	Ni*	Cu*	Zn*			As	Se*		
					Mo*					Ag	Cd		Sn*	Sb	Te		
	Ba								Pt	Au	Hg	Tl	Pb	Bi			

Figure 1.1 Periodic table showing biologically interesting trace elements. * indicates essential elements.

clinical interest (essential elements are also indicated). As more and more research is carried out on the role of trace elements in biological systems the clearer it becomes that there is often no way to make a definite distinction between essential and nonessential trace elements. Arsenic has recently been found to be essential. Even elements such as cobalt, iron, zinc, and copper, which are well established as essential, become toxic at high levels. Other elements such as nickel, selenium, floruine, and vanadium are now known to be essential, but at very low levels. The presence of these elements at only slightly higher levels results in toxicity. Table 1.2 is a chronology of the discovery of the essentiality of trace elements to animals (2). Note that the metaloids, for example, arsenic, selenium, tellurium, bismuth, and antimony, are included under the term *metal* in this book.

The well-known toxic elements include arsenic, berylium, cadmium, mercury, lead, antimony, tellurium, and thallium. Of these there are certainly some data to suggest that arsenic, cadmium, and lead may be essential at very low levels. In fact, one might hypothesize that, because living systems developed in the sea (which contains all elements at some level), most if not all elements may ultimately be shown to be essential. Indeed, here is a challenge for the trace metal analytical chemist, to improve detection limits so that it is possible to determine the less common elements at the very low concentrations in which they occur naturally in living systems. This will then allow research into their biological behavior.

A further complicating factor in trace element toxicity is the symbiotic effect of coexisting metal species. Thus, for example, research has shown that the presence of selenium results in some protection against heavy metal (e.g., mercury and cadmium) toxic effects. This problem emphasizes the importance of being able to do multielement analysis. Therefore multielement techniques such as plasma–optical emission spectrometry take on added importance.

It must be stressed that it is the form of the metal that determines its biological and environmental effects. A striking example is the case of chromium.

Table 1.2 Discovery of Trace Element Essentiality for Animals[a]

Element	Year	Investigators
Fe	17th century	
I	1850	Chatin
Cu	1928	Hart, Steenbock, Waddell, and Elvehjem
Mn	1931	Kemmerer, Todd, Elvehjem, and Hart
Zn	1934	Todd, Elvehjem, and Hart
Co	1935	Underwood and Filmer; Marston; Lines
Mo	1953	DeRenzo, Kaleita, Heytler, Oleson, Hutchings and Williams; Richert and Westerfield
Se	1957	Schwarz and Foltz; Patterson, Milstrey, and Stockstad
Cr	1959	Schwarz and Merz
Sn	1970	Schwarz, Milne, and Vinyard
V	1971	Schwarz and Milne; Hopkins and Mohr
F	1972	Schwarz and Milne; Messer, Armstrong, and Singer
Si	1972	Schwarz and Milne; Carlisle
Ni	1973	Nielsen

[a] Reprinted with permission from (2) G. H. Morrison, *Critical Reviews Anal. Chem.* **8,** 287 (1979). Copyright CRC Press, Inc., Boca Raton, FL.

Trivalent chromium is an essential element and yet extremely low levels of hexavalent chromium are known to be carcinogenic. Thus it will be very important analytically to be able to determine the chemical forms of elements in biological and environmental samples.

1.1.2 Background Levels

Difficulties in accurately determining the levels of trace elements in real samples abound. This is a particularly serious problem in the clinical field because of the importance of identifying and treating patients with deficiencies or excesses. The chaotic state of our capabilities in this area is emphasized by the following statement by Versieck and Cornelis in 1980 (3):

> Some investigators (4) consider a serum Cr level of 0.5 ng/ml as the upper level in normal individuals; others (5) accept a value of 5 ng/ml as definite proof of Cr deficiency.

It must be a high-priority goal of trace element analytical researchers to develop methodology that will help obviate problems of this type.

This brings us to the question of determining the background levels of trace elements in biological and environmental samples. If one charts (e.g., for chromium in Table 1.3) (6) the base levels of trace elements that were thought to

Table 1.3 Reported Chromium Concentrations in Blood[a]

Author	Year	Concentration (μg/L)
Grushko, Y. M. (*Biokhimiya* **13:**124)	1948	35
Urone, P. F., et al. (*Anal. Chem.* **22:**1317)	1950	50
Monacelli, R., et al. (*Clin. Chim. Acta.* **1:**577)	1956	180
Volod'ko, L. V., et al. (*vestn. Akad. Nauk. Belarusk* **1:**107)	1962	200
Schroeder, H. A.; et al. (*J. Chronic Dis.* **15:**941)	1962	520; 170
Wolstenholme, N. A. (*Nature* **203:**1284)	1964	1000[a]
Feldman, F. J., et al. (*Anal. Chim. Acta* **38:**489)	1967	29
Hambidge, K. M. (in *Newer Trace Elements in Nutrition,* Dekker, New York)	1971	7
Cary, E. E., et al. (*J. Agr. Food Chem.* **19:**398)	1971	7
Davidson, I. W. F., et al. (*Amer. J. Obstet. Gynecol.* **116:**601)	1973	4.7
Pekarek, R. S., et al. (*Anal. Biochem.* **59:**293)	1974	0.72

[a] From Mertz (6).

occur in samples at various times up to the present, there is a distinct recent trend toward lower and lower values. This is undoubtably the result of improvements in analytical methodology and equipment that allow trace element determinations to be done with better and better detection limits. This problem also emphasizes the importance of good background correction.

In the case of atomic absorption spectrometry (AAS) little background correction was done until the paper of Koirtyohann and Pickett (7) in which a method was described employing a continuum source (this was ten years after Walsh's classic paper (8) that introduced analytical atomic absorption spectrometry). Subsequently many instrument manufacturers supplied such a source in their instruments. After this date atomic absorption data appearing in the literature for elements present at near background levels were much improved. However, with furnace atomizers continuum-source background correction is sometimes not entirely satisfactory (i.e., especially when the background absorption is large and shows fine structure). Fortunately the recent introduction of background correction based on the Zeeman effect improves this situation. A similar scenario can be told for other trace element analysis techniques.

1.2 THE TECHNIQUES

Historically, gravimetric and titrimetric methods were first employed for trace element analysis. However, it was necessary to employ a number of time-consuming and error prone steps of separation and preconcentration to isolate and bring the level of the trace constituent to a detectable level prior to determination by these techniques. It was not, therefore, until the advent of instrumentation that practical and accurate trace element analysis became possible.

In the 1930s optical emission spectrometric methods of analysis were developed. Initially flames were employed as atomization–excitation sources, followed by furnaces and arcs and sparks. Colorimetric–molecular-spectrophotometric methods were introduced. These approaches include the related techniques of nephelometry, turbidimetry, and molecular fluorometry. Practical electrochemical methods were introduced in the late 1930s and early 1940s. The most commonly used electrochemical approaches are polarography and anodic stripping voltametry.

It was not, however, until the relatively recent introduction of atomic absorption spectrometry and modern neutron activation and related methods that routine trace element analysis became a fast developing science. More recently the introduction of the plasma atomizer–excitation sources for use with optical emission and mass spectrometric methods promises to expedite greatly trace element analysis methodology.

Table 1.4 lists the techniques of trace metal analysis showing the detection limits available for the metals of interest biologically and environmentally (2). Levels of trace metals thought to be present in normal serum are included for comparison.

A cursory survey of the literature suggests that between 65 and 75% of all trace element analyses are done by the techniques of valence electron atomic spectrometry with atomic absorption (to this date) responsible for the largest fraction of this work. Emission spectrometry, with the advent of the inductively coupled plasma (ICP) source, is presently enjoying a renaissance. In the larger laboratories this technique will probably largely replace flame atomic absorption in those applications in which more than about three elements must be done per sample. Furnace atomic absorption, however, will remain preeminent in the foreseeable future when best detection limits are necessary. Most trace element analysis laboratories do not have access to neutron activation equipment— a technique that is competitive with respect to detection limits with furnace atomic aborption. X-ray fluorescence can be employed in some applications for trace element analysis, but is better used for determining major and minor elements (i.e., from 0.01 to 100%). Electrochemical techniques have been used over the years for trace element analysis. Interferences, when complex samples are to be analyzed, are usually too severe for this approach to have general usefulness in biological or environmental trace element analysis. Molecular spectrophotometric methods, widely used in the past for trace element analysis, have been largely supplanted in most countries by atomic absorption and inductively coupled plasma emission spectrometries.

Table 1.4 Sensitivities and Detection Limits of Analytical Techniques[a]

Element	Normal Serum Expected Levels (ppb)	ASV[b] Detection Limits (ppb)	Flame AAS[c] Detection Limits (ppb)	ET–AAS[d] Detection Limits (ppb)	ICP–ES[e] Detection Limits (ppb)	NAA[f] Sensitivity (ppb)
Trace Essential Metals						
Co	0.3	—	5	0.6	2	10
Cr	0.28	—	2	0.4	2	300
Cu	1140	0.5	2	0.4	2	2
Fe	1150	—	1	0.8	2	2000
Mn	1.4	—	3	0.1	0.5	1
Mo	4	—	10	0.2	5	1000
Ni	30	—	8	0.1	5	700
Se	11	—	100	0.1	30	10
Si	80	—	500	7.0	10	—
Sn	33	0.2	1000	0.4	3	30
V	1	—	20	0.5	2	2
Zn	1000	0.4	0.6	0.2	1	100
Trace Toxic Metals						
As	190	—	100	0.2	20	50
Ba	79	—	200	0.06	5	20
Be	4	—	20	0.001	3	—
Bi	20	0.05	50	0.3	50	—
Cd	5	0.005	1	0.003	1	5
Hg	3	1	2200	0.5	50	3
Li	31	—	20	0.6	—	—
Pb	46	0.01	10	0.05	20	500
Sb	54	0.1	30	0.2	200	7
Tl	—	0.01	300	0.07	200	—

[a] Reprinted with permission from G. H. Morrison, *Critical Reviews Anal. Chem.* **8,** 287 (1979). Copyright CRC Press, Inc., Boca Raton, FL.

[b] Anodic stripping volametry.

[c] Flame atomic absorption spectrometry.

[d] Electrothermal atomic absorption spectrometry.

[e] Inductively coupled plasma emission spectrometry (ultrasonic nebulizer).

[f] Neutron activation analysis.

Table 1.4 also includes the levels of trace metals to be expected in serum samples. The detection limits quoted were not obtained on serum samples and are usually better than would be obtained with the evaluation of serum samples. However, in most instances flame and furnace atomic absorption cover the concentration levels of elements in serum. Separations will often be needed. Inductively coupled plasma emission spectrometry detection limits are frequently too poor for these determinations directly. Because of the inherent multielement characteristics of ICP emission (particularly with a quantometer) the time per sample analysis is small relative to AAS when more than about three elements are to be done. Thus sample preconcentration becomes a practical process when ICP emission is employed.

1.2.1 Detection Limits

The above discussion is based on instrument capability to reach detection limits required for serum analysis. A comparison of detection limits required for other biological and environmental samples shows that for many plant tissues, animal tissues, and other body fluids similar or better detection limits must be obtained. In the case of sediments, soils, and similar samples, detection limits often need not be so good. However, with soils, the matrix is very complex and the sample contains high levels of interfering substances. Thus detection limits in these situations for the techniques will be much poorer than those given in Table 1.4.

Obtaining good detection limits is important, particularly when samples with background levels of the trace elements are to be analyzed. The steady decrease in these background values over the years attests to the fact that improved detection limits are still required if accurate values of trace element levels in biological and clinical samples are to be obtained.

The term trace in trace analysis does not have commonly agreed upon concentration limits. For this monograph trace means any concentration below about 0.01% (i.e., 100 μg/mL or μg/g). Ultratrace is an expression sometimes used for levels below nanogram per millileter (or nanograms per gram). The term is not employed in this work.

The most commonly used concentration expressions in trace metal analysis are summarized in Table 1.5:

Sometimes, particularly when referring to quantities detectable by furnace atomic absorption, the absolute value in micrograms, nanograms, or picograms is given.

1.2.2 Separation and Preconcentration

Despite recent rapid advances in analytical instrumentation, it is still often necessary to use separation and preconcentration methods prior to the determinative step. These are time consuming and error prone (losses and/or contamina-

Table 1.5 Concentration Terms Used in Trace Metal Analysis

	Parts per Million (ppm)	Parts per Billion (ppb)
mg	mg/L, mg/kg	—
μg	μg/mL, μg/g	μg/L, μg/kg
ng	—	ng/mL, ng/g

tion) and should be used only if essential. The reasons for doing a separation and preconcentration step are to bring the concentration of a trace element to a detectable level and/or to separate it from interfering substances (usually high concentration elements of the sample matrix). Rarely is it necessary to separate the individual trace elements from one another. A section on separation is given in Chapter 3. The reader is also referred to a good critical review by Bachmann on separation and concentration (9).

Solvent extraction and ion exchange chromatography are the most commonly used separational approaches. Historically, solvent extraction was the method most commonly employed for separation and preconcentration of trace metals. Depending on the requirements either single or group metal extraction can be performed. Thus specificity of the reagent is a point of importance. In both cases adjustment of pH and its control during the separation are usually crucial.

More recently the many techniques of chromatography have assumed an important role in separation and preconcentration of trace metals. In this regard ion exchange chromatography is preeminent. At present rapid developments are occurring in the production of chelating resins. Generally their advantage over ion-exchange resins is having better specificity and often better capacity. For the separation and determination of the chemical forms of the elements, chromatography is without peer. In solution applications reverse phase high performance liquid chromatography and ion exchange chromatography are often the techniques of choice. Gas chromatography is employed for gaseous samples.

1.2.3 Blanks

The need for running blanks with each set of determinations cannot be overemphasized. In trace element analysis the detectable amount is often set by the concentration of elements in the reagents and by the contamination levels from the laboratory environment. Table 1.6 shows the level of contamination of human blood by heparin in 1966 (10).

To keep blanks low it may be necessary to use specially purified acids and double distilled or specially deionized water. With the latter the singly distilled

Table 1.6 Trace Element Contamination of Human Blood by Heparin (μg/mL)[a]

Element	Heparin	Blood
Ba	2.5–12	0.069
Ca	300–2900	62.0
Cu	0.65	1.1
Mn	3.6	0.026
Sr	5–92	0.039
Zn	28	6.5

[a] From Bowen (10).

water in my laboratory was found to contain 0.1 ppb lead, a level too high when dealing with lake waters. Double distillation of the water lowered the lead level by one order of magnitude, thus making the determination feasible.

Considerable imparting of impurities to the blank (and hence the sample being analyzed) can come even from analytical grade reagents. Murphy (11) analyzed various grades of hydrochloric acid for trace elements; and his results are summarized in Table 1.7. Thus it may be necessary to employ a subboiling still when very low blanks are essential. Commercial high-purity acids are very expensive.

1.3 THE LABORATORY

Trace metal analysis is carried out throughout the world in a variety of laboratories. Most of these, including my own, have many serious deficiencies. Ideally, a specially designed "clean" laboratory which can be isolated from all sources of external contamination, should be employed. Unfortunately, such facilities are extremely expensive and are not available to most workers.

Table 1.7 Some Impurities in Hydrochloric Acid (ng/g)[a]

Element	ACS Reagent	Commercial High Purity	Subboiling Distillation
Pb	0.5	<1	0.07
Cd	0.03	0.5	0.02
Cu	4	1	0.1
Ni	6	3	0.2
Cr	2	0.3	0.3

[a] From Murphy (11).

Contamination from metal-bearing ingredients in dusts is often a crippling problem in trace metal analysis. These dusts originate from the outside, adjacent rooms and hallways, and from the air in the laboratory. It is difficult and expensive to eliminate such dusts from the laboratory. However, it is important that methods for restricting exposure of the samples to dusts are available in a trace metal analysis laboratory.

Much conventional laboratory hardware is constructed of metal. It is important in a trace metal analysis laboratory to minimize use of such fixtures. In addition to the consideration of metal taps and other plumbing, this also means such precautions as elimination of metal parts in fume hoods (even fasteners), unprotected metal in light fixtures, and metal-containing paints.

The laboratory should be easy to keep clean. This means materials used in floors, walls, and ceilings should not collect or trap dust readily. Projections that are dust trappers should be minimized and where they do exist should be easily reached and cleaned. Therefore a vacuum cleaner is essential equipment in a trace metal analysis laboratory.

1.4 THE SAMPLES

Biological and environmental sample types conjure up in one's mind an almost all-encompassing group of samples. For example, under environmental samples one might include geological substances such as rocks and minerals and also industrial materials because these samples are a source of trace elements to the environment. However, it was important in structuring the coverage of this book to leave out such distantly related sample types. Specifically then the substances covered under environmental samples are air, water, soils, and sediments; under biological samples are botanical and zoological fluids and tissues. These latter include clinical samples and foods as well as the many other tissues and fluids of plants and animals.

Collection, preservation, and physical sample preparation are very important aspects of the chemical analysis process. Methodology depends on the sample type and the reason for the analysis. For most pruposes sediments and soils require little sample preservation. They must be sieved to give the desired particle size prior to chemical treatment. Water samples must usually be filtered and then acidified. This should be done as soon as possible after collection. Water samples must be frozen or kept very cool prior to the filtration step. Metals in air are collected by passing the air through an impinger, a filter, or an adsorbent. The sample thus produced is usually ready for chemical treatment. Biological samples are difficult to preserve and physically prepare and the methodology depends greatly on sample type. Preservation is generally essential to prevent adverse biological changes from occurring. Sampling, preservation, and physical sample preparation are, however, largely beyond the scope of this book.

1.5 CHEMICAL SAMPLE PREPARATION

Chemical sample preparation is a very important aspect of the trace metal analysis of biological and environmental samples. Choice of method depends to a large extent on the sample type and the purpose of the analysis.

Basically, the two approaches to sample decomposition that are employed are acid attack or fusion. In the case of acid treatment, mineral acids or mixtures of mineral acids are commonly utilized. Fusions are accomplished by using a large excess of fusing agent (e.g., an alkali metal carbonate or hydroxide) over sample.

If an acid attack is to be used for the total decomposition of a sample containing silicon, it is essential to employ hydrofluoric acid in the reaction mixture. Likewise, if organic matter is abundant, it is necessary to include a strong oxidizing agent to release the trace metals. In this regard perchloric acid together with nitric acid are often used. Nitric acid by itself or hydrochloric and nitric acid together do not usually extract trace metals totally from samples. Sulfuric acid is a dehydrating agent and may result in strongly reducing conditions occurring. This is indicated when an organic mixture turns brown or black. To maintain oxidizing conditions, hydrogen peroxide may be added periodically to the sulfuric acid solution. Sulfuric acid has the highest boiling point of the mineral acids and is used to expel the remaining traces of fluoride from acid mixtures.

Fusing agents greatly boost the dissolved salt content of a sample solution and hence are seldom favored over acid decompositions when the latter are applicable. When abundant organic matter is present, an oxidizing fusion is often necessary. For this purpose sodium peroxide can be added to the fusion mixture. Sodium or potassium hydroxide are commonly used fusing agents, being particularly useful when arsenic is to be determined. Lithium metaborate has in recent times attained widespread acceptance as a fusing agent when silicate or oxide material must be decomposed.

1.6 QUALITATIVE ANALYSIS

Qualitative analysis is the means by which a substance is identified in a sample. Very little attention is now given in the teaching of analytical chemistry to the subject of qualitative analysis. Yet it is essential to be convinced of the presence of a substance before undertaking its determination. (Basically, qualitative analysis is beyond the scope of this book.)

Trace metals historically were detected in samples by the spot test. In this approach the sample solution is reacted with a chemical reagent that is relatively specific for the metal of interest. As a result of the reaction a colored solution or precipitate was formed. Sometimes instead of a color a fluorescence occurred. The specificity of the reagent for forming a given color with the metal under consideration is important.

With the advent of instrumentation qualitative analysis could be readily accomplished by the techniques of spectroscopy. Flame tests, whereby the metal of interest imparts a characteristic color to a flame, are an early example of spectroscopic qualitative analysis. Flame emission and arc emission spectrometry with photographic recording of the sample spectra are important methods of trace metal detection. Colorimetric methods (molecular solution spectrophotometry) are also commonly employed in qualitative analysis. In this instance it is often possible to identify the form of a metal in solution. Nuclear analytical methods are also very valuable in qualitative analysis. Mass spectrometry occupies the ultimate position in qualitative analysis. When all else fails, mass spectrometry with due precautions can be relied upon to prove the presence of a metal or metal compound.

1.7 ERROR

Error relating to a trace element analysis may occur at all points from sampling through to the determinative step. The magnitude of error encountered in my experience is sampling > sample preparation > determination.

Because sampling is a science in itself it cannot be covered adequately in this book. It is important, however, for the analyst to be convinced that the sample as presented is representative and hence worth analyzing.

Sample preparation often includes both physical and chemical preparation steps. Physical sample preparation including grinding and sieving if necessary can result in severe contamination if metal implements are employed. Losses of trace metals may result if any fraction of the sample is rejected. Chemical sample preparation is crucially important and no unanimity of opinion exists on the proper treatment for each sample type. A later chapter is wholly dedicated to this topic. The determinative step, during which the sample is introduced to an instrument, is usually relatively interference free. Yet great care is essential in calibration and eliminating remaining interference effects.

In this era of analytical instrument sophistication one might assume that trace metal data in the literature would be on the whole quite reliable. This is far from the case. It is only in the past five to ten years that laboratories in general have employed useful checks on accuracy and precision. Thus there is frequently no method of judging the validity of trace metal data. This is certainly true for most data obtained prior to the above years.

1.7.1 Accuracy Assessment

It is not always necessary to have the highest accuracy in trace metal analysis, but it is important to be able to access the accuracy of an analysis. This is particularly crucial with clinical analyses. An excellent two-volume treatise on *Accuracy in Trace Analysis* was published in 1974 by the U.S. National Bureau of Standards (11). Generally three approaches to accuracy assessment are employed:

1. Analysis of standard reference samples.
2. Recovery studies.
3. Use of an absolute method, for example, isotope dilution.

The use of standard reference samples is much preferred. However, it is not always possible to obtain such samples for the sample type and elements desired. Fortunately a number of agencies, notably the National Bureau of Standards (NBS) in Washington, D.C., are greatly expanding their offerings in this area. Table 3.2 lists some of the most useful standard reference samples known to me; the supplier and the supplier's address are also given.

Standard reference samples are relatively expensive and are often in short supply. Hence they may not be available for routine day-by-day control of analytical data. For this purpose in-house laboratory standards may be prepared and employed together with an occasional standard reference sample. However, great care is essential in preparation of such standards to ensure homogeneity, freedom from contamination, and a long useful life. Organizations that have long experience in such endeavors should be consulted regarding proper methodology.

Another indispensable tool for controlling trace metal data is the round robin or interlaboratory comparison study. Fortunately these have become quite popular in the past five years. Although such studies may not necessarily ensure accuracy, they give some basis for comparison of data being recorded in the literature. Again, great care is essential in planning and executing such studies. Proper sample preparation, preservation, and dispersal are essential. Consultation with mathematicians, particularly statisticians, is also very important in the design, data processing, and assessment and presentation of data.

The following dramatically demonstrates the need for accuracy assessment in routine trace metal analysis. Lead is a very important element in biological and environmental studies. Yet lead results on such samples remain among the least satisfactory in trace element analysis. In 1975 Loescher (13) conducted an interlaboratory comparison study among laboratories in Ontario, Canada, on the determination of lead in water sludge and sediment samples. Table 1.8 summarizes the results. As can be seen the large range of results is unsatisfactory for most samples.

In the clinical field Lauwerys et al. (14) reported the results of an interlaboratory comparison carried out in Europe of lead determinations in blood, urine, and aqueous solutions. These results are summarized in Table 1.9.

An interlaboratory study of the determination of lead in seawater (by three

Table 1.8 Lead in Water Sludge and Sediment

Samples	Water[a] (μg/L)			Sludge (μg/g)	Sediment (μg/g)	
	1	2	3		1	2
Mean of labs	6	4	334	1690	14	14
Standard deviation	4	3	56	120	5	6
Range	2–130	2–170	280–420	1–2500[b]	5–20	4–25

[a]Water 1 and 2 are splits of same sample.

[b]One lab used H_2SO_4, thus precipitating most Pb.

Table 1.9 Lead in Clinical Samples[a]

Sample		Units	*n*	Mean	Range	% of Labs with RSD < 10%
Blood C			51	15	3–49	47
D		μg/100mL	52	24	10–87	65
E			55	23	1–115	—
Urine A		μg/L	33	63	5–159	42
B			34	84	8–185	47
Aqueous solution	1	μg/L	51	209	25–670	65
	2[b]		50	194	18–50	—

[a]From Lauwreys et al. (14).

[b]Dilution by ten of aqueous solution 1.

laboratories) was reported in 1982 by Lamathe et al. (15). At levels of 3 and 50 μg/L, a positive bias of 30 and 17%, respectively, was obtained. Electrothermal atomic absorption spectrometry was employed subsequent to selective extraction on Chelex-100 resin.

REFERENCES

1. R. A. Zingaro, *Environ. Sci. Technol.* **21,** 282 (1979).
2. G. H. Morrison, *Critical Reviews Anal. Chem.* **8,** 287 (1979).
3. J. Versieck and R. Cornelis, *Anal. Chim. Acta* **116,** 217 (1980).
4. W. Niedermeier, E. E. Creitz, and H. L. Holley, *Arthritis Rheum.* **5,** 496 (1962).
5. W. Niedermeier and J. H. Griggs, *J. Chronic Dis.*, **23,** 527 (1971).
6. W. Mertz, *Clin. Chem.* **21,** 468 (1975).
7. S. R. Koirtyohann and E. E. Pickett, *Anal. Chem.* **37,** 601 (1965).
8. A. Walsh, *Spectrochim. Acta* 7, **108** (1955).

9. K. Bachmann, *Critical Reviews Anal. Chem.* **12,** 1 (1981).
10. H. M. Bowen, *Trace Elements in Biochemistry,* Academic Press, London, 1966.
11. T. J. Murphy, (P. D. Lafleur, Ed.), *Accuracy in Trace Analysis: Sampling, Sample Handling and Analysis,* National Bureau of Standards Special Publ. 422, 1974, p. 117.
12. P. D. Lafleur, *Accuracy in Trace Analysis,* Vol. 2, Proceedings of the Seventh Materials Research Symposium at Gaithersburg, MD, October 7–11, 1974, U.S. Department of Commerce, 1976.
13. B. Loescher, Report to Participants in the Toronto Pb Intercomparison Study, 1975, 20 pp.
14. R. Lauwreys, J. P. Buchet, H. Roels, A. Berlin, and J. Smeets, Intercomparison Program of Pb, Hg, and Cd Analyses in Blood, Urine and Aqueous Solutions, *Clin. Chem.* **21,** 551–557(1975).
15. J. Lamathe, C. Maqurno, and J. C. Equel, *Anal. Chim. Acta* **142,** 183(1982).

CHAPTER

2

TECHNIQUES AND INSTRUMENTATION

2.1 ATOMIC ABSORPTION SPECTROMETRY

Atomic absorption spectrometry is still the most widely used technique for the determination of trace metals. The equipment is generally available. It can be found in analytical trace metal laboratories and in a variety of biological, clinical, and environmental research and routine analytical establishments. The approach is relatively simple. For these reasons atomic absorption methods are emphasized throughout this book.

Atomic absorption detection limits depend on the type of atomizer used and the sample matrix. Generally these are of the order of micrograms per milliliter to below nanograms per milliliter, the latter being the case when a furnace atomizer is employed. These detection limits are usually adequate for determining trace metals in a wide variety of samples such as soils, sediments, sludges, and rocks. However, for waters and biological tissues it may sometimes be necessary to employ some method of preconcentration prior to the determinative step. In addition, when determining background levels it may be necessary to employ preconcentration techniques for almost all samples prior to atomic absorption analysis. The analyst must beware of the many tables of elemental detection limits that have been obtained by atomic absorption analysis of distilled water containing very low concentrations of the elements in acid solutions. Such values do not apply to complex real sample solutions because in these solutions detection limits are usually much poorer.

Generally it is necessary to dissolve a sample prior to atomic absorption ysis, which is a distinct disadvantage of the technique. Wet chemical manipulations are, of course, fraught with potential errors resulting from contamination or loss of analyte. Much work is currently being done to overcome this problem, but at this writing no generally acceptable approach exists for direct solid sample atomic absorption analysis.

2.1.1 Historical Introduction

The phenomenon of the absorption of radiation by atoms has been used for investigations in physics since the early part of the nineteenth century when

Fraunhofer observed a number of dark lines in the sun's spectrum. The first analytical application of atomic absorption was to the determination of mercury by Muller (1). It was not until 1955, when Walsh (2) outlined the general usefulness of the approach to elemental analysis, that real analytical atomic absorption spectroscopy was born. In the relatively short period of three decades since this development, atomic absorption spectroscopy has become one of the most important techniques for trace element analysis.

During the early commercialization of atomic absorption spectroscopy, extravagant claims were made concerning the general lack of interferences encountered with the technique. This is peculiar in that Walsh in his original paper limited his claims for superiority to a simpler pattern of atomic spectral interference and a greater tolerance to thermal fluctuations in the atomizer as compared to emission spectroscopy.

Between 1955 and 1962 there was little useful commercial instrumentation available. The first commercial offering, an unmodulated spectrophotometer, had serious deficiencies. During the early years, in response to many requests for information, Walsh and his coworkers published papers describing in detail instrumentation suitable for routine atomic absorption work, for example, Box and Walsh (3) (see Fig. 2.1). This equipment is remarkably similar in principle to modern atomic absorption units. In fact, except for the development of electrothermal atomizers and devices for automatic background correction, little of fundamental importance has been introduced into commercial units to this day. Double-beam or dual-channel units, advanced electronics, microprocessors, and so on, have only resulted in equipment that is used more easily. Modern equipment, of course, despite its push-button, black-box technology, operates on the

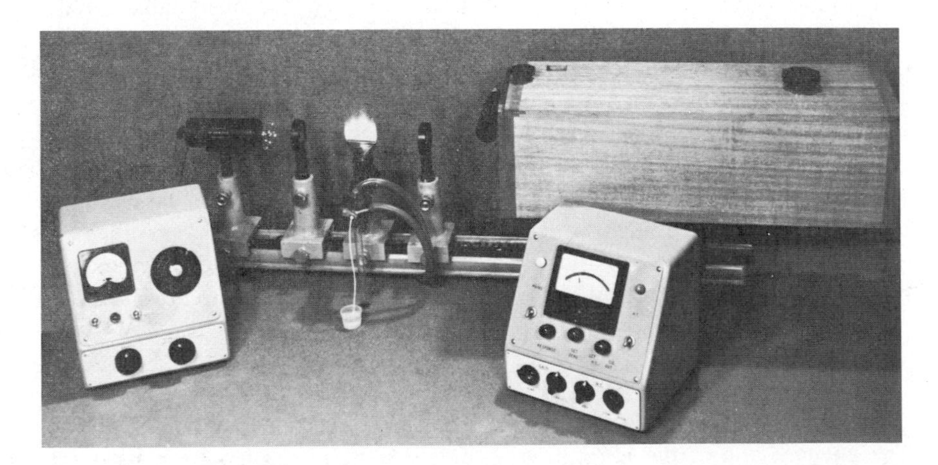

Figure 2.1 Atomic absorption spectrometer developed by Box and Walsh (3).

same physics and chemistry as the original equipment. Therefore, it is important to stress the need for analysts to master these principles.

2.1.2 Principles of an Atomic Absorption Spectrometer

A block diagram of an atomic absorption unit is given in Fig. 2.2. Atoms absorb radiation only at discrete wavelengths characteristic of the absorbing species. Thus radiation from the source, produced from a vapor of the metal of interest, is absorbed at a discrete wavelength (s) by atoms of that element in the atomizer. As a result, the radiation beam intensity is attenuated by an amount that is proportional to the concentration of the element under consideration in the atomizer. The source radiation is modulated (chopped or electronically modulated) so that selective amplification of the lamp signal is possible. This effectively eliminates amplification of flame emission. The latter, however, is a source of noise.

The function of the atomizer is to produce free atoms from the introduced sample. In atomic absorption the atomizer is usually a flame- or furnace-type device. Radiation of a characteristic wavelength is most frequently produced by a hollow-cathode or electrodeless discharge lamp. The desired wavelength is isolated from other absorbing or nonabsorbing lines by a monochromator. The transducer (detector) is commonly a photomultiplier tube.

Absorbance, the quantity usually measured in atomic absorption spectrometry, can be expressed as follows:

$$A = \log I_0/I = abc$$

where I_0 is the intensity of the incident beam, I the intensity of the transmitted beam, a a constant (characteristic of the particular system), b the path length of the optical beam (which can be kept constant), and c the concentration of the element of interest in the atomizer. This equation predicts a linear relationship between the absorbance and the concentration of the considered element. In practice, this occurs from the detection limit over two to three orders of magnitude.

An atomic absorption working curve is depicted in Fig. 2.3. An estimate of the precision obtainable is also shown. As can be seen, best precision is obtained in the middle straight line portion of the curve. At the upper end of the graph, curvature is experienced and precision decreases markedly.

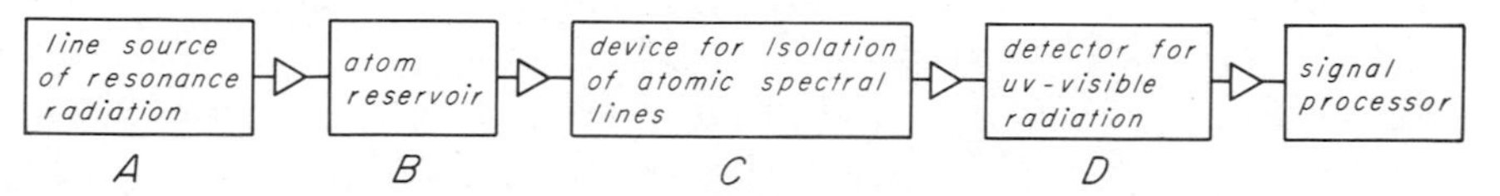

Figure 2.2 Block diagram of an atomic absorption spectrometer.

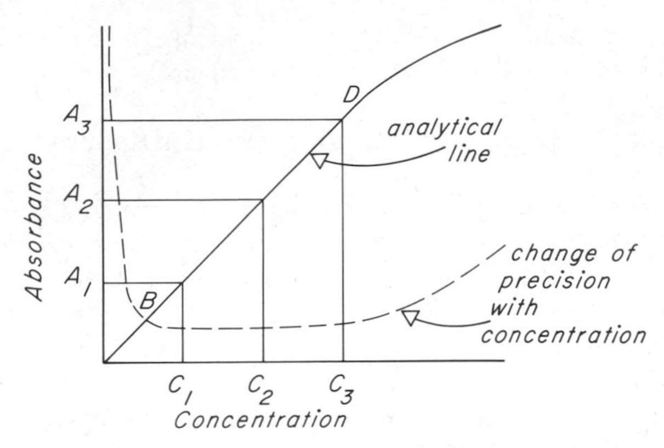

Figure 2.3 Calibration graph showing the shape of analytical line and the change of precision with concentration.

2.1.3 Elements Covered

Figure 2.4 is a periodic table showing only those elements that can be analyzed by atomic absorption. Elements most easily determined by this technique are Cu, Zn, Cd, Sb, Bi, Mn, Fe, Co, Ni, Ag, Au, Pb, Mg, and Ca. Hydride generation is the favored technique for As, Sn, Ge, Se, and Te, although furnace atomic absorptioin is gaining in popularity for these elements. Mercury is de-

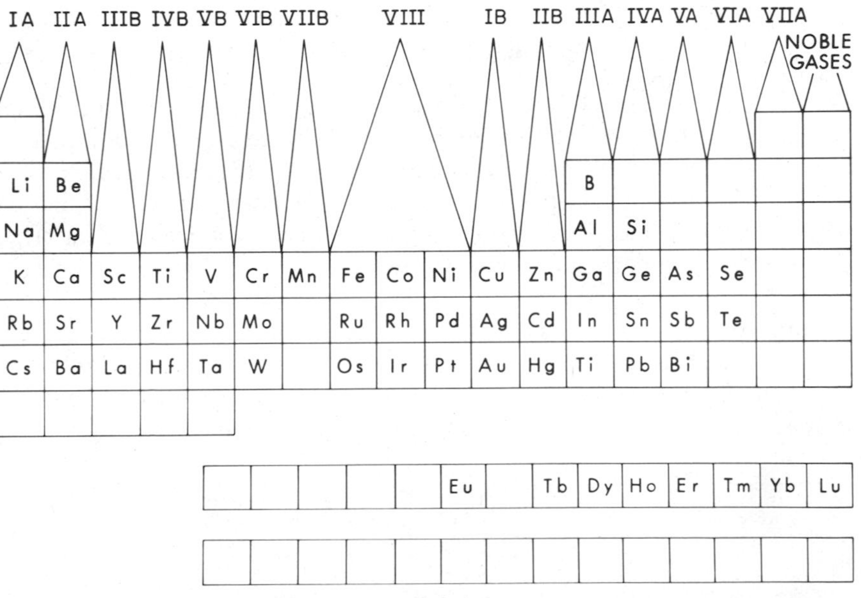

Figure 2.4 Periodic Table showing elements determinable by atomic absorption spectrometry.

termined using cold vapor atomic absorption. A nitrous oxide–acetylene flame is used for the elements in groups IIIB, IVB, VB, and VIB, for Ru, Os, Ir, Al, Si, Be, Sr, and Ba, and for the rare earths.

2.1.4 Radiation Sources

A line source of radiation is needed for absorption by the atoms in the atomizer. For best sensitivity a narrow line width emission source should be used. This can be produced by a hollow-cathode lamp such as the one shown in Fig. 2.5.

The hollow-cathode lamp is filled with about 2 torr of an inert gas, usually argon or neon. A high voltage is generated between the anode and cathode (the latter is made from metal of the analyte element) that results in ionization of the inert gas. The positively charged inert gas ions bombard the cathode and eject atoms of the analyte element. These are excited by bombardment with ions and electrons and emission then occurs.

Hollow-cathode lamps can be made from most of the elements that are conventionally determined by atomic absorption spectrometry. Single-element lamps are best. However, it is possible to buy multielement lamps for some elements. These generally give lower intensities for the constituent elements that single-element lamps and for the most part should be avoided when possible.

Conventional hollow-cathode lamps for elements such as arsenic, tellurium, selenium, lead, and mercury are generally unsatisfactory. In these instances electrodeless discharge lamps are available. These lamps require a separate power supply, but are recommended despite this additional cost. Recently, controlled temperature gradient lamps have become commercially available for elements such as arsenic. These show excellent intensities and appear to have sensitivity (narrow line width) advantages over electrodeless discharge lamps.

All lamps have a finite operating lifetime. In hollow-cathode lamps this lifetime is limited by loss of metal from the cathode, which is redeposited elsewhere in the lamp, and by the leakage of gases into the lamp from the surroundings. The leakage occurs whether the lamp is used or not. Loss of metal from the cathode can be a particularly serious problem for some elements in a multielement lamp.

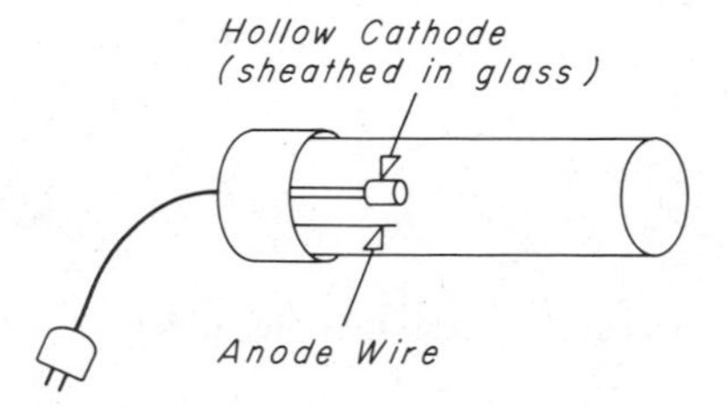

Figure 2.5 Drawing of a hollow cathode lamp.

2.1.5 Flame Atomizers

As a general rule, a flame atomizer should be used when applicable. In comparison with electrothermal atomizers, better precision is obtainable. Interference problems are complex with electrothermal devices. A typical burner nebulizer–premix chamber is shown in Fig. 2.6.

Each element, run at a given analytical line, has its characteristic atomic absorption sensitivity. Table 2.1 lists flame detection limits obtained using a Perkin–Elmer Model 603 spectrometer for water solutions that are very low in total dissolved salts. Detection limits, it must be remembered, are matrix dependent. Therefore such a list can only be used in a relative sense.

There has been little new in analytical atomic absorption spectroscopy (flame technology) since the introduction of the nitrous oxide–acetylene flame by Amos and Willis (4). The three flames in common use today are air–hydrogen, nitrous oxide–acetylene, and air–acetylene. Of these the last is most commonly employed. Choice of flame is basically dictated by the temperature required to

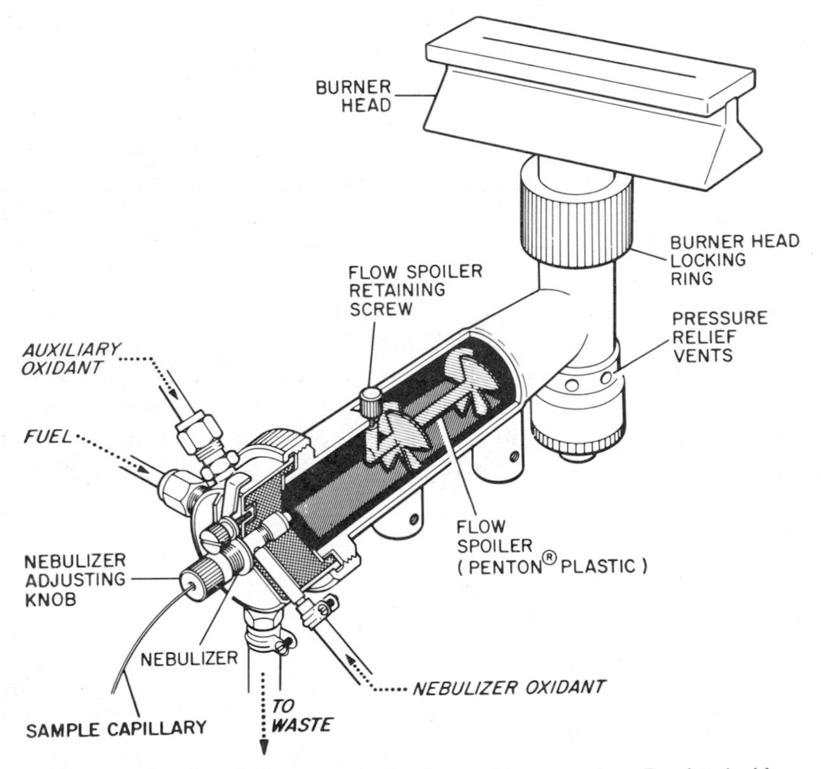

Figure 2.6 Drawing of a nebulizer, premix chamber, and burner system. Reprinted with premission of *Atomic Spectrometry*.

Table 2.1 Flame Detection Limits[a] for Perkin-Elmer[b] 603

Element	Detection Limit	Element	Detection Limit	Element	Detection Limit	Element	Detection Limit
Ag	0.002	Co	0.01	Mg	0.001	Ru	0.1
Al	0.05	Cr	0.006	Mn	0.05	Sb	0.08
As[c]	0.5	Cu	0.002	Mo	0.1	Se[c]	0.05
Au	0.02	Fe	0.01	Na	0.005	Si	0.05
B	0.7	Ga	0.08	Ni	0.01	Sn	0.1
Ba	0.01	Ge	0.10	Os	0.1	Sr	0.005
Be	0.01	Hg[c]	0.3	Pb[c]	0.05	Te[c]	0.1
Bi	0.025	Ir	0.6	Pd	0.02	Ti	0.1
Ca	0.001	K	0.01	Pt	0.1	Tl	0.05
Cd	0.001	Li	0.01	Rh	0.01	V	0.1
						Zn	0.001

[a]Nitrous oxide-acetylene flame used for refractory elements.

[b]Use of distilled-water standards and conditions recommended by the manufacturer (μg/ml.).

[c]Electrodeless discharge lamp recommended.

atomize the analyte in the presence of potential interferences. Little use is made of the relatively cool air–hydrogen flame except for elements such as arsenic and selenium, whose principal analytical wavelengths are at 193.7 and 196.0 nm, respectively. In the wavelength region below 200 nm, absorption of radiation by other flames is prohibitively high.

Interferences in flame atomic absorption can be classified as atomic spectral, physical, chemical, ionization, and nonspecific (background). The most troublesome types are chemical and nonspecific interferences.

Atomic spectral interferences are in most cases negligible. Because of the very narrow line emission characteristic of hollow-cathode and electrodeless discharge lamps and the simplicity of absorption spectra, there is little problem from spectral line overlap with interfering lines. Thus few problems are to be expected from this type of interference.

Physical interference results when solutions being aspirated have different viscosities. For example, standard solutions often have low viscosities characteristic of very dilute solutions. These should not be used with relatively viscous highly acid solutions that result when samples are dissolved and then diluted only slightly with distilled water. In this instance standards should be prepared with a similar acid content to the samples.

Appreciable fractions of easily ionizable elements are ionized at the temperatures of air–acetylene and particularly nitrous oxide–acetylene flames. Thus to ensure the same degree of ionization between standards and samples for easily ionizable elements and hence prevent ionization interference the samples and

Table 2.2 The Elements, Flame Type, and Interference

Element	Flame Type[a]	Comments
Ag	A,l	Cationic interferences minimal; avoid halides.
Al	N	Fe above 0.2% depresses, Ti enhances, H_2SO_4 depresses. Some ionization occurs.
As	H	Interferences severe; better to use method.
Au	A,l	Use a very lean flame; cationic interferences minimal.
Bi	A,l	Significant interference from base metals; can use hydride method as alternative for complex solutions.
Cd	A,l	Cationic interferences slight.
Co	A,l	Cationic interferences slight.
Cr	A,r	Gives best sensitivity, but extensive cation interference occurs.
	A,l	Poorer sensitivity, but most cationic interference suppressed.
	N	Cationic interferences minimal.
Cu	A,l	Cationic interferences minimal.
Fe	A,l	Cationic interferences minimal.
Hg	–	Use cold absorption tube method.
Ir	A,r	Complex interference patterns, interference minimized; use a Na–Cu mixture.
Mn	A,l	Cationic interference minimal.
Mo	N,r	A variety of cations interfere; use Al to suppress these.
Ni	A,l	Some interference from base cations; use very lean flame to minimize.
	N,r	Cationic interferences minimal.
Pb	A,l	Use 2833-Å line. Cationic interferences minimal. Avoid SO_4.
Pd	A,l	Several cationic interferences; use La to suppress these.
Pt	A,l	Complex cationic interferences; use La to suppress these.
	N,r	Minimal cationic interferences.
Rh	A,l	Complex cationic interferences; use La to suppress.
Sb	A,l	Spectral interferences occur in presence of lead using 2176 Å. Some cationic interferences; use hydride or other electrothermal method in complex solutions.
Se	H	Interferences severe; use hydride or other electrothermal method.

Table 2.2 (*Continued*)

Element	Flame Type[a]	Comments
Sn	N,r	Sensitivity best in hydrogen flame, but interferences very severe. Poor sensitivity in air-acetylene flames. Use hydride or other electrothermal method if applicable.
Te	A,l	Some cationic interference; use hydride or electrothermal methods where applicable.
V	N,r	Overcome cationic interferences with Al.
Zn	A,l	Minimal cationic interferences.

[a]A = air–acetylene, l = lean, N = nitrous oxide–acetylene, H = hydrogen–air, r = reducing.

standards should contain an ionization buffer. The elements of concern in this case are sodium, potassium, rubidium, and cesium. If an nitrous oxide–acetylene flame is used, a few other elements (e.g., alkaline earth and rare earths) may present a problem.

A common difficulty in working with the more refractory elements is a failure to completely dissociate compounds of these elements in a flame, resulting in chemical interference. This problem is particularly severe when working with air–acetylene or lower temperature flames. For example, when phosphate is present a significant fraction of the calcium remains bound as a phosphate at air–acetylene flame temperatures. Thus a standard made by dissolving calcium carbonate with hydrochloric acid and dilution with distilled water cannot be used to measure the calcium content of a phosphate-containing sample solution. The remedy commonly employed is to add a releasing agent at about 1% to the sample and the standards to prevent the formation of calcium phosphate molecules and to give roughly similar major ion contents.

Nonspecific or background interference is caused by absorption of radiation at the analyte wavelength by molecular species and/or scatter of incident radiation in the atomizer. This interference is generally most significant with work near the detection limit. Background interference problems are minimal in flame atomic absorption.

Table 2.2 gives guidance on choice of flame and on interference problems. It lists the elements with the flame type commonly employed. Comments are made concerning interferences. The comment "cationic interferences are minimal" refers to solutions with less than a few hundred micrograms per milliliter of the individual cations. In more concentrated solutions, any cation or anion may be considered a potential interference and may have to be added to the standard solutions. Ionization interferences can be overcome with an easily ionizable alkali metal such as potassium.

As can be seen, refractory elements require the high-temperature nitrous oxide–acetylene flame. Other elements, for example, chromium and nickel, may benefit from the use of this flame in overcoming persistent interferences. On the negative side, the nitrous oxide–acetylene flame results in a high incidence of ionization interference, can yield appreciable shot-noise problems, and is generally more difficult and less desirable to work with.

Nonspecific interferences resulting from molecular absorption and/or scatter are much less of a problem with flame atomizers than with electrothermal devices. The following section discusses electrothermal equipment. However, when working near the detection limit with any element, the need for background correction should be evaluated. As with flames, little is new in nebulizer–premix chamber technology. Ultrasonic nebulization, capable of yielding high atomization efficiencies, has yet to become a commercial success in atomic absorption.

2.1.6 Electrothermal Atomizers

L'vov (5) introduced electrothermal atomizers into atomic absorption analysis. The term *electrothermal atomizer* will be used in this chapter to refer to electrically heated devices such as graphite furnaces and rods. The common designation *flameless atomizer* will not be employed, to avoid implications that other equipment such as cold vapor absorption tubes are being discussed in this section.

Most early commercial atomic absorption equipment was less than optimal for electrothermal work. Electronic systems failed to follow the very fast transient signals obtained with electrothermal atomization. Installation of baffling was essential in many optical systems to cut down on stray light. Heating cycles were not well controlled and hence were not reproducible.

The main aim in electrothermal work is to obtain atomization under isothermal conditions. This will minimize interference problems. In this regard temperature control, use of a L'vov platform and a rapid heating ramp to the atomizing temperature are essential. Because such features have only recently been developed, the literature through the early years is often very contradictory as to optimum experimental parameters and interference patterns.

Instrument improvements that have recently been made that are important in making electrothermal AAS a generally useful tool are:

1. Most furnace power supplies now allow more than six steps of temperature programming with ramping facilities for each step. This aids greatly in obtaining adequate matrix modification to overcome interferences.
2. The rate of heating during the atomize step has been improved so that atomization under isothermal conditions can be anticipated for most elements.

3. The L'vov platform approach can be used to help further the delay of atomization until isothermal conditions are obtained.
4. Temperature control exists based on actual temperature measurement.
5. Zeeman background correction is available.
6. Good automatic samplers are now available.

I believe that the graphite furnace with Zeeman background correction capability is the best combination for routine electrothermal work. Recently a new approach to background correction, the Smith–Heiffje system, has been introduced (6). There is at this time, little practical experience with this method.

2.1.6.1 L'vov Platform

L'vov (7) pointed out the importance of separately controlling the appearance temperature and temperature of the gas phase in a furnace. One method of doing this is to place the sample on a small graphite platform that rests on the bottom of the tube. In this way the sample is in large part heated by radiation from the tube walls. This delays the appearance time of the analyte signal so that atomization occurs under nearly isothermal conditions. As a result there is a minimal difference in the width and height of the signal obtained from pure solutions and those obtained in more complex matrices. Figure 2.7 from Slavin et al. (8) shows a comparison of the magnesium chloride signal obtained for 1 ng of thallium atomized from the wall with that from the platform. This dramatically illustrates the advantage of using the L'vov platform for thallium. In general the L'vov platform is recommended for work with the more volatile elements. (I also find it good for more refractory elements such as aluminum.)

Background correction is very important in electrothermal work. Currently best background correction available commercially is based on use of the Zeeman effect. However, equipment containing such a capability is generally very expensive. Thus at this writing most instruments are equipped with the less expensive deuterium arc for background correction.

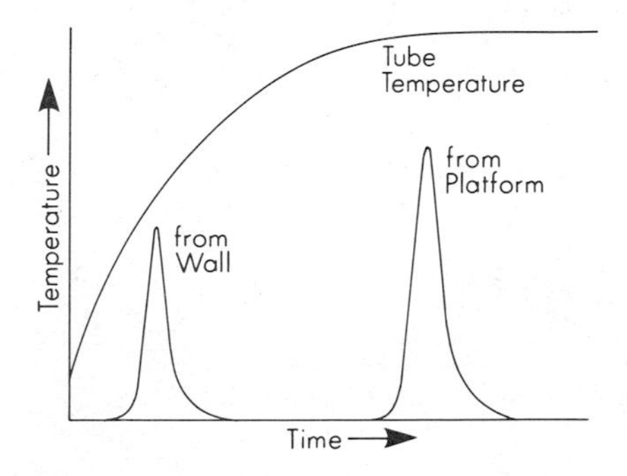

Figure 2.7 Comparison of thallium signal obtained by atomization from the wall and from a platform (8).

Electrothermal atomizers have a small sample capacity. It is difficult to inject manually the microliter volumes, which are applicable to these devices, reproducibly into the atomizer. Automatic samplers are now available for most equipment and I highly recommend them.

Detection limits with furnaces are potentially up to three orders of magnitude better than with flames. Table 2.3 lists the elements often done by electrothermal atomization and their detection limits in low-salt-content solutions. Detection limits are, of course, matrix dependent. Cruz and Van Loon (9) found that in rock sample solutions the detection limits for most elements given in Table 2.3 were seriously degraded. It must be stressed, however, that some of this effect resulted from using nonoptimal early equipment.

2.1.6.2 Interences Using Electrothermal Atomization

Interferences in electrothermal work can be very complicated. This is particularly true with samples containing large amounts of a complex matrix. Many interferences can be minimized by using isothermal atomization. In spite of this the method of standard additions must often still be employed. This approach, however, does not overcome all matrix interferences.

Physical interference problems can arise in pipetting of standards and samples when the viscosities of each are very different. With some of the highly acid or high-salt-content geological sample solutions, this source of error can be significant. The method of standard additions can be used to overcome this problem.

Any phenomenon that alters the rate of volitalization of the analyte from a

Table 2.3 Detection Limits for Analyte in Low-Salt-Content Water[a]

Element	Detection Limit (pg)	Element	Detection Limit (pg)	Element	Detection Limit (pg)
Ag	0.5	Ir	1000	Se	200
As	25	Mn	0.8	Sn	200
Au	25	Mo	15	Te	100
Bi	30	Ni	30	Ti	500
Cd	0.5	Pb	5	V	200
Co	10	Pd	30	Zn	0.2
Cr	20	Pt	500		
Cu	5	Rh	50		
Fe	10	Sb	10		

[a]Perkin-Elmer 603 and HGA-2100 were used with the conditions recommended by the manufacturers.

sample compared with a standard is an interference. Coexisting cations and anions in sample solutions can alter the rate of volatilization. Interferences of this type are as yet very poorly quantified. To minimize interference it may be possible to change the composition of the matrix, either by reagent addition or by selective volatilization of matrix components prior to atomization. Signal integration or standard addition is also employed. For the latter approach to be valid, the added analyte must assume the same chemical form in the sample as the sample analyte. Standard addition also requires that the calibration graph be linear over the concentration range employed. Calibration graphs have very short linear regions in electrothermal work. Of course, using standard addition is time consuming, but it may be essential for the most accurate work. Recently designed autosamplers containing a standard additions program are a great boon to the use of this important technique.

With some samples, for example, seawater, there is a danger that during the heating cycle the analyte can be lost before the atomization step. This is a particularly serious potential source of error when metal chlorides are formed. It is therefore advisable to convert all chlorides to a less volatile form, such as nitrate or sulfate, before the ashing step. Use of the L'vov platform is also highly recommended. A number of elements such as selenium, zinc, cadmium, and lead can become volatile at relatively low temperatures. It may be possible in these instances to vaporize the analyte prior to the matrix. When this approach is not valid, recent tellurium research has shown that an element such as nickel or silver (in the case of arsenic, selenium, or tellurium) can be added to stabilize the element in the atomizer during the ashing step.

By far the most important interference in using electrothermal atomizers results from signal caused by background (light scatter or molecular adsorption). This problem, most severe at ultraviolet wavelengths, also must be considered in the visible region of the spectrum. The effect can be so acute that the backgrounnd correction capability of the instrument is exceeded. Under this circumstance much of the older equipment will still generate values without any indication of error. The consequences of such a phenomenon are obvious. If there are no warnings indicating background correction overload but the equipment has an energy meter, observation that the energy meter needle drops drastically toward zero can be evidence of the problem. Most commercial atomic absorption units cannot correct for background when the interference produces a signal greater than an absorbance of one. Zeeman background correction, however, is useful well above this value.

Matrix modification and/or background correction methods are used to overcome nonspecific interferences. In seawater samples and body fluids, compounds such as NaCl and organic materials come off at relatively low temperatures. Their generation can interfere with the determination of low-boiling-point elements (Cd, Pb, Zn, etc.). For high-boiling-point elements, compounds such as NaCl and KCl can be reduced to tolerable levels by selective volatilization. A common remedy is to add NH_4NO_3 and to volatilize chloride as NH_4Cl. The presence or formation of refractory oxides (CaO, MgO, Al_2O_3, etc.) can interfere severely with the determination of high-boiling-point metals. Selective volatilization is not possible in this case. It must be emphasized that the method of standard additions is not applicable to overcoming nonspecific interferences.

2.1.6.3 Background Correction

There are a number of approaches to background correction. With most instruments an automatic, continuum-source background correction is employed. In this method the signals obtained with a continuum and line source are monitored separately. Their difference can give a background correction value. If narrow slit widths are used, significant absorption of the continuum radiation at the analyte wavelength may occur. This effect leads to a loss of sensitivity. Instruments that give dual-beam background correction or the equivalent are to be preferred because of drift problems inherent with continuum sources.

Nonabsorbing lines have been employed for background correction. However, this method, frequently used in the early years of atomic absorption, has been largely replaced by the continuum approach. When background correction is needed in the visible region, the nonabsorbing line method is still employed. Some recent equipment provides a tungsten iodide continuum source for automatic correction in the visible region of the spectrum.

When there is direct spectral overlap between the fine structure of a molecular absorption band and the analyte line, continuum-source background compensation will give erroneous reseults. Failure to ensure by adjustment each time that the two optical beams are coincident and fill the same fraction of the optical aperture gives erroneous "corrected values." Again, Zeeman background correction obviates problems of these types.

An intense magnetic field can be used to split atomic lines into their Zeeman components. If an atom exhibits the normal Zeeman effect, a central component (π) and two components on the wings ($\pm\sigma$) are produced. Using polarizers, the signals for analyte plus background and for background can be monitored separately. The background signal can then be subtracted.

A new method of background correction has been recently introduced by Instrumentation Laboratories. In this approach the analyte plus background signal is measured in the conventional way with a hollow-cathode lamp running normally. The lamp is then pulsed momentarily with a high current. Under this condition the lamp emission becomes severely broadened and analyte sensitivity drops drastically. Thus a measurement made under the latter condition represents mainly background, which can then be subtracted. Little experience at the time of writing is available on the performance characteristics of this method.

Slavin et al. (10) have published a generally very useful paper in which they outline their experiences with the stabilized temperature platform furnace and Zeeman background correction. This paper contains two tables, combined and reproduced here as Table 2.4, that lists recommended experimental conditions from their experience and the literature. Their tables assume use of pyrolitically coated tubes, maximum power heating, gas interrupt, integrated absorption, and Zeeman background correction. They feel, however, that the conditions given for most elements apply as well with continuum-source background correction.

2.1.7 Optical System

A double-beam optical system for an atomic absorption system is illustrated in Fig. 2.8. This approach differs from a single-beam system in that the light from the radiation source is split into two components: (1) the sample beam passes through the atomizer and (2) the reference beam passes around this device. Then electronically these two beams can be ratioed to eliminate instabilities in the radiation source. This prevents variations in radiation source intensity from being recorded as signal change. Thus radiation sources can be turned on and used immediately without need for warm-up. This adds to convenience and increases the useful lifetime of a lamp.

To guide the radiation beam through the instrument mirrors and/or lenses can be employed. In the case of lenses the focusing point is somewhat wavelength dependent, which requires corrective elements to be present with the lens. It is important to minimize light loss in the spectrometer and thus high quality optical components must be used. However, use of atomic absorption spectrometers in areas of a laboratory containing chemical fumes can result in rapid deterioration of optical surfaces and a serious degradation in optical throughput.

The monochromator as shown in Fig. 2.9 is the device in which the optical beam is dispersed into its wavelength components. In modern equipment this is performed by a reflecting grating. Light from the source passes through the entrance slit and is focused onto the grating surface. The desired wavelength component of the dispersed radiation is then focused out of the monochromator through the exit slit onto the photomultiplier.

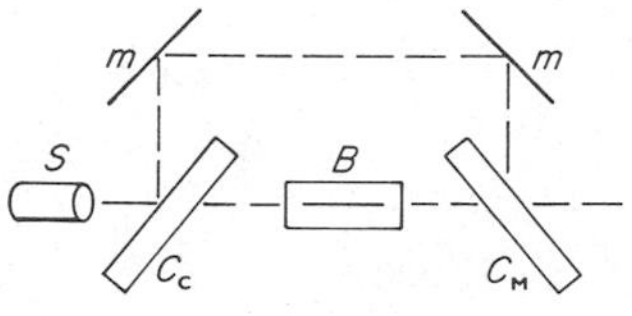

Figure 2.8 Diagram of a double-beam optical system: m is the mirror; C_C, chopper; C_M, half-silvered mirror; S, source; and B, burner.

Table 2.4 Furnace Conditions[a]

	Zeeman/PE5000							
Element	Wavelength (nm)	Slit Width (nm)	Sensitivity Loss[h] (%)	Modifier	Char Temp. (°C)	Atom Temp. (°C)	C^b	Footnote
Ag	328.1	0.7	6	200 μg PO_4	650	1600	1.4	*d*
Al	309.3	0.7	10	50 μg $Mg(NO_3)_2$	1700	2400	10	
As	193.7	0.7	11	20 μg Ni	1500	2500	17	
Au	242.8	0.7	16	50 μg Ni	1000	2200	13	
B	249.7	0.7	40	5 μg Ca	1000	2700	700	*e*
Ba	553.6	0.4	4		1200	2700	6.5	*e*
Be	234.9	0.7	50	50 μg $Mg(NO_3)_2$	1500	2500	1.2	
Bi	223.1	2.0	35	20 μg Ni	900	1900	23	
Ca	422.7	0.7	5		1100	2600	0.8	*e*
Cd	228.8	0.7	2	200 μg PO_4 + 10 μg $Mg(NO_3)_2$	900	1600	0.35	*d*
Co	240.7	0.2	15	50 μg $Mg(NO_3)_2$	1400	2400	6	
Cr	357.9	0.7	12	50 μg $Mg(NO_3)_2$	1650	2500	3.5	
Cu	324.7	0.7	45		1200	2300	8	
Cs	852.1	0.4			900	1900	5	*f*
Fe	248.3	0.2	8	50 μg $Mg(NO_3)_2$	1400	2400	2	
Hg	253.7	0.7	40	$K_2Cr_2O_7$	250	1000		
K	766.5	0.4	40					
Li	670.8	0.4	15		900	2600	1.4	*f*
Mg	285.2	0.7	9		900	1700	0.35	*f*
Mn	279.5	2.0	9	50 μg $Mg(NO_3)_2$	1400	2200	2	
Mo	313.3	0.7	5		1800	2700	9	*e*
Na	589.5		5					
Ni	232.0	0.2	9	50 μg $Mg(NO_3)_2$	1400	2400	9	

P	213.6	0.7	30	50 μg $Mg(NO_3)_2$	1400	2700	3000	
Pb	283.3	0.7	17	200 μg PO_4 + 10 μg $Mg(NO_3)_2$	600	900	11	*d*
Pd	247.6	0.7	9		900	2300	24	*e,f*
Pt	265.9	0.7	20		1300	2500	17	*e*
Rb	780.0	0.4			800	1900	2.3	*f*
Sb	217.6	2.0	5	20 μg Ni	1100	2400	38	
Se	196.0	2.0	12	20 μg Ni + 25 μg 10 $Mg(NO_3)_2$	900	2000	30	
Si	251.6	0.2	2		1400	2700	40	*f*
Sn	286.4	0.7	6	200 μg PO_4 + 20 μg $Mg(NO_3)_2$	1000	2100	23	*d,g*
Sr	460.7	1.4	2		1300	2600	1.4	*e,f*
Te	214.3	2.0	7	20 μg Ni	900	2000	15	
Ti	365.3	0.2	4		1400	2700	43	*e*
Tl	276.8	0.7	35	1% H_2SO_4	600	1500	10	
V	318.4	0.7	25		1500	2700	30	*e*
Zn	213.9	0.7	12	6 μg $Mg(NO_3)_2$	600	1800	0.04	*d*

[a] Reprinted with permission from Slavin et. al. (10). The sample size is typically 20 μL. The drying time is 60 sec at 160°C with 1-sec ramp. For platform char, the hold time is 45 sec after 1-sec ramp. The tube is cleaned out with a final 6-sec firing at 2600°C using 1-sec ramp. The data apply to the HGA-500 and HGA-400 and to the Zeeman/5000, Models 5000, 4000, or 3030. New pyrolytically coated tubes are assumed. EDLs are assumed for all the elements where they are available.

[b] *C* is the characteristic amount in pg/0.0044 A sec for the Zeeman/5000. Integrated absorbance signals are assumed.

[c] In a few cases the detection limits have been improved from more recent work, see note *g*.

[d] PO_4 is either $(NH_4)_2HPO_4$ or $NH_4H_2PO_4$, whichever is less contaminated with the analyte metal.

[e] Atomization from the wall.

[f] In certain cases we have added approximate detection limit data (±100%) where the data were not previously available.

[g] If deuterium background correction is used, a different wavelength is optimum and the *C* will be different.

[h] Due to anomalous Zeeman splitting and/or incomplete splitting.

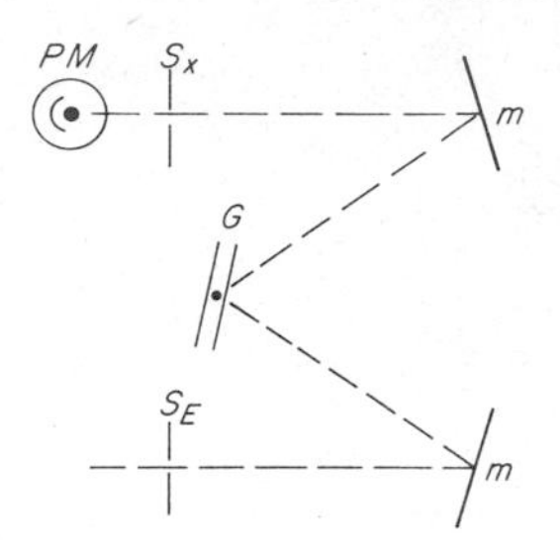

Figure 2.9 Diagram of a monochromator: m is the mirror; G, grating; S_E, entrance slit; S_x, exit slit; and PM, photomultiplier.

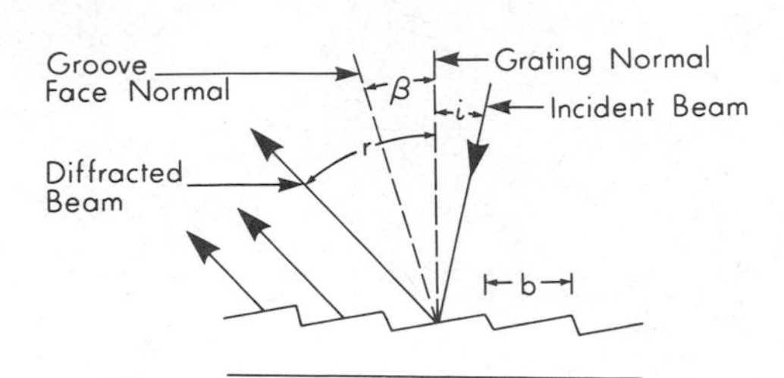

Figure 2.10 Diagram of a grating surface.

Part of a grating surface is diagramed in Fig. 2.10 The surface consists of closely spaced parallel lines ruled into the surface at a given angle (blaze angle).

Incident radiation is dispersed at the surface and the individual components are reflected back. Different wavelengths diverge from the surface at different angles. The wavelength with the greatest intensity is the one whose angle of reflection equals the angle of incidence.

The blaze angle (angle at which the grating is cut) affects the reflected wavelength intensity distribution. That is, a grating is blazed to give the greatest spectral intensity in a chosen wavelength region. The further a wavelength is from this region the greater the light loss at that wavelength. For this reason some instruments have two gratings—one blazed for the ultraviolet and the other for the visible regions of the spectrum.

2.2 PLASMA EMISSION SPECTROMETRY

Inductively coupled plasma emission spectrometry is rapidly becoming one of the most important techniques of trace element analysis. One of its distinct advantages over atomic absorption is capability for simultaneous multielement analysis (quantometer). In addition, calibration curves are linear over at least five orders of magnitude (compared with two or three orders with atomic absorption). On the negative side, detection limits are poorer by one or two orders of magnitude for most elements as compared to electrothermal atomic absorption. Work is going on to improve detection limits. Plasma emission spectrometry requires the attention of a highly trained spectroscopist and instrumentation is very expensive. Hence this technique will for the most part be available only in the larger analytical laboratories.

2.2.1 Historical Introduction

Bapat (11) in 1942 was the first to maintain an inductively coupled plasma at atmospheric pressure. Then Reed (12) in 1961 first produced a stable, inductively coupled atmospheric pressure plasma in a flowing system. But it was Wendt and Fassel (13) and Greenfield et al. (14) who simultaneously but independently studied the analytical usefulness of this plasma.

Sample introduction into the plasma as conceived by Greenfield (14) was through an axial pathway up the center of the plasma. A patent obtained by Greenfield et al. (15) contained the drawing of a torch that had this property and that is similar in many respects to commercial offerings today. With the exception of some debate over the radio frequency powers that should be employed (high in the United Kingdom and low in the United States), the development of commercial instruments was relatively straightforward. (Conversions using older spectrometers, however, often produced poor equipment because of stray light problems.) At present commercial offerings of inductively coupled plasma emission equipment are basically of two types: instruments capable of simultaneous multielement analysis (polychromator) and sequential devices (a computer-controlled slew scan monochromator).

During the same period development of a commercial dc arc plasma was occurring. One instrument (Spectrametrics Ltd.) employs an Eschelle spectrometer with a cassette of slits for multielement analysis. The dc arc plasma has a more complex pattern of interferences.

The Eschelle spectrometer is known to possess excellent resolving power. Thus when used with an inductively coupled plasma (relatively very rich in spectral lines), excellent analytical performance is obtained. An inductively coupled plasma is now available with the Eschelle spectrometer. In this book the descriptions and procedural material presented are for an inductively coupled plasma.

Figure 2.11 shows a block diagram of an inductively coupled plasma emission spectrometer (ARL Model 34,000) for simultaneous multielement analysis (typical of quantometers). The components are (1) a radio-frequency generator; (2) a torch, the upper portion of which is encircled with copper induction coils; (3) a spray chamber and nebulizer; (4) an argon supply and a flow system; (5) a spectrometer with an entrance slit and a family of exit slits; (6) photomultipliers; and (7) measuring electronics and computer.

2.2.2 Plasma

Plasmas are very hot gases in which atoms are present in large part as ions. A torch for the production of an inductively coupled plasma consists of three concentric rings of quartz as shown in Fig. 2.12. Three argon flows pass upward

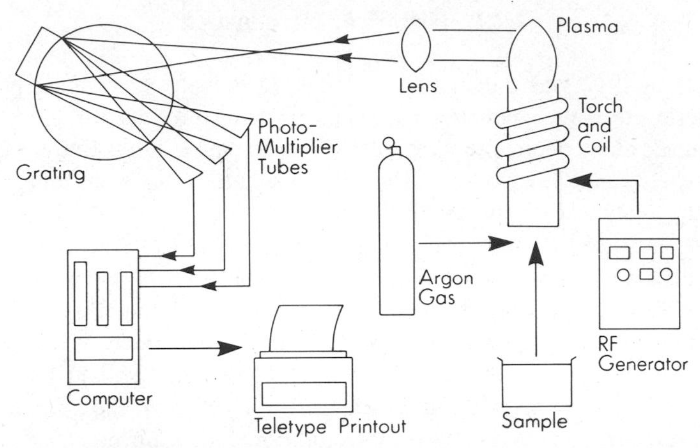

Figure 2.11 Block diagram of an inductively coupled plasma emission system. Reprinted with permission of Applied Research Laboratories, Sunland, California.

through these three channels. The flows are (from inside out) the sample carrier gas flow, the plasma gas flow, and the coolant gas flow.

Argon gas is nonconducting. Thus it is necessary to obtain argon ions and electrons to interact with the magnetic field in the radio-frequency coil. These are produced using a Tesla coil. The electrons and ions accelerate on each half cycle. These species experience resistance to their flow and heating occurs. The result is the ignition of a plasma in the coil region that extends above the torch as shown in Fig. 2.11.

The temperature of the plasma in the region of the coil is 8000 to 10,000 K. For emission spectrochemistry purposes the plasma is viewed about 15 mm above

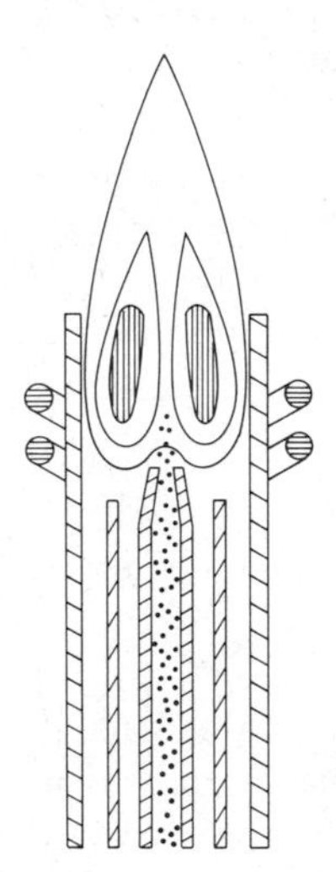

Figure 2.12 Inductively coupled plasma and torch system.

the coil region. Here the temperature is about 6500 K. The flow of argon in the outer part of the torch is introduced tangentially at about 10 L/min. This coolant flow prevents the plasma from melting the torch. The sample carrier gas flow in the central part of the torch is about 1 L/min. This gas passes through the nebulizer/spray chamber system and carries the sample up through the center of the plasma. The third argon flow is the plasma gas flow.

The residence time of the atoms in the viewing point of the plasma is considerably longer than residence time for atoms in a flame. Atoms are emitting in an optically thin layer in the center of the plasma. This means that self-absorption of radiation is minimal. Thus linearity of emission signal over several orders of magnitude of concentration can be expected. This property and the fact that multielement analysis is easy with plasma emission spectrometry are distinct advantages for using this technique over flame atomic absorption. In addition, it is my view that repeatability with plasma emission is generally better than with flame atomic absorption.

Sample uptake is one of the most important remaining problems in inductively coupled plasma emission spectrometry. The two types of nebulizers now commonly in use are concentric and cross-flow nebulizers. In both types continuous smooth flow of the sample to the plasma cannot be guaranteed over prolonged periods. At the time of writing nebulizer problems are far from solved. Using a peristaltic pump with nebulizers helps produce a more constant liquid flow. When samples contain high solids, the above nebulizers are particularly troublesome. But it is now possible to purchase a Babington-type nebulizer that can be recommended for such samples.

The emission signal for an element in a plasma depends on observation height, plasma power, and sample uptake argon flow rate. Because these parameters will be different for each element, it will be necessary in multielement analysis to choose the best compromise conditions.

2.2.3 Spectrometer

Because of the very high temperature of plasmas, a very rich spectrum is generated from the contained atoms. Thus good resolution is essential. Many types of spectrometers have been used with inductively coupled plasmas. Little success has, however, been reported in adapting older spectrometers designed for use with dc arc and ac spark sources. The main reason is that such spectrometers were not meant to be used with the very high radiation fluxes experienced with plasmas. Thus stray light problems abound. Spectrometers designed to handle this problem must contain light traps, baffles, filters, and minimal secondary optics.

Grating imperfections are a serious problem in plasma spectrometers. For this reason holographically produced gratings are generally preferred.

For simultaneous multielement analysis two types of spectrometers are common: (1) the Eschelle and (2) the type in which grating and primary and secondary slits are positioned on the Rowland circle. In the latter an exit slit is placed at each position on the Rowland circle where a wavelength of the element under consideration focuses. The radiation passes through the slit onto a photomultiplier. When large number of lines must be accommodated, close packing of photomultiplier tubes is essential. It may thus be necessary to use a mirror at the exit slit to reflect the radiation above or below the plane of the slit when adjacent lines are too close to place the tubes side by side.

A person purchasing a quantometer must decide before the instrument is manufactured what elements and what lines are to be employed. The spectrometer is then constructed with exit slits and photomultipliers fixed in the appropriate positions. Furthermore, choice of analyte wavelength will depend on what main types of samples are to be analyzed. This is because lines must be chosen that minimize spectral interference for the particular sample matrix. A compromise may also be necessary with detection limit requirements when spectral interferences are a problem at the most sensitive wavelength of an element.

The Spectrametrics Eschelle spectrometer provides very high resolution, thus making it very useful in plasma work. In this particular offering a prism is used in combination with the Eschelle grating. The prism is employed as an order sorter in which the orders are dispersed at right angles to the grating dispersion. This gives a two-dimensional arrangement of the spectrum (i.e., orders are arranged vertically and the individual lines within an order horizontally). A two-dimensional removable cassette with slits is then designed to the user's specifications. Each cassette can contain 20 or so elements.

Sequential multielement instruments have a computer-controlled slew scan monochromator. These instruments are slower when multielement analysis is necessary. However, the instrument is much more flexible than simultaneous multielement instruments. Any set of elements and lines can be chosen for a given set of samples. (The user is not locked into a fixed set of lines and elements as with the quantometer.) Sequential instruments are therefore to be preferred when used for research or when samples being analyzed vary greatly in the matrix. It is also possible to have a scanning monochromator on one of the channels of the fixed slit instruments, but such an arrangement involves compromises.

2.2.4 Electronics and Computer

Because of the large amount of data generated by a multichannel instrument and because of the complexity of instrument control, a computer must be incorporated into the design. With the rapid developments now going on in the computer field it is possible to incorporate a relatively low cost computer that can be programmed to control the instrument, do background correction, and accumulate and make calculations on the data in a "user friendly" way. This means that the user need know very little about computers to operate this aspect of the instrument. It must, however, be emphasized that no degree of sophistication in this aspect of instrument design will overcome the fact that spectrochemical analysis depends on certain spectroscopic and chemical principles. The analyst must therefore be forever diligent in learning these principles and not be lulled to sleep by instrument salespeople who try to gloss over these problems!

2.2.5 Interferences

Classically, spectrochemical interferences can be grouped into three categories—physical, chemical, and spectral. Proponents of plasma emission spectrometry are quick to point out the relative freedom from chemical interferences enjoyed by the plasma compared with flame atomizers. This is because plasmas have temperatures that are about twice those of the hottest generally used flames. On the other hand it is also important to stress that because of these high temperatures, relatively complex line spectra are to be anticipated compared with those obtained with flames. Regarding continua spectra both atomizers are to some degree unique. Flames generate continua due to molecules resulting from combustion products and entrained air. Plasmas are generated in argon and therefore plasma gases generate little problem in this regard. However, continua are produced by ion recombination and to a small extent by entrainment of molecules in the air in the observation zone. Both effects can be minimized by proper torch design and careful choice of observation height above the torch.

Interference correction for spectral overlap and/or baseline shift is frequently essential in plasma emission analysis. This is true in spite of minimizing such problems by choosing the best available lines for the elements being determined in the samples of interest. Such manipulations often require a good portion of the computer's capability. Thus with older instruments that incorporated computers with a relatively low capacity, serious sacrifices were often essential in instrument operation and data accumulation and calculation. Modern instruments do not generally suffer greatly from such problems.

2.2.5.1 Correction for Spectral Interferences and Baseline Shift

Modern instruments allow for spectral scans to be made in the region of the lines for each element. Obviously, this is very easily done with slew scan monochromators. For simultaneous multielement instruments a movable primary slit may be employed for this purpose. A unique computer-based correction can then be made for each line at which a problem is detected.

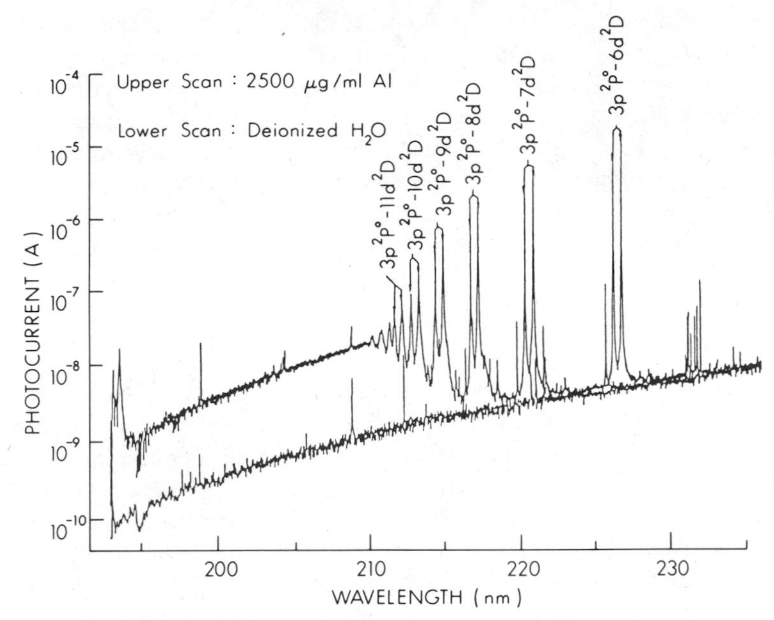

Figure 2.13 Tracings of ICP spectrum for Al and H_2O. Reprinted with permission of Applied Research Laboratories, Sunland, California.

Figure 2.13 shows a wavelength versus signal obtained for deionized water (bottom curve) and 2500 μg/mL aluminum (upper curve). In the wavelength range 195–220 nm there is a very significant baseline shift exhibited for aluminum. In the 210–235 nm range aluminum line spectra can be seen. Thus when high levels of aluminum are present in a sample solution, it would be necessary to employ background correction when analyte lines occur in this spectral region. Other commonly occurring elements causing similar problems are iron, calcium, and magnesium.

Figure 2.14 shows a wavelength scan around the zinc 202.55-nm line. The B's are the baseline obtained for deionized water; S's are the signal obtained for micrograms per milliliter zinc; and the *'s represent the signal for 1000 μg/mL chromium. This illustrates a spectral overlap of the chromium line on the zinc line. A slight baseline shift (S's versus B's) is also present. A correction factor for chromium can be obtained by running different concentrations of chromium. These can be used to obtain a chromium calibration graph that can be used to correct for any chromium interference level on zinc.

A significant baseline shift and spectral overlap interference problem are illustrated in Fig. 2.15. This arises for the molybdenum 284.82-nm line when magnesium is present. The 1000 μg/mL solution (*'s) illustrates a rather flat

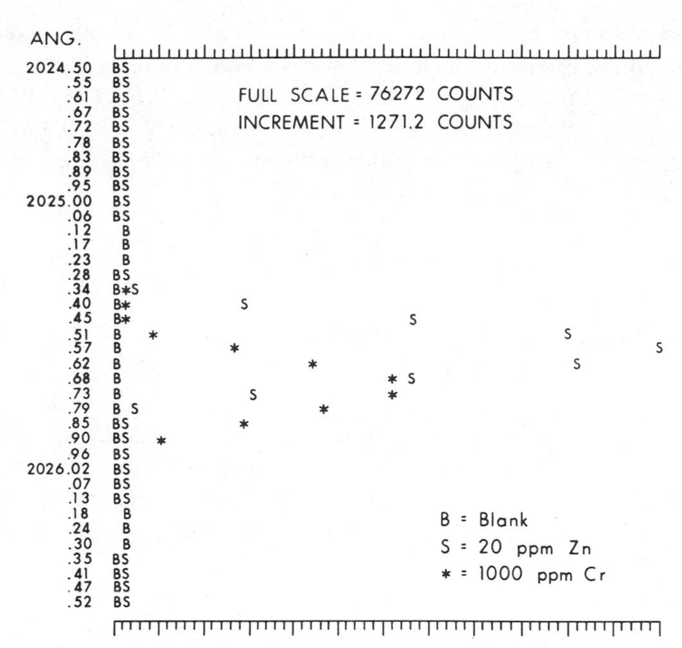

Figure 2.14 Spectral scan around the 2025.6-Å Zn line. Reprinted with permission of Applied Research Laboratories, Sunland, California.

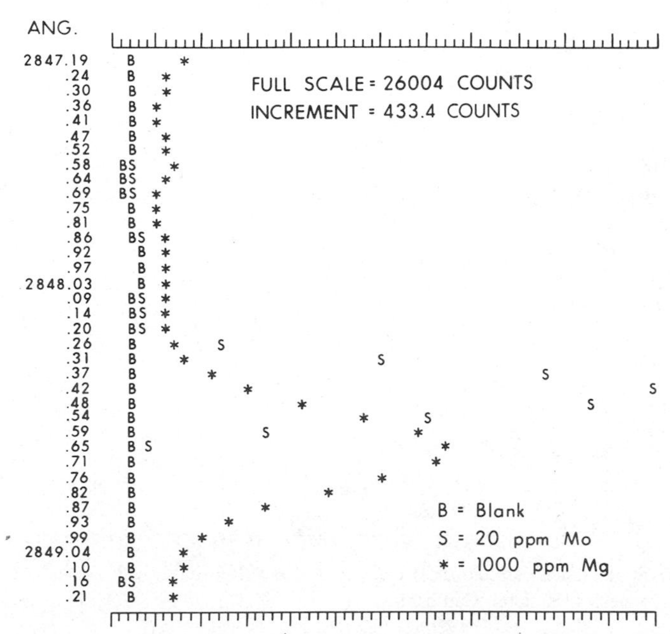

Figure 2.15 Spectral scan around the 2848.4-Å Mo line. Reprinted with permission of Applied Research Laboratories, Sunland, California.

baseline shift in this spectral region. The S's are the 20 μg/mL molybdenum spectrum and the B's represent a deionized water solution.

A severely sloping background shift is shown in Fig. 2.16. The *'s represent the output obtained around the 193.76-nm arsenic line (S's) from 1000 μg/mL aluminum. Here a two-point correction (one on either side of the arsenic line) would be necessary.

2.2.5.2 *Chemical Interference*

Chemical interferences (e.g., the interference of phosphate on calcium through failure to break down the calcium phosphate molecules) are minimal with plasmas. This is because at the high temperatures of the plasma, the breakdown of such molecules should be complete. However, there is some evidence of chemical interference occurring for some of the more refractory oxides. Generally such problems are expected to be greater with the DCP compared with the ICP.

Typical detection limits for inductively coupled plasma emission using real

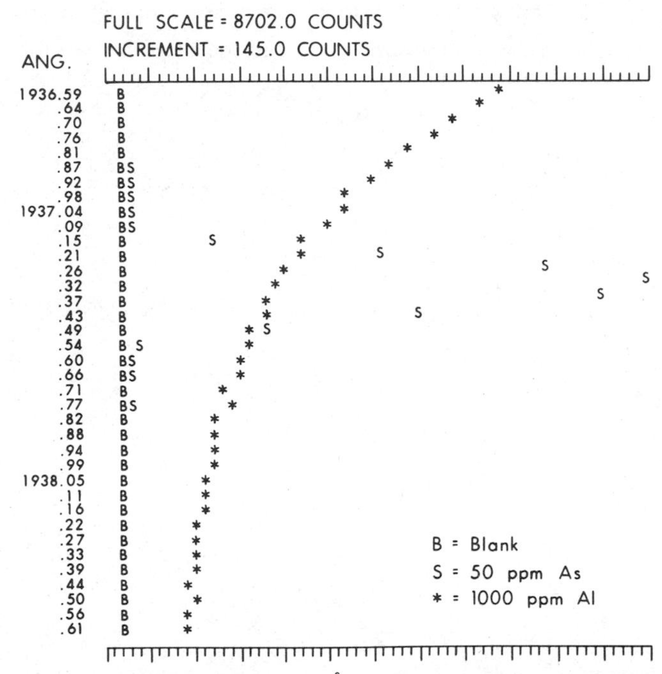

Figure 2.16 Spectral scan around the 1937.3-Å As line. Reprinted with permission of Applied Research Laboratories, Sunland, California.

samples are illustrated in Table 2.5. Elements are grouped in detection limit ranges.

2.3 SPECIAL EQUIPMENT

2.3.1 Hydride Generation

The elements arsenic, antimony, bismuth, selenium, tellurium, tin, germanium, and lead form volatile covalent hydrides. These can be produced in solution and

Table 2.5 Typical ICP Detection Limits[a]

0.1–1 g/L	1–10 g/L	10–100 g/L	100–1000 g/L
Mg	Ag	Al	K
Ca	B	As	Bi
Mn	Ba	An	Ge
Sr	Be	Ga	I
	Ce	Gd	In
	Co	Hf	Pr
	Cr	Ag	Te
	Cu	In	
	Fe	Ir	
	Li	La	
	Ni	Mo	
	V	Nb	
	Yb	P	
	Zn	Pb	
		Pd	
		Rh	
		Se	
		Si	
		Ta	
		Ti	
		Tl	
		W	
		Y	
		Zr	
		Sb	
		S	
		Na	

[a]Limits for most sensitive lines.

volatilized into an atomizer for determination by atomic absorption or plasma emission spectrometry. Using hydride generation virtually all of the elements in a sample solution enter the atomizer at one time and thus a large increase in sensitivity is obtained as compared to direct aspiration into a flame or plasma. Of the elements listed above, arsenic, selenium, tellurium, and tin are commonly determined by the hydride approach. Furnace atomic absorption is a competitive method for these elements.

A variety of equipment has been proposed for hydride generation. This ranges from simple inexpensive manual devices such as a disposable syringe to complex automated systems. Several commercial systems exist. Typically a hydride generation system consists of a generation vessel, a method of separating the liquid from the gas, and the atomizer.

For plasma emission the hydride is injected directly into the plasma atomizer. The two approaches that exist are continuous hydride generation systems and injection of discrete samples. The former are presently favored. There is a distinct trend toward automated systems for hydride generation. Such systems are dominated by the segmented stream approach typical of the Technicon autoanalyzer. Recently Liversage et al. (16) reported the use of flow injection for semiautomation of hydride generation for determining arsenic by inductively coupled plasma emission. A block diagram of the apparatus is shown in Fig. 2.17. This is a very simple approach to hydride generation for the inductively coupled plasma.

In atomic absorption spectrometry, atomization is accomplished by a flame or a heated quartz or Vycor tube. The tube is heated with resistance wire or by an air–acetylene flame. Heated tube atomizers give significant improvements in sensitivity over flames and are highly recommended. Exposure to flames causes slow devitrification of the quartz tube atomizer, which eventually crumbles.

2.3.2 Mercury Cold Vapor Method

Mercury is determined almost entirely by atomic absorption spectrometry. Elemental mercury has an appreciable vapor pressure at room temperature. Thus if the mercury in a sample can be obtained in the elemental form, the vapor thus produced can be swept into the optical beam and no heated atomizer is necessary. This approach was popularized by Hatch and Ott (17). In their approach the sample is decomposed in the presence of acids and oxidizing reagents (to prevent loss of mercury). The excess of oxidizing agent is prereduced and then a reducing agent is added to reduce mercury to the elemental form.

The basic types of apparatus that have been used are closed and open systems. In a closed system the mercury vapor is recirculated and the cold vapor

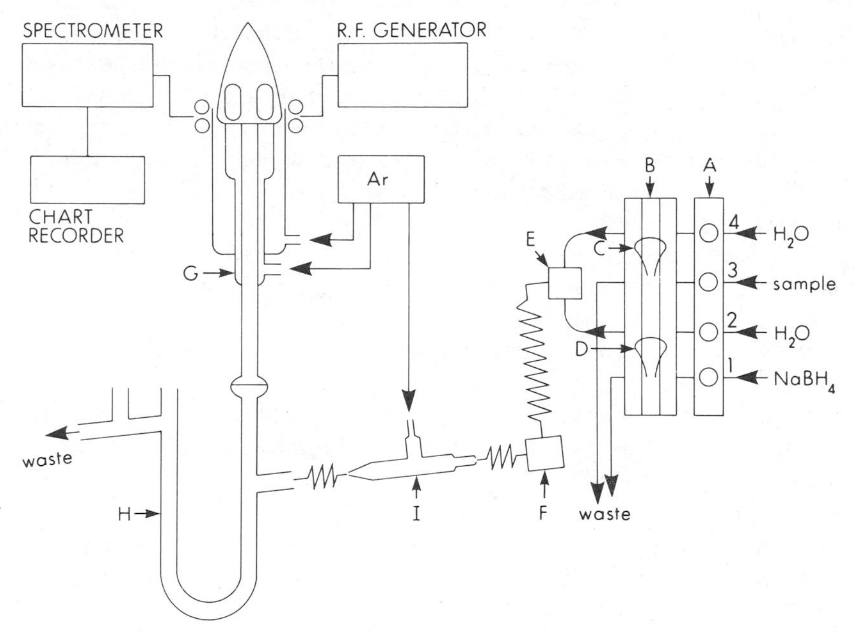

Figure 2.17 Diagram of a hydride flow injection hydride-ICP system.

absorption cell, placed in the optical beam, has windows on its ends. The cold vapor absorption cell in the open system is windowless. Mercury is generated in a glass vessel fitted with a ground glass cap containing a fritted bubbler tube.

2.4 CHOOSING INSTRUMENTATION

This section covers some of the considerations that are important in choosing atomic spectrometry instrumentation. It is important that equipment obtained be suitable for trace element analysis.

2.4.1 Atomic Absorption Equipment

It is interesting to note that the atomic absorption instruments today are in all important respects similar to the first instrument exhibited by Walsh (the inventor of analytical atomic absorption spectrometry) in 1955. The main innovations since then are in electronics where microprocessor control and data processing have taken over, and in the introduction of furnace atomizers by L'vov in 1959. This speaks very highly for Walsh's insight into the method he was developing.

The prospective buyer of atomic absorption equipment is faced with a bewildering array of conflicting claims and optional extras. Customers must carefully consider the type of work they wish to do. They must be certain that this is best served by atomic absorption spectroscopy. Analyses can be categorized as follows: (1) major-element, high-accuracy analyses; (2) accurate trace analyses; (3) high-precision trace analysis; and (4) combinations of these. For the work discussed in this book, (2) and (3) will be of great interest. Those wishing to work in category (1) require a flame instrument; there will likely be no need for background correction accessories and a furnace is most certainly unnecessary. Workers contemplating analyses in categories (2) and (3) need background correction and may or may not require an electrothermal atomizer depending on the sample type and elements to be analyzed. In most instances accessories can be added at a later stage. The customer must be careful not to be oversold.

In choosing flame accessories the customer should not skimp, particularly when equipment is to be used for routine operations. It is best to have highly automated gas control with all available safety features. It is also useful to purchase a nitrous oxide as well as an air–acetylene burner. Although many of the metals can be analyzed using air–acetylene flames, for a few, for example, vanadium, silicon, aluminum, osmium, and molybdenum, the hotter nitrous oxide flame is essential. This flame gives better chemical interference suppression and is even often used for elements such as calcium and chromium.

The graphite furnace is the most satisfactory of electrothermal atomizers. It maximizes sample size and suffers least from reaction with the sample constituents. Reproducibility is generally best with the graphite furnace. The furnace heating rate should be in excess of 2000°C/sec and accurate temperature control is important.

Graphite tubes should be internally grooved to restrict sample spreading when organic samples are to be handled. A pyrolitic coating of the graphite gives better results for a large number of elements. The aquisition of L'vov-type platforms is important.

Autosampler accessories can be obtained for both furnaces and rods. These are highly recommended for improving reproducibility. A recorder is essential for work with electrothermal devices.

A number of atomic absorption instruments are being offered with CRT output either built in or as an accessory. This is an important feature in furnace atomizer work. Peak shape and appearance time can be analyzed using such a device, and this is important in diagnosing potential interferences.

Hollow-cathode lamps are now very stable, thus negating the necessity of double-beam optics to overcome this source of instability. Electrodeless discharge lamps on the other hand require 30-min warm-up periods and therefore can be used more quickly with double-beam equipment. When background correction is necessary, equipment with double-beam background correction is desirable. The deuterium arc tends to vary in intensity, as do electrodeless discharge lamps. The double-beam instrument is probably a better buy when large numbers of routine analyses are to be done, particularly when nonskilled personnel are involved.

At present, Zeeman-effect-based background correction is best. It is very desirable to have this capability for furnace work even though the correction is expensive.

Conventional hollow-cathode lamps are still the most useful line sources for most elements. Boosted (high-intensity) hollow-cathode lamps may be introduced in the future for some of the more refractory elements.

Electrodeless discharge lamps (EDL) are now offered commercially. These are highly recommended for elements such as arsenic, selenium, phosphorus, lead, and tin. Lamp warm-up time is about 30 min, but this disadvantage is greatly overweighed by long lifetime and higher spectral output.

Rapid developments are now being made in the area of microprocessor utilization in atomic absorption instrumentation. Some manufacturers offer completely microprocessor-controlled instrumentation—a very desirable step forward. One should note, however, that atomic absorption spectroscopy is a technique based on various laws of physics and chemistry. These fundamentals dictate problems and interferences that will always be present. Although automation and microprocessor control can ease the operation of instrumentation, they cannot negate spectroscopic principles. Analysts must therefore continue to be diligent in their efforts to understand basic atomic absorption theory.

2.4.2 Plasma Emission Spectrometry

Although there are several types of plasmas available for use analytically, it is my opinion that at the moment the inductively coupled plasma is superior for analysis of most of the substances covered in this volume. An important exception is the gas chromatographic analyses of volatile metal species (covered in Chapter 10). In this application a microwave-induced plasma appears to be superior when used as the gas chromatography detector. Another readily available offering, the dc plasma, appears to suffer more strongly from problems of chemical interference than the ICP.

There are basically two types of ICP atomic emission instrumentation available—the sequential and quantometer designs. The quantometer allows simultaneous multielement analysis on 40 or more separate channels. These are fixed wavelength channels (fixed at time of manufacture) for the elements that the user has specified. Because sample matrix and sensitivity requirement determine which element lines can be considered, it is important that manufacturer and purchaser discuss the selection of analytical wavelengths carefully so that the least amount of problems will subsequently be experienced.

Once the wavelengths have been chosen and the exit slits fixed at the proper position in the instrument, it is a major job (usually requiring manufacturer's personnel) to change the wavelength of a channel. This means that the purchaser is virtually locked into the elements chosen at the time of purchase.

It is possible with most quantometer equipment to install a scanning monochromator in one of the channels. This is highly recommended because of the increased flexibility such a device gives to the user.

The sequential ICP analyzes one element at a time in sequence. Although this process is now relatively rapid, the time per sample for multielement analysis is much greater than with the quantometer-type system.

Why then buy a sequential system? The sequential ICP is more flexible. This is important to the purchaser who must do many different combinations of elements on different samples. A selection may be made in the wavelength between different lines of the same element as matrix and sensitivity may require. Scans can be performed very easily to determine interference problems. In a fixed-channel multielement instrument, compromise conditions (e.g., argon flow rates, viewing point in the plasma, etc.) must be used. With a sequential instrument the best conditions for each element can be used.

Wavelength scanning is an important feature of any ICP emission instrument. This can be done readily using the sequential type of instrument. However, limited (in wavelength span) scans can be made around each line with quantometer equipment if the background correction accessory is ordered. Background correction is essential in much of the work done near the detection limit of the element, and the ability to do wavelength scans gives great flexibility in this operation.

Because of increased problems (compared with atomic absorption) of atomic spectral interferences, it is important to purchase a spectrometer of good resolution for emission work. This is particularly important in work with ICPs because of the very rich spectrum obtained for many elements. When 15 or so elements are present in appreciable amounts in a sample (as is common), a very complex spectral pattern is obtained.

Excellent resolution can be obtained with an Eschelle spectrometer. With this device a high order (> 100th) of emission radiation is viewed. Because dispersion increases with increasing spectral order, this would seem to be an excellent approach for use with the ICP. One minor problem with an Eschelle is that the light throughput of the spectrometer is reduced compared with conventional units, because the radiation intensity is reduced as higher orders are employed. However, the gratings used in modern Eschelle spectrometers are specifically blazed for use in the higher orders, thus obviating this problem somewhat.

REFERENCES

1. K. Muller, *Z. Phys.* **65,** 739 (1930).
2. A. Walsh, *Spectrochim. Acta* **7,** 108 (1955).
3. G. F. Box and A. Walsh, *Spectrochim. Acta* **16,** 255 (1960).
4. M. D. Amos and J. B. Willis, *Spectrochim. Acta* **22,** 1325 (1965).
5. B. V. L'vov, *Inzh. Fix Zh.* **2,** 44, (1959).
6. S. B. Smith and G. M. Hieftje, *Appl. Spectrosc.* **37,** 419 (1983).
7. B. V. L'vov, *Spectrochim. Acta* **36B,** 153 (1978).
8. W. Slavin, D. C. Manning, and G. R. Carnrick, *At. Spectros.* **2,** 137 (1981).
9. R. B. Cruz and J. C. VanLoon, *Anal. Chim. Acta* **72,** 231 (1974).
10. W. Slavin, G. R. Carnrick, D. C. Manning, and E. Prusekowska, *At. Spectros.* **3,** 69 (1983).
11. G. I. Bapat, *Vestn. Elektroprom.* **2,** 1 (1942).
12. T. B. Reed, *J. Appl. Phys.* **32,** 821 (1961).
13. R. H. Wendt and V. A. Fassel, *Anal. Chem.* **37,** 920 (1965).
14. S. Greenfield, I. L. W. Jones, and C. T. Berry, *Analyst* **89,** 713 (1964).
15. S. Greenfield, I. L. W. Jones, and C. T. Berry, *U.S. Patent* **3,** 467, 471 (September 16, 1969).
16. R. Liversage, J. C. Van Loon, and J. C. de Andrade, *Anal. Chem. Acta* **161,** 275 (1984).
17. W. R. Hatch and W. L. Ott, *Anal. Chem.* **40,** 2085 (1968).

CHAPTER

3

BASIC MATERIALS

In trace metal analysis it is important to prevent contamination and loss of analyte from samples. Problems from contamination are particularly severe when levels of ppb and below are encountered. Such concentrations are common when furnace atomic absorption is employed. In this case detection limits are often determined by impurity levels in reagents and contamination during sample preparation. The following elements are ubiquitous in laboratories: zinc, calcium, potassium, sodium, and iron; and restricting contamination from these elements is challenging. The fact that generally accepted background levels of trace elements in botanical, zoological, and clinical samples have been steadily decreasing in recent times is a testimony to improvements in contamination prevention.

To minimize problems of analyte loss and contamination, special precautions are essential in preparation of high-purity water, preparation of standard and calibration solutions, choice of reagents, and choice of bottles for storage of samples and standard solutions. The following sections will be helpful to the analyst in ensuring that loss of analyte and contamination are minimized.

3.1 IMPURITIES IN REAGENTS

3.1.1 Water

Water of high purity is essential in sample preparation and for preparation of standard and calibrating solutions. The purity of water is usually expressed in terms of resistivity and total solids. For trace element analysis in the micrograms per liter range and below it is essential to have finished water with a resistivity of better than 10 MΩ/cm and total solids below 1 μg/mL. Methods of purification include distillation, deionization, and a combination of both.

3.1.1.1 Deionization System

Deionizing systems are faster than distillation systems (i.e., liters per minute rather than liters per hour). For example, in a popular system the speed is 3*L*/

min through three stages of purification. The water has a resistivity of about 10 MΩ/cm. This system contains two stages of demineralization followed by a filtration stage to remove down to submicron-sized particles including bacteria. The system (pump conductivity meter but without cartridges) costs about $1400.*

3.1.1.2 Distillation System

All-glass distillation apparatus (as opposed to stills with metal parts) is essential for use in trace metal analysis laboratories. Pyrex brand glass stills are available commercially from most scientific apparatus suppliers. One company advertises a Pyrex glass still with one stage of distillation for less than $2000.* The water specifications are 2.0–1.2 MΩ/cm resistivity and 3.0–0.2 ppm total solids, depending on the speed of distillation. Thus the still that operates at 1.4 L/hr (the slowest) gives water with the best specifications, and the fastest still, 6.0 L/hr, gives water with the worst specification. Two stages of distillation can be employed if water of higher purity is essential.

When the raw water supply is high in dissolved ionic constituents, a demineralizer can be employed as a first stage of purification. The cost of such an apparatus as advertised is about $4800.*

Organic contamination of water used for trace metal analysis can sometimes be a problem. Organic materials can be removed if one stage of distillation is done in the presence of alkaline permanganate solutions.

When the ultimate in water purity is essential, a subboiling still can be used. Distillation from such an apparatus is very slow (< 1 L/day; therefore this water should only be used when absolutely essential). Such a still is diagramed in Fig. 3.1—a design by Murphy (1).

Storage of high-purity water can result in contamination. Even carefully washed glass and plastic containers will result in unacceptable levels of contamination if the water is stored over extended periods. Therefore when highest purity is essential, it is important to use only freshly prepared water.

3.1.2 Reagents

It is essential to use high-purity chemicals throughout for trace metal analysis. A variety of designations are used for high-purity chemicals. For example, Fisher Scientific Co. Canada Ltd. lists three grade designations that could generally be considered suitable for trace metal analysis: Fisher certified ACS, Fisher certified, and Fisher reagent grade. The first two grades have an actual lot analysis on their labels and the product meets or exceeds these specifications.

*United States dollars, 1982.

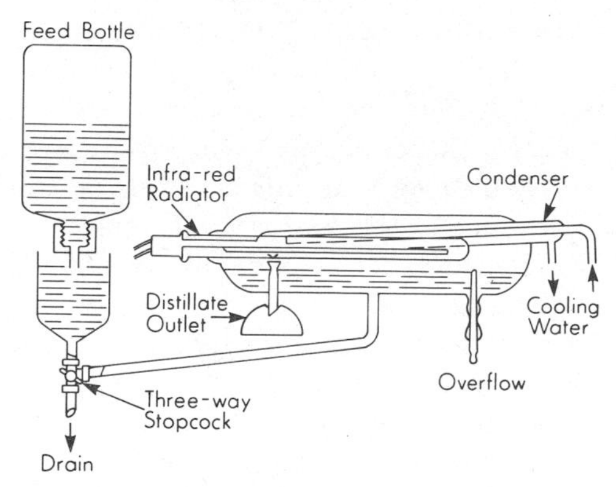

Figure 3.1 Quartz subboiling still (1).

For some purposes the reagent grade chemicals such as these may not be sufficiently pure. This is frequenlty true when furnace atomic absorption analysis is required. For this reason some chemical companies offer even higher purity chemicals. For example, the British Drug Houses, Chemicals, Canadian catalogue lists Suprapur and Aristar grades, the latter to be used when the highest level of purity is required. For hydrochloric acid the Suprapur and Aristar grades are approximately ten and twenty times more expensive, respectively. It goes without saying that these reagents should be used only when absolutely essential.

The high cost of ultrapure reagents may be a problem for some laboratories. This is particularly true of acids that are required in relatively large amounts in a trace metal analysis. To overcome this problem Kuehner et al. (2) used a commercially available (Quartz Products Corp., Plainfield, NJ) quartz subboiling still for the production of ultrapure HCl, HNO_3, H_2SO_4, and $HClO_4$ at reasonable cost. A similar design of a Teflon subboiling still was used for purification of HF. Production of other high-purity reagents has been discussed by Zief and Horvath (3). In this paper they cover strategies for purification of inorganic salts, organic reagents, bases, and water. This reference provides excellent information for analysts requiring any high-purity reagents.

Storage generally degrades the quality of chemicals; thus older chemicals should be discarded and replenished by a fresh supply.

Reagent blanks should always be run in trace metal analysis. It is impossible to overemphasize the importance of this point. This is true even if the highest grade reagents, such as Aristar, are used. The reagent blank represents not only the level of impurity in the reagents but also will take account of any other contamination that occurs. Thus in my laboratory, it is customary to run up to three blanks with each sample set. The level of analyte in the blank is often the factor that determines the limit of detection in an analysis.

3.2 STANDARD SOLUTION PREPARATION

It is common to prepare 100–1000 mL of stock (standard) solution of a metal at a concentration of 1000 to 10,000 μg/mL from which calibration standard solutions are prepared. Calibration standards are made up from time to time as required. These concentration levels for stock solutions are chosen because the weight of the chemical required for preparing the solution is sufficiently large that an accurate weighing can be done. Also solutions of this concentration can usually be made to be stable for a period up to one year. It is important to stress, however, that solutions of some elements such as iron, copper, zinc, and manganese will usually be much more stable than solutions of elements such as mercury, antimony, bismuth, and tantalum. To stabilize most metal solutions the final solution should contain at least 1% acid. Usually nitric acid or hydrochloric acid is employed for this purpose.

The solid-metal-containing substance for production of stock solution is usually the metal or a simple metal salt. These are weighed and then dissolved in acid and the solution diluted to the appropriate volume. It is often sufficient to use reagent grade material for this purpose because of the large dilutions usually required in preparing calibration solutions. If, however, these stock solutions are to be used directly for interference studies such as those frequently required in plasma emission spectrometry interference investigations, only the highest purity materials should be employed. For this purpose there are a few chemical companies that specialize in very high-purity-chemicals.* It is now possible to purchase commercially prepared stock metal solutions. In the early years of the production of such solutions I experienced some difficulties with concentration, purity, and stability. However, at present, stock metal solutions purchased from reputable chemical companies are useful for many purposes.

Table 3.1 list methods recommended for the preparation of single-metal 1000 μg/mL stock solutions (4).

I prepare only single-metal stock solutions. However, particularly with plasma emission spectrometry, it may be desirable to produce mixed-metal solutions. It is recommended that in these cases an intermediate mixed-metal solution be prepared from a dilution of the single-metal stock solutions. This additional step is recommended because mixed-metal solutions tend to be less stable than those made to contain one metal. On this point it is important to emphasize that the metal solutions that are mixed together should form a compatible mixture. Suggestions for such mixtures have been given by McQuaker et al. (5).

It is important to emphasize that whereas many metal solutions of concentrations 1000 μg/mL and above in 1% or stronger acid are stable up to one year, dilutions of these may be very much less stable. As a general rule, the lower the

*Two of these are Spex Industries, Box 798, Metuchen, NJ, and Johnson Mattey chemicals, 74 Hutton Garden, London.

Table 3.1 Standard Stock Solutions

Element	Procedure
Aluminum	Dissolve 1.000 g Al wire in minimum amount of 2 *M* HCl; dilute to volume.
Antimony	Dissolve 1.000 g Sb in (1) 10 mL HNO_3 plus 5 mL HCl, and dilute to volume when dissolution is complete; or (2) 18 mL HBr plus 2 mL liquid Br_2; when dissolution is complete, add 10 mL $HClO_4$, heat in a well-ventilated hood while swirling until white fumes appear, and continue for several minutes to expel all HBr; then cool and dilute to volume.
Arsenic	Dissolve 1.3203 g of As_2O_3 in 3 mL 8 *M* HCl and dilute to volume; or treat the oxide with 2 g NaOH and 20 mL water; after dissolution dilute to 200 mL, neutralize with HCl (pH meter), and dilute to volume.
Beryllium	(1) Dissolve 19.655 $BeSO_4 \cdot 4H_2O$ in water, add 5 mL HCl (or HNO_3), and dilute to volume. (2) Dissolve 1.000 g Be in 25 mL of 2 *M* HCl and dilute to volume.
Boron	Dissolve 1.000 g B in 8 mL 10 *M* HNO_3; boil gently to expel brown fumes and dilute to volume.
Chromium	(1) Dissolve 2.829 g $K_2Cr_2O_7$ in water and dilute to volume. (2) Dissolve 1.000 g Cr in 10 mL 2 *M* HCl and dilute to volume.
Cobalt	Dissolve 1.000 g Co in 10 mL 2 *M* HCl and dilute to volume.
Copper	(1) Dissolve 3.929 g fresh crystals of $CuSO_4 \cdot 5H_2O$ and dilute to volume. (2) Dissolve 1.000 g Cu in 10 mL HCl plus 5 mL water to which HNO_3 (or 30% H_2O_2) is added dropwise until dissolution is complete. Boil to expel oxides of nitrogen and chlorine and dilute to volume.
Gold	Dissolve 1.000 g Au in 10 mL hot HNO_3 by dropwise addition of HCl, boil to expel oxides of nitrogen and chlorine, and dilute to volume. Store in amber container away from light.
Iron	Dissolve 1.000 g Fe wire in 20 mL 5 *M* HCl; dilute to volume.
Lead	(1) Dissolve 1.5985 g $Pb(NO_3)_2$ in water plus 10 mL HNO_3 and dilute to volume. (2) Dissolve 1.000 g Pb in 10 mL HNO_3 and dilute to volume.
Lithium	Dissolve a slurry of 5.3228 g Li_2CO_3 in 300 mL of water by addition of 15 mL HCl; after release of CO_2 by swirling, dilute to volume.
Manganese	(1) Dissolve 1.000 g Mn in 10 mL HCl plus 1 mL HNO_3 and dilute to volume. (2) Dissolve 3.0764 g $MnSO_4 \cdot H_2O$ (dried at 105°C for 4 h) in water and dilute to volume. (3) Dissolve 1.5824 g MnO_2 in 10 mL HCl in a good hood, evaporate gently to dryness, dissolve residue in water, and dilute to volume.

Table 3.1 (*Continued*)

Element	Procedure
Mercury	Dissolve 1.000 g Hg in 10 mL 5 *M* HNO_3 and dilute to volume.
Molybdenum	(1) Dissolve 2.00425 g $(NH_4)_2MoO_4$ in water and dilute to volume. (2) Dissolve 1.5503 g MoO_3 in 100 mL 2 *M* NH_4OH_2 and dilute to volume.
Palladium	Dissolve 1.000 g Pd in 10 mL HNO_3 by dropwise addition of HCl to hot solution and dilute to volume.
Platinum	Dissolve 1.000 g Pt in 40 mL hot aqua regia, evaporate to incipient dryness, add 10 mL HCl, and again evaporate to moist residue. Add 10 mL HCl and dilute to volume.
Selenium	Dissolve 1.4050 g SeO_2 in water and dilute to volume or dissolve 1.000 g Se in 5 mL of HNO_3; then dilute to volume.
Silicon	Fuse 2.1393 g SiO_2 with 4.60 g Na_2CO_3, maintaining melt for 15 min in Pt crucible. Cool, dissolve in warm water, and dilute to volume. Solution contains also 2000 μg/mL sodium.
Silver	(1) Dissolve 1.5748 g $AgNO_3$ in water and dilute to volume. (2) Dissolve 1.000 g Ag in 10 mL HNO_3; dilute to volume. Store in amber glass container away from light.
Tellurium	(1) Dissolve 1.2508 g TeO_2 in 10 mL HCl; dilute to volume. (2) Dissolve 1.000 g Te in 10 mL warm HCl wth dropwise addition of HNO_3 and then dilute to volume.
Tin	Dissolve 1.000 g Sn in 15 mL warm HCl; dilute to volume.
Vanadium	Dissolve 2.2963 g NH_4VO_3 in 100 mL water plus 10 mL HNO_3; dilute to volume.
Zinc	Dissolve 1.000 g Zn in 10 mL HCl; dilute to volume.

[a]Reprinted from Dean and Rains (4), p. 332–335, by courtesy of Marcel Dekker, Inc. 1000 μg/mL as the element in a final volume of 1 L unless stated otherwise.

concentration of a metal solution, the less stable the solution is likely to be. I have found that the micrograms per liter solutions used for calibration in furnace atomic absorption may be so unstable that they should be made up fresh each day of use.

3.3 STANDARD REFERENCE SAMPLES

The importance of standard reference samples to accuracy assurance in trace metal analysis cannot be overemphasized. These are samples that have been analyzed by a number of methods and in several laboratories and to which certified values have been given for various constituents. Any laboratory engaged in trace metal analyses must acquire such relevant standards as are available. These samples should form the basis of accuracy assessments in the laboratories. Of course, spiking, use of internal control samples, and other methods commonly used for quality control of results can be useful. However, these approaches have severe limitations and must be supplemented by analysis of standard reference samples. To ensure that the standard reference samples remain in good condition, the supplier's instructions should be carefully heeded. Instructions for proper use of these materials are also often provided and must be followed. Table 3.2 lists samples and supplying agencies of which I have knowledge.

Table 3.2 Standard Reference Samples (trace metals)[a]

Sample	Supplier
SRM 1566 oyster tissue	National Bureau of Standards, Washington, DC 20234
SRM 1567 wheat flour	National Bureau of Standards, Washington, DC 20234
SRM 1568 rice flour	National Bureau of Standards, Washington, DC 20234
SRM 1569 brewers yeast	National Bureau of Standards, Washington, DC 20234
SRM 1571 orchard leaves	National Bureau of Standards, Washington, DC 20234
SRM 1572 citrus leaves	National Bureau of Standards, Washington, DC 20234
SRM 1573 tomato leaves	National Bureau of Standards, Washington, DC 20234
SRM 1575a pine needles	National Bureau of Standards, Washington, DC 20234
SRM 1577 bovine liver	National Bureau of Standards, Washington, DC 20234
SRM 1632a coal (bituminous)	National Bureau of Standards, Washington, DC 20234
SRM 1633a coal fly ash	National Bureau of Standards, Washington, DC 20234
SRM 165 river sediment	National Bureau of Standards, Washington, DC 20234
SRM 1648 urban particulate	National Bureau of Standards, Washington, DC 20234
SRM 1963a water	National Bureau of Standards, Washington, DC 20234
SRM 50 albacore tuna	National Bureau of Standards, Washington, DC 20234
Furnace dust (L.D.)	Bureau of Analysed Samples Ltd., Newby Middlesbrough, Cleveland, England, TS8 9EA
Furnace dust (electric)	Institute de Recherches de la Siderurqie Francaise B.P. 129, 78104-Sainte Germain, enLaye, France
TEG-50-A Gelatin Multitrace element standard	Dr. D.H. Anderson, Industrial Laboratory, Kodak Park Divisions, Eastman Kodak Co., Rochester, NY 14650
Water	Environmental Protection Agency, Environmental Monitoring and Support Lab, Cincinnati, OH

Table 3.2 **Standard Reference Samples (trace metals)**[a]

Sample		Supplier
Pepper bush	CRM	The National Institute for Environmental Studies, Yatake-Machi, P.O. Yatabe, Tsukuba, Ibaraki 300-21, Japan
Pond sediment	CRM	The National Institute for Environmental Studies, Yatake-Machi, P.O. Yatabe, Tsukuba, Ibaraki 300-21, Japan
Chlorella	CRM	The National Institute for Environmental Studies, Yatake-Machi, P.O. Yatabe, Tsukuba, Ibaraki 300-21, Japan
Freeze-dried serum	CRM	The National Institute for Environmental Studies, Yatake-Machi, P.O. Yatabe, Tsukuba, Ibaraki 300-21, Japan
Hair	CRM	The National Institute for Environmental Studies, Yatake-Machi, P.O. Yatabe, Tsukuba, Ibaraki 300-21, Japan
Mussel	CRM	The National Institute for Environmental Studies, Yatake-Machi, P.O. Yatabe, Tsukuba, Ibaraki 300-21, Japan
Sewage sludge A, Sewage sludge B, Sewage sludge C, Sewage sludge D, Incinerator sludge		Waste Water Technology Centre, Fisheries and Environment Canada, P.O. Box 5050, Burlington, Ontario
GXR2 soil GXR5 soil SXR6 soil		U.S. Department of the Interior, Geological Survey, Denver CO
Kale		H.J.M. Bowen, Department of Chemistry, Reading University, Reading, U.K.
Serum		Center for Disease Control, Atlanta, GA
Blood serum		International Atomic Energy Agency, Vienna, Austria
Animal bone		International Atomic Energy Agency, Vienna, Austria
Fish soluble		International Atomic Energy Agency, Vienna, Austria
Milk powder		International Atomic Energy Agency, Vienna, Austria
Corn flour		International Atomic Energy Agency, Vienna, Austria
Wheat flour		International Atomic Energy Agency, Vienna, Austria
Potatoes		International Atomic Energy Agency, Vienna, Austria
Pig muscle		International Atomic Energy Agency, Vienna, Austria
Sea plant		International Atomic Energy Agency, Vienna, Austria
Copepod		International Atomic Energy Agency, Vienna, Austria

[a]Other groups that can be consulted for standard reference samples are the Bureau of Analysed Standards, Newham Hall, Middlesborough, Yorkshire, England (a wide range of metallurgicals) and the Horticultural Department, Michigan State University, East Lansing, MI (apple, cherry, peach, citrus leaves).

3.4 CONTAINERS

Much has been written on the suitability of different container materials for storage of samples and standards in trace metals analysis. There seems to be lack of general agreement on the relative merits of borosilicate glass and a variety of plastics for sample storage. This is due to differences in behavior during storage of the various metal ions.

3.4.1 Behavior of Ag, As, Cd, Se and Zn

Massee et al. (6) studied the loss of silver, arsenic, cadmium, selenium, and zinc at trace levels from distilled and artificial seawater samples stored in various containers at different pHs. A good survey of the literature describing similar studies for 41 elements was also given. Particular attention in this study was paid to the ratio R of inner container surface (contacting the solution) to solution volume. The study is particularly relevant to furnace atomic absorption work. Radioactive tracers of the elements studied were employed.

The authors classify sorption losses into four categories as follows:

1. Sorption loss dependent on chemical form and concentration of the analyte element.
2. The effect of a solution parameter such as pH and presence of dissolved salts, complexing agents, dissolved gases, suspended material, and microorganisms on sorption losses.
3. The effect of the type of container on sorption losses. This includes chemical composition of walls, surface roughness, surface cleanliness (recommends cleaning in 8 M HNO_3), and surface area exposed to the solution.
4. Sorption loss dependent on external factors such as length of storage, temperature, light exposure, and agitation.

3.4.1.1 Zinc

Table 3.3 contains typical sorption loss data obtained for zinc. It is interesting to note the sorption dependence on inner container surface to solution volume ratio (R value), which is not always the worst for high R (e.g., borosilicate glass containers, pH 4, after seven days storage).

3.4.1.2 Silver

It should be noted that Massee et al. do not state whether brown bottles were employed with this element. Thus the relevance of the results for silver must be questioned.

Table 3.3 Sorption Behavior of Zinc[a]

Matrix	Distilled Water												Artificial Seawater											
Material	Polyethylene				Borosilicate				PTFE				Polyethylene				Borosilicate Glass				PTFE			
pH	4		8.5		4		8.5		4		8.5		4		8.5		4		8.5		4		8.5	
R/cm^{-1} [b]	1.4	3.4	1.4	3.4	1.0	4.2	1.0	4.2	1.0	5.5	1.0	5.5	1.4	3.4	1.4	3.5	1.0	4.2	1.0	4.2	1.0	5.5	1.0	5.5
Contact time											Sorption %													
1 min							21	20																
30 min				66			24	21			5	12						12	31					
1 h				65			23	22			3	16						9	31					
2 h			3	64			24	21			5	22							10	29	5			
4 h			3	60			25	21			4	28							9	30	4			
8 h			5	58			25	23			3	33							9	28	5			
24 h			8	56			26	22			5	27							5	26	4			
2 d			9	52			25	20			3	25							4	21	4			
3 d			6	53			20	22			3	19							4	18	5			
7 d			11	57							4	20					10	4	3	9	4			
14 d			11	57							5	20					27	19	–	10	5			
21 d			10	55							4	20					25	17	3	9	5			
28 d			12	56							6	20					20	19	4	9	5			

[a] Reprinted with permission from Massee et al. (6).
[b] R = inner container surface to solution volume ratio.

Of the elements studied silver showed the greatest problem due to sorption. No significant loss of silver from solution was noted for pH 1 or 2 with either type of water and for any of the containers and *R* values investigated. Polyethylene proved to be the worst choice for silver at pH 4 or higher. Results for artificial seawater showed less sorption than for distilled water samples for most conditions used.

3.4.1.3 Cadmium

No significant loss of cadmium from solution occurred at pH 1 or 2. At pH 4 no significant loss of cadmium was recorded for either of the waters and for all containers except for artificial seawater in borosilicate glass containers after seven days.

3.4.1.4. Arsenic and Selenium

No significant loss of arsenic and selenium was occasioned under any of the conditions studied.

3.4.2 Behavior of Chromium

Arnand and Ducharmne (7) investigated the stability of chromium ions at low concentration in aqueous and biological matrices stored in glass, polyethylene, and polycarbonate containers. They investigated storage at room temperature, +4 and −10°C. Borosilicate glass containers were shown to be best for storage of solutions up to six months. In addition, storage at low temperatures was found to minimize losses by adsorption and contamination due to leaching of impurities from container walls. Evaporation of solutions during storage, a severe problem with plastic containers, was also minimized at low storage temperatures. Polyethylene containers were found to be particularly poor for storage of aqueous solutions. However, serum and plasma samples could be satisfactorily stored in polyethylene containers if they were frozen at below −10°C.

3.4.3 Other Studies

Because the subject of loss of metals from solution during sampling and storage is such an important aspect affecting the analytical result, it is important to give some guidance for other metals. Thus Table 3.4, taken from Massee et al. (6), is presented so workers might refer to a study relevant to their applications.

Table 3.4 Survey on Sorption of Trace elements by Aqueous Solution from Various Materials[a]

Element	Matrix	pH Range	Material[b]	Ref.
Ag	Distilled water	5–11	B, M	13
	Distilled water	2–6	B, PE, PP	14
	Distilled water	7	B, PP, T, PS	15,16[c]
	Distilled water	1.5–4	B, PE, T, B + desicote	17
	Distilled water	7	B, PE, T, B + desicote	18
	Seawater		B	19
	Seawater	1.5;8	B, PE	20
Al	Synthetic and natural water	1.5–8	B, PE	25
	Natural water	6.7	PE	33
	0.5% NaCl	1.5–12	B	24
Au	0.5% NaCl	1.5–11	B	21
	Seawater		PE	52
Ba	Distilled water	1–7	B, PE	22[c]
	Hardwater	4–7	B, PP	30
	0.5% NaCl	1.5–11	B	21
Be	Distilled water	1.7	B, PE	22[c]
Ca	0.5% NaCl	1.5–12	B	24
Cd	Distilled water	2–6	B, PE, PP	14
	Distilled water	1–7	B, PE	21[c]
	Distilled water	1–10	B, PE	32[a]
	Distilled water	6–10	B, PE, PP, PVC	31
	Synthetic and natural water	1.5–8	B, PE	25
	River water	7.5–8	PE	23
	0.5% NaCl	1.5–11	B	21
Co	Synthetic and natural water	1.5–8	B, PE	25
	Natural water	6.7	PE	33
	0.5% NaCl	1.5–12	B	22
	Seawater		B	19
	Seawater	1.5;8	B, PE	20
Cr	Distilled water		B, PE	34[c]
	Synthetic and natural water	1.5–8	B, PE	25
	Natural water	6.7	PE	33
	Seawater		B	19
	Seawater		B, PE, PP	35[c]
Cs	Hardwater	4–7	B, PP	30
	Seawater	1.5;8	B, PE	20
	Seawater		B	19
Cu	Distilled water	7.5–8	PE	23
	Synthetic and natural water	1.5-8	B, PE	25
	0.5% NaCl	1.5–12	B	24
	Estuarine water		PE, T	36[c]

Table 3.4 (*Continued*)

Element	Matrix	pH Range	Material[b]	Ref.
Fe	Distilled water	0–13	B	38
	Distilled water	0–13	PE	40
	Synthetic and natural water	1.5–8	B, PE	25
	Natural water	6.7	PE	33
	0.5% NaCl	1.5–12	B	24
	Estuarine water		PE, T	36[c]
	Seawater	1.5;8	B, PE	20
Hg	Distilled water	0–14	PE	40
	Distilled water	0–14	B	41
	Distilled water	0–7	PE	42
	Distilled water		PE, PP	43
	Distilled water		B, PE	44
	Creek water		PE	45
	Natural water	2;7	B, PE, PVC	46
	Ice and river water		PE	47
	River water	5–8	PE	48
	River water	7.5–8	PE	23
	Seawater		PE	26
In	0.5% NaCl	1.5–11	B	21
	Seawater	1.5;8	B, PE	20
La	Hard water	4–7	B, PP	30
	Natural water	6.7	PE	33
Li	0.5% NaCl	1.5–11	B	21
Mg	0.5% NaCl	1.5–12	B	24
Mn	Distilled water	4–14	B	49
	Distilled water	1–7	B, PE	22[c]
	Synthetic and natural water	1.5–8	B, PE	25
	Natural water	6.7	PE	33
	0.5% NaCl	1.5–12	B	24
Ni	Distilled water	2–6	B, PE, PP	14
	Synthetic and natural water	1.5–8	B, PE	25
	0.5% NaCl	1.5–12	B	24
Pb	Distilled water	2–6	B, PE, PP	22[c]
	Distilled water	1–7	B, PE	32[c]
	Distilled water	1–7	B, PE	50
	HNO_3 and H_2O_2		B, PE	25
	Synthetic and natural water	1.5–8	B, PE	24
	0.5% NaCl	1.5–12	B	
Pd	0.5% NaCl	1.5–11	B	21
Pt	0.5% NaCl	1.5–11	B	21
Sb	0.5% NaCl	1.5–11	B	21
	Seawater		B	19
	Seawater	1.5;8	B, PE	20

Table 3.4 (*Continued*)

Element	Matrix	pH Range	Material[b]	Ref.
Se	Distilled water	0–7	B, PE	28
	Natural water	1.5–7.2	B, PE	29
	Seawater		B	19
Sn	0.5% NaCl	1.5–11	B	21
Sr	Synthetic and natural water	1.5–8	B, PE	25
	Hard water	4–7	B, PP	30
	0.5% NaCl	1.5–12	B	24
	Seawater	1.5;8	B, PE	20
Ti	0.5% NaCl	1.5–12	B	24
Tl	0.5% NaCl	1.5–11	B	21
U	Seawater	1.5;8	B, PE	20
V	0.5% NaCl	1.5–12	B	24
Zn	Distilled water	2–6	B, PE, PP	14
	Distilled water	1–7	B, PE	22[c]
	Synthetic and natural water	1.5–8	T	25
	0.5% NaCl	1.5–12	B	24
	Seawater		B	19
	Seawater	1.5;8	B, PE	20
	Seawater		PE	26
	Seawater	5.0n8.6	B, PE	27

[a] Reprinted with permission from Massee et al. (6).

[b] B = borosilicate glass, PE = polyethylene, T = polyfluoroethylene, M = metal surfaces, PS = polystyrene, PVC = polyvinylchloride.

[c] Complexing agents were also considered.

3.4.4 Container Composition

3.4.4.1 Glass

Soda–lime soft glasses are not suitable for storage of trace metal solutions because ion exchange of metals with ions on the glass surface occurs readily. Borosilicate glass is used by many workers for storage of trace metal solutions. Hetherington and Bell (8) demonstrated that vitreous silica containers were best for trace metal solution storage. However, these are fragile and very expensive.

3.4.4.2 Plastic

Most workers now choose plastic containers for storage of trace metal solutions. It is important to realize, however, that plastics have pores through which evaporation can occur. In this regard Teflon containers show the least problem. At first glance, Teflon containers might seem to be the ultimate for trace metal solution storage. However, Zief and Horvath (3) report the presence of fine particulates embedded in Teflon container walls. These particulates result in contamination from manganese, zinc, aluminum, iron, and copper.

Other plastics may contain metal-bearing particulate depending on the production method. For example, polyethylene has been shown to be contaminated with chromium, aluminum, antimony, cobalt, manganese, copper, aluminum, and vanadium during the molding process (9). Plastics may also have metal contamination resulting from additives used during production for polymer stabilization. Erickson (10) has presented a very useful table summarizing the trace metal content of materials used in storage of solutions. Table 3.5 compiled from Robertson (11) is very useful.

For most purposes I recommend high-density linear polyethylene for storage of solutions for trace metal analysis. However, in the case of mercury, borosilicite glass containers may be superior. Solutions must be acidified. Storage at reduced temperature (e.g., +4°C) is also recommended.

Some care is also essential in choosing sample containers for storage of solid samples. It is common to employ plastic bottles or plastic-lined bags. It is important to realize that plastic bags may contain a zinc-bearing powder to make them easy to open. I suggest that clean, wide-mouth, high-density linear polyethylene bottles be used for solids.

3.4.4.3 Cleaning Container Surfaces

It is important to clean container surfaces before using for storage of trace metal solutions. Again, there is much variation in opinion among different workers on the recommended approach. I recommend the following for cleaning:

1. *Teflon.* Clean with a detergent followed by aqua regia (3:1 $HCl:HNO_3$) treatment at about 50°C for several hours. To remove organic matter, Teflon can be soaked in 1% $KMnO_4$ followed by HCl rinsings.
2. *Linear high density polyethylene.* Clean with detergent followed by a 1:1 HNO_3 soaking for two to three days.
3. *Glass.* Do *not* use chromic acid solutions. Treat with aqua regia at elevated temperature for several hours.

Water rinses should be done after the above treatments. It is important to use high-purity water for this purpose.

Table 3.5 Trace Metal Content (ppb) of Some Materials

	Metals				
Materials	Sc or Se	Cr	Mn	Fe	Co
Teflon PVC	<0.004	<30		35	1.7
(i) Structural	4.5	2		2.7×10^5	45
(ii) Tygon					
(a)		6000	2000	5×10^4	
(b)		< 40		3000	2.6
Nylon (structural)					1.4×10^6
Plexiglass	<0.002	<10		<140	<0.05
Polyethylene					
High pressure		15–300	<10	600–2100	5
Low pressure		180–1500			10–370
Process unknown	.008–.36	19–76	7.2–1900	1.1×10^4	0.07–0.31
Polyethylene hose	11	254		7.4	140
Surgical (rubber) tubing					
(c)	<8				<30
(d)	185	4.2×10^5		<100	7500
(e)					
Neoprene rubber	3090				2300
Quartz tubing					
(f)	0.03	6.5		395	0.44
(g)	0.18	230			0.64
(h)	0.10	225			1.7
Borosilicate glass	106			2.8×10^5	81
Vycor glass					
Millipore (HA) filters (i)	0.8	1.8×10^4		330	13
Nuclepore filters	3	2000	130	2.8×10^4	25
Kimwipe	–	500		1000	24

[a] Reprinted from (11) p. 290 by courtesy of Marcel Dekker Inc.

3.5 LABORATORY REQUIREMENTS

3.5.1 Contamination

The prevention of contamination during trace metal analysis is a constant problem. Thiers (12) identifies three cases relating to contamination.

1. Positive contamination results from addition of analyte contaminants to the sample.

Commonly used in Trace Metal Sampling and Storage[a]

Metals						
Ni	Cu	Zn	Ag	Sb	Pb	Hg
	22	9.3	<0.3	0.4		
	630	7100	<5	2700		
2×10^5	1×10^4	5000	600		2×10^5	
	6.5	6.2×10^4	<10	<5		66
	<10	<10	<0.03	<0.01		
	4	90	20	<5	200	
		300	<10	<10		
	6.6–17	25–28	<0.1	0.18–0.83		
		55	<300	9000		
	<6	3.08×10^6	1240	<100		
		4.1×10^7	<700	360		
		5.35×10^6				
		1.84×10^7	<1000	290		
	2.0	1.5	0.05	0.05		
	0.05	21	<0.1	43		
	0.16	20	<0.1	58		
		730	<0.01	2900		
				1.1×10^6		
		2400	<0.5	39		<0.015
	1800	2300		<20		
		4.9×10^4	<0.8	16		

2. Negative contamination occurs when losses of analyte are experienced.
3. Pseudocontamination is an error occurring because of the presence of a substance other than analyte.

There are analytical personnel who maintain that it is necessary to use a specially designed "clean" laboratory for trace metal analysis. I agree that for best detection limits and when sub-ppm analyses are done that a clean laboratory greatly expedites analytical work. However, there are numerous analysts who are unable to make use of such a laboratory and must make do with a facility that is less than ideal for trace metal analysis. I am in this category. I experienced contamination of samples during trace lead analysis that necessitated discarding all the results obtained over a three-month period.

The amount of analyte detected in the analytical blank is often the factor that establishes the lowest level of the analyte that can determined by a given method. Murphy (1) defines the analytical blank as "the contamination by the element or compound being determined from all sources external to the sample." This definition excludes the instrument blank or noise level. The following, section discusses the laboratory problems affecting the level of analyte in the blank as defined above.

3.5.2 "Clean" Laboratory

My laboratory is situated in downtown Toronto (a light industrial city of 2.5 million). Despite the use of conventional filters on incoming air, it is impossible to obtain blank values for many metals (e.g., zinc, lead, copper) that are below 1 μg/L. This makes furnace atomic absorption work difficult at best. It can be shown that this problem is caused by contamination of the samples by air particulates.

At the National Bureau of Standards (NBS) at Gaithersburg, MD a class 100 laboratory has been built that is used for trace element analysis (1). Class 100 means that there are less than 100 particles larger than 0.5 μm per Cubic foot of air. This requires that special filters be employed on the incoming air. These filters have an efficiency of 99.97% for 0.3-μm particles (1). Air flows through prefilters, then through the high efficiency filters, and then blows down from the ceiling. The laboratory is kept at a positive pressure. Figure 3.2 shows a cutaway of the clean laboratory facility used at NBS. This laboratory also has class 100 fume hoods for work on the dissolution of samples. Murphy (1) demonstrated that contamination due to lead was reduced by a factor of 1000 in the class 100 laboratory over that obtained in a conventional one.

3.5.3 Conventional Laboratories

A number of things can be done to conventional laboratories to reduce the contamination problem. Evaporation chambers as shown in Fig. 3.3 can be used to reduce airborne contamination. These have filtered air and are used to isolate samples being heated, from the laboratory air. In my laboratory decompositions are usually carried out in closed Teflon reaction vessels. This minimizes contamination during the decomposition stage. However, if evaporations are necessary, contamination from the air can still be a problem.

In a conventional laboratory trace metal contamination can come from exposed metal surfaces. Thus exposed metal in fume hoods and throughout the laboratory should be minimized. Any remaining exposed metal can be painted

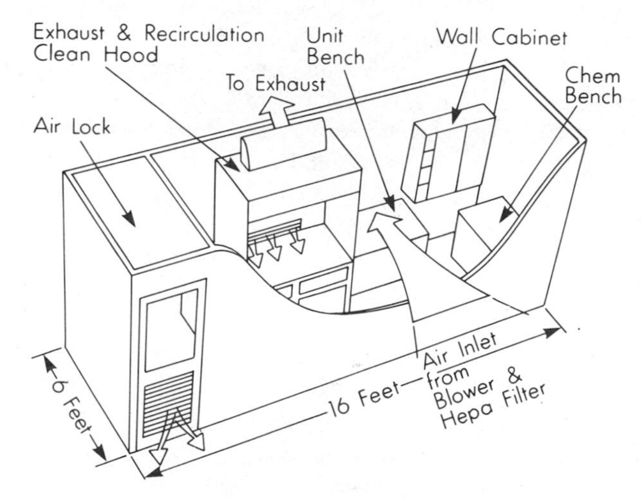

Figure 3.2 Schematic diagram of a positive pressure filtered air laboratory (1).

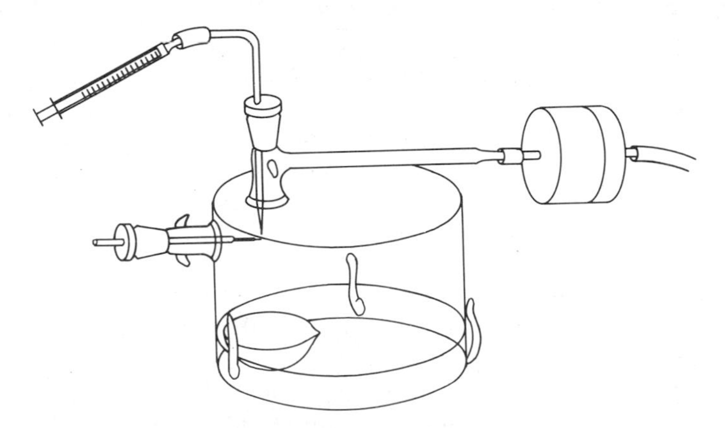

Figure 3.3 Evaporation chamber.

with an epoxy-based paint. Another useful precaution that can be taken is to seal all windows and to filter air coming in from hallways and so on. In my laboratory, despite all these precautions plus a slight positive laboratory pressure, contamination from some metals still cannot be reduced to acceptable levels.

3.5.4 Laboratory Layout

A trace metal analytical laboratory should have the following components: a wet chemical laboratory area, balance room, chemical/sample storage room, and instrument room. The instrument and balance rooms should be isolated from the air of the wet chemical laboratory area.

The wet chemical laboratory is used for sample preparation. This includes both dry and wet ashing of samples. The basic requirements are fume hoods, benches with electrical outlets, sinks, natural gas, compressed air, hot and cold water, and, if possible, distilled water services. The benches should have chemical-resistant tops. A variety of drawers and cupboards for equipment should be available. Lighting must be excellent.

Basic equipment should include hot plates with aluminum or graphite digestion blocks, hot water bath, drying oven, ashing and muffle furnaces, sand bath, water purification system, closed Teflon decomposition vessels and pressure cooker, vacuum pumps, refrigerator (explosion proof), and freezer. A vacuum cleaner should also be available at all times.

In addition, the wet chemical laboratory should contain a minimum of exposed metal. This means that plumbing as far as possible should be plastic with specially protected metals used when necessary. Metal abounds in most hardware used to suspend roofs, and hang lighting, in most fasteners, and in lighting fixtures themselves. It is therefore necessary to obtain nonmetallic substitutes or to treat the metallic ones with special acid resistant sealers and protective coatings.

To minimize problems caused by dust, laboratory fixtures should be readily cleanable by vacuum cleaner. Dust-catching surfaces should be kept to a minimum and must be easily accessible. Windows, if present, should be sealed. Filtered air should be supplied to the room to replace air used by fume hoods.

Fume hoods are used for evaporation of acids and other processes involving hazardous fumes. Thus they must have a satisfactory air flow (face velocities above 0.5 m/s) and the air flow should be checked periodically. The interior surfaces and fixtures inside the fume hood must not contain exposed metal. This means that special attention is needed to assure that metal fasteners, light fixtures, and so on, are treated with acid resistant coatings. Lining materials in fume hoods can be made from chemical resistant painted or asbestos cement board, PVC, or fiberglass-reinforced plastic. In the case of the last material, exposure to hydrofluoric acid fumes could be a problem. The fume hood sash should be easily adjustable and be constructed with safety glass.

A special fume hood for perchloric acid digestions is necessary. This is to reduce the potential hazard of fire and explosion during use of perchloric acid. These hoods are constructed of type 316 stainless steel and would be expected to have good resistance to chemicals; however, I found that my hoods rapidly became tarnished with rust and the metallic dusts thus produced were a contamination problem. Subsequently all surfaces were treated with chemical-resistant sealers. Perchloric acid fume hoods have special wash-down systems that allow the duct work and the inside of the hood to be completely flushed with water periodically. This minimizes build-up of perchloric acid residues and the fire and explosion hazard thus produced.

The instrument room should ideally have controlled temperature and humidity. The air in this room must be free from acid and other chemical fumes. In my laboratory acid fumes in the instrument room air have seriously degraded optical and electronic systems in the instruments. This has resulted in very costly repairs and replacement. Adjustable-height, efficient fume hood canopies must be installed over instrument atomizers (flames, plasmas, furnaces, etc.). The flow up the canopies must be checked periodically. The benches on which the instruments sit should be heavy and be suspended to minimize vibration.

3.5.5 Major Instrumentation in a Trace Metal Analysis Laboratory

Several major items of equipment are essential in a trace metal analysis laboratory. These are semimicro balance, expanded scale pH meter, ultraviolet–visible spectrophotometer, chart recorder, atomic absorption spectrometer, and/or plasma emission spectrometer (and related equipment and accessories).

It is an advantage to have access to both atomic absorption and plasma emission equipment. If this is not possible, however, atomic absorption spectrometry is the best choice of a single technique at present because, for very low levels of metals furnance atomic absorption is essential. Plasma emission spectrometry detection limits are not adequate for determinations in the sub-ppb range.

3.6 SAFETY IN THE TRACE METAL ANALYSIS LABORATORY

This topic is basically beyond the scope of this book. Nevertheless a few comments on safety should be included.

Fortunately the subject of laboratory safety is receiving much greater attention now than it was five years ago. This is largely the result of the scrutinization by government regulatory agencies. Much more is known now about laboratory safety and a satisfactory range of safety equipment presently exists for most purposes.

The trace metal analysis laboratory has numerous potential hazards and workers should acquaint themselves with safety problems. Each person should know the location of safety equipment and be acquainted with its operation. The following describes some of the more important aspects of safety, but the outline is by no means complete. Breathing apparatus for both dusts and hazardous vapors must be available. Plastic gloves and aprons to be used in conjunction with safety glasses and masks or visors for protection from exposure to hazardous liquids or solids should be included. In case of accidental exposure to chemicals, eye wash stations and showers must be present. Recepticles for different types of waste (solid chemicals, liquids, glass, etc.) should be properly labeled. There should be posters throughout the laboratory giving safety information. Data sheets on hazardous chemicals should be posted in the appropriate locations. Compressed gas cylinders must be securely fastened to benches and walls. When not in use the pressure regulators should be removed from cylinders and the cylinders capped. Flash-back arrestors should be employed with flammable gases in compressed gas cylinders.

It is common to see chemicals in a chemical storage area that are more than ten years old. For safety reasons old chemicals should be checked and hazardous onces disposed of periodically. Spilled chemicals can sometimes present a serious hazard. Thus neutralizing compounds and special instructions for cleanup should be available.

Fire precautions are an important aspect of safety in a trace metal analysis laboratory. Therefore a protocol for leaving the laboratory in case of fire should be established. Fire safety equipment should include CO_2 and dry chemical fire extinguishers and a fire blanket. To minimize the fire hazard safety cans must be used for the storage and dispensing of flammable liquids and the liquids stored in approved cabinets. Large-scale storage of flammable solvents should not be permitted in the laboratory.

REFERENCES

1. T.J. Murphy (P.D. Lafleur, Ed.), *Accuracy in Trace Analysis: Sampling, Sample Handling and Analysis,* National Bureau of Standards Special Publ. 422, 1974, p. 117.
2. E.C. Kuehner, R. Alvarez, P.J. Paulsen, and T.J. Murphy, *Anal. Chem.* **44,** 2051, (1972).
3. M. Zief and J. Horvath (P.D. Lafleur, Ed.), *Accuracy in Trace Analysis: Sampling, Sample Handling and Analysis,* National Bureau of Standards Special Publ. 424, 1974, p. 363.
4. J.A. Dean and T.C. Rains (R. Mavrodineau, Ed.), *Procedures Used at the National Bureau of Standards to Determine Selected Trace Elements in Biological and Botanical Materials,* NBS Special Publ. 492, 1977.
5. N.R. McQuaker, P.D. Kluckner, and Gok N. Chang *Anal. Chem.* **51,** 888 (1979).
6. R. Massee, F.J.M.J. Maessen, and J.J.M DeGoeij, *Anal. Chim. Acta* **127,** 181 (1981).
7. V.P. Arand and D.M. Ducharme (P.D. Lafleur, Ed.), *Accuracy in Trace Analysis: Sampling, Sample Handling and Analysis,* National Bureau of Standards Special Publ. 422, 1974, p. 611.
8. G. Hetherington and L.W. Bell, (M. Zief and R. Speights, Eds.), *Ultrapurity Methods and Techniques.* Dekker, New York, 1976, p. 353–400.
9. R.W. Karin, J.A. Buono, and J.L. Fasching, *Anal. Chem.* **47,** 2296 (1975).
10. P. Erickson, *Marine Trace Metals Sampling and Storage,* National Research Council of Canada, No. 16472, Report (1) (1977).
11. D.E. Robertson, (M. Zeif and R. Speights, Eds.), *Contamination Problems in Trace Element Analysis and Ultra Purification, Ultrapurity Methods* and *Techniques.* Dekker, New York, 1972, pp. 207–253.

12. R.E. Thiers *Methods Biochem. Anal.* **5,** 273 (1975).
13. J.W. Hensley, A.O. Long, and J.E. Willard, *Ind. Eng. Chem.* **41,** 1415 (1949).
14. A.W. Streumpler, *Anal. Chem.* **45,** 2251 (1973).
15. F.K. West, P.W. West, and F.A. Iddings, *Anal. Chem.* **38,** 1566. (1966).
16. F.K. West, P. W. West, and F.A. Iddings, *Anal. Chim. Acta* **37,** 112 (1967).
17. W. Dyck, *Anal. Chem.* **40,** 454 (1968).
18. R.A. Durst and B.T. Duhart, *Anal. Chem.* **42,** 1002 (1970).
19. D.F. Schutz and K.K. Turekian, *Geochim. Cosmochim. Acta.* **29,** 259 (1965).
20. D.E. Robertson, *Anal. Chim Acta.* **42,** 533 (1968).
21. A.E. Smith, *Analyst* **98,** 209 (1973).
22. A.D. Shendrikar, V. Dharmarajan, H. Walker-Merrick, and P.W. West, *Anal. Chim. Acta* **84,** 409 (1979).
23. J. Gardiner, *Water Res.* **8,** 157 (1974).
24. A.E. Smith, *Analyst,* **98,** 65 (1973).
25. K.S. Subramanian, C.L. Chakrabarti, J.E. Sueiras, and I.S. Maines, *Anal. Chem.* **50,** 444 (1978).
26. Y. Dokiya, H. Ashikawa, S. Yamazaki, and K. Fuwa, *Spectrosc. Lett.* **7,** 551 (1974).
27. Y.M. Petrov, *Zh. Anal.* Khim. **29,** 686 (1974).
28. A.D. Shendrikar and P.W. West, *Anal. Chim. Acta* **74,** 189 (1975).
29. V. Cheam and H. Agemian, *Anal. Chim. Acta* **113,** 237 (1980).
30. G.G. Eichholz, A.E. Nagel, and R.B. Hughes, *Anal. Chem.* **37,** 863 (1965).
31. W.G. King, J.M. Rodriguez, and C.M. Way, *Anal. Chem.* **46,** 771 (1974).
32. W.C. Hoyle and A. Atkinson, *Appl. Spectrosc.* **33,** 37 (1979).
33. P. Benes and E. Steines, *Water Res.* **9,** 741 (1975).
34. A.D. Shendrikar and P.W. West, *Anal. Chim. Acta* **74,** 91 (1974).
35. T.R. Gilbert and A.M. Clay, *Anal. Chim. Acta* **67,** 289 (1973).
36. R.E. Pellenbarg and T.M. Church, *Anal. Chim. Acta* **97,** 81 (1978).
37. F. Ichikawa and T. Sato, *Radiochim. Acta* **12,** 89 (1969).
38. P. Benes, J. Smetana, and V. Majer, *Collect. Czech. Chem. Commun.* **33,** 3410 (1968).
39. P. Benes and J. Smetana, *Collect. Czech. Chem. Commun.* **34,** 1360 (1969).
40. P. Benes and I. Rajman, *Collect. Czech. Chem. Commun.* **34,** 1375 (1969).
41. P. Benes, *Collect. Czech. Chem. Commun.* **35,** 1349 (1970).
42. J.H. Lo and C.M. Wai, *Anal. Chem.* **47,** 1869 (1975).
43. R.W. Heiden and D.A. Aiken, *Anal. Chem.* **41,** 151 (1979).
44. C. Feldman, *Anal. Chem.* **46,** 99 (1974).
45. R.V. Coyne and J.A. Collins, *Anal. Chem.* **44,** 1093 (1972).
46. R.M. Rosain and C.M. Wai, *Anal. Chim. Acta.* **65,** 279 (1973).
47. H.V. Weiss, W.H. Shipman, and M.A. Gutman, *Anal. Chim. Acta.* **81,** 211 (1976).
48. K.I. Mahan and S.E. Mahan, *Anal. Chem.* **49,** 662 (1977).
49. P. Benes and A. Garbe, *Radiochim. Acta* **5,** 99 (1966).
50. H.J. Issaq and W.L. Zielinski, *Anal. Chem.* **46,** 1328 (1974).
51. T.M. Florence and G.E. Bailey, *Critical, Reviews Anal. Chem.* **219,** (August 1980).
52. R.W. Karin, J.A. Buone, and J.L. Fasching, *Anal. Chem.* **47,** 2296 (1975).

CHAPTER

4

SAMPLE DECOMPOSITION

4.1 INTRODUCTION

Most determinative methods currently used in trace metal analysis require that the sample be presented in liquid form. Thus it is important to have methods that are appropriate for converting a largely organic matrix into a solution suitable for analysis.

Biological samples consist of many different types of organic and inorganic compounds, which often vary greatly in their behavior toward the reagents used for decomposition. For example, fats are much more difficult to decompose than starches and carbohydrates. Also, the trace elements are present in different forms in samples, such as inorganic salts (food additives) or as a part of an organic molecule (as iron is in hemoglobin). Because of this variation it is important to realize that no one method will be suitable for decomposition of biological samples.

Basically there are three approaches to the decomposition of organic samples: (1) wet ashing, (2) dry ashing, and (3) fusion. Wet ashing involves treatment of the organic sample with a mixture of acids including an oxidizing agent, which may be either an acid or a salt. Dry ashing is the treatment of the sample (1) at elevated temperature (usually above 450°C) in air to remove the organic matter as oxides of carbon or (2) at 50–100°C under reduced pressure in an oxygen plasma discharge. In fusions the sample is mixed with a reagent capable of fluxing the sample and this approach is generally reserved for samples high in inorganic constituents.

The potential advantages and disadvantages of dry and wet ashing can be summarized as follows:

Wet ashing involves more operator time, but generally takes less overall time. A small sample must generally be used. Because a relatively large volume of reagents must be employed, contamination due to reagent impurities can be a problem.

Dry ashing involves a longer overall time, but operator involvement is minimal. Relatively large samples can be treated if high-temperature ashing is employed. (Plasma ashing at low temperature is applicable to only very small samples.) In high-temperature ashing, because containers must be left open to the atmosphere, contamination caused by airborne substances and by components of the ashing apparatus may be a problem. Losses may also occur because of high temperature incorporation of analyte into the walls of the container.

In all approaches, volatility loss of analytes may be a problem. The analyte may also be lost to nondecomposible residues.

4.1.1 Volatility, Residue, and Container Wall Losses

4.1.1.1 Volatility

There is considerable confusion in the literature concerning loss of elements caused by the presence and the formation of volatile compounds during sample ashing. Hence in carrying out a volatility loss study it is essential to be sure that measured losses actually result from volatilization and not the incorporation of analyte into the container wall or an insoluble residue. I believe that such problems have invalidated a number of published studies.

Mercury and many of its compounds are volatile at relatively low temperatures. Thus when mercury is to be determined it is important to take special precautions during sample decomposition. Therefore it is essential to ensure oxidizing conditions throughout the decomposition procedure.

The covalent-hydride-forming elements arsenic, antimony, selenium, tin, and tellurium are readily lost as the hydride at relatively low temperatures. Melting and boiling points of these compounds are given in Table 4.1. Thus as with mercury, oxidizing conditions should be maintained throughout the decomposition step.

The presence or formation of metal halides can be a problem in sample decomposition. Many metal halides are volatile at relatively low temperatures. Table 4.2 lists some of the more commonly determined metals with the melting and/or boiling points of the metal halides and other relatively high-volatility compounds.

It is evident from Table 4.2 that many metals could be lost at the temperatures

Table 4.1 Melting and Boiling Points of Covalent Hydrides

Hydrides	Melting Point (°C)	Boiling Point (°C)
ASH_3	−116	−55
SbH_3	−88	−17
H_2Se	−60	—
SnH_4	−150	−52
H_2Te	−49	−2

Table 4.2 Melting and Boiling Points of Metal Compounds

Element	Compound	Melting Point (°C)	Boiling Point (°C)
Ag	Ag	960	2210
	AgBr	432	—
	AgCl	455	1550
	AgF	690	—
	AgI	558	1506
Al	Al	660	2467
	$AlBr_3$	98	264
	$AlCl_3$	190	—
	AlF_3	1290	—
	AlI_3	190	360
	Al_2O_3	2045	2985
	Al_2S_3	1100	—
As	$AsBr_3$	33	220
	$AsCl_3$	−85	63
	AsF_3	−9	−63
	AsF_5	−80	−53
	AsI_3	146	403
	AsI_5	76	—
	As_2O_3	315	—
	As_2S_3	300	707
B	B	2300	2550
	BBr_3	−46	92
	BCl_3	−147	13
	BF_3	−127	−100
	BI_3	50	210
	B_2O_3	460	1860
	B_2S_3	310	—
Be	Be	1278	2980
	$BeBr_2$	490	520
	$BeCl_2$	405	510
	BeI_2	510	590
	BeO	2530	—
Bi	Bi	271	1560
	$BiBr_3$	218	453
	$BiCl_3$	230	448
	BiF_3	727	—
	BiI_3	408	508
	Bi_2O_3	860	—

Table 4.2 (*Continued*)

Element	Compound	Melting Point (°C)	Boiling Point (°C)
Cd	Cd	321	766
	$CdBr_2$	568	862
	$CdCl_2$	568	960
	CdF_2	1100	1758
	CdI_2	387	796
Co	Co	1495	2900
	$CoBr_2$	678	—
	$CoCl_2$	724	1049
	CoF_2	1200	1400
	CoI_2	515	—
	CoO	1935	—
Cr	Cr	1890	2482
	$CrBr_2$	840	—
	$CrCl_2$	824	—
	$CrCl_3$	1150	—
	CrO_2Cl_2	−97	117
	Cr_2O_3	2400	—
Cu	Cu	1083	2596
	CuBr	492	1345
	$CuBr_2$	498	—
	CuCl	430	1490
	$CuCl_2$	620	—
	CuF	909	—
	CuI	605	1290
	Cu_2O	1235	—
	CuO	1326	—
	Cu_2S	1100	—
Fe	Fe	1535	3000
	$FeCl_2$	670	—
	$FeCl_3$	306	—
	FeO	1420	—
	Fe_2O_3	1565	—
	FeS_2	1171	—
	FeS	1190	—
Hg	Hg	−38.9	356
	$HgBr_2$	236	322
	$HgCl_2$	276	302
	Hg_2F_2	570	—
	HgI_2	259	354
	HgS	584	—

Table 4.2 (*Continued*)

Element	Compound	Melting Point (°C)	Boiling Point (°C)
Mo	Mo	2610	—
	$MoCl_5$	194	268
	MoF_6	18	35
	MoO_3	795	—
Mn	Mn	1244	2096
	$MnCl_2$	650	1190
	MnF_2	856	—
	MnI_2	639	—
	MnO_2	535	—
	Mn_3O_4	1704	—
Ni	Ni	1453	2732
	$NiBr_2$	63	—
	$Ni(CO)_4$	−25	43
	$NiCl_2$	1001	—
	NiI_2	797	—
	NiO	1990	—
	NiS	797	—
Pb	Pb	327	—
	$PbBr_2$	373	—
	$PbCl_2$	500	—
	PbF_2	855	1290
	PbI_2	402	955
	PbO_2	888	—
	PbS	1114	—
Sb	Sb	631	1380
	$SbBr_3$	97	280
	$SbCl_3$	73	282
	$SbCl_5$	79	—
	SbF_3	292	—
	SbF_5	7	150
	SbI_3	170	400
	SbI_5	79	400
	Sb_2O_3	656	1550
	Sb_2S_3	550	1150
Se	Se	217	285
	Se_2Cl_2	−85	—
	H_2Se	−60	—
	Se_2I_2	68	—
	SeO_2	320	—
	SeO_3	118	—
	$SeOCl_2$	9	176

Table 4.2 (*Continued*)

Element	Compound	Melting Point (°C)	Boiling Point (°C)
Si	Si	1410	2360
	$SiBr_4$		154
	$SiCl_4$	−70	58
	SiF_4	−90	—
	SiH_4	−185	—
	SiI_4	121	288
	SiO_2	1713	2300
Sn	Sn	232	2260
	$SnBr_2$	216	620
	$SnCl_2$	246	652
	$SnCl_4$	−33	114
	$SnBr_4$	31	202
	SnI_2	320	717
	SnI_4	145	365
	SnO_2	1127	—
	SnS	882	1230
Te	Te	452	1390
	$TeBr_4$	210	339
	$TeBr_4$	381	327
	$TeCl_2$	210	—
	$TeCl_4$	224	380
	TeF_6	−36	36
	TeI_4	280	—
	TeO_2	733	1245
Tl	Tl	304	1458
	TlBr	480	815
	TlCl	430	720
	TlF	327	655
	TlI	440	823
	Tl_2O	300	1080
	Tl_2S	450	—
V	V	1890	—
	VCl_4	−28	148
	VI_2	760	—
	V_2O_5	690	—
Zn	Zn	420	907
	$ZnBr_2$	394	650
	$ZnCl_2$	283	732
	ZnF_2	872	—
	ZnI_2	446	—
	ZnO	1975	—
	ZnS	1850	—

commonly used for dry ashing (i.e., 450–850°). It is also important to realize that many biological and environmental samples contain these compounds.

4.1.1.2 Loss to Residues and Container Walls

Analyte can be incorporated into container walls during ashing at high temperature. The mechanism is not clear, but the reaction between a metal oxide and silicate can form glasses that are not easily attacked by mineral acids. The presence of sodium chloride, a common constituent of organic samples, is believed to add to the severity of this problem.

Many organic materials contain appreciable quantities of silicon. During ashing procedures silicon often remains in a residue as silicon dioxide or a silicate. Such residues can readily trap analyte metals.

4.1.2 Ashing

4.1.2.1 Dry Ashing

Although dry ashing generally requires a long period of time, it is attractive because operator involvement is low, large sample sizes can be employed, and unless a reagent addition is necessary, contamination due to reagents is low.

Oxidizing conditions should be maintained as much as possible throughout the ashing periods. The temperature employed depends upon the presence or absence of volatile compounds and the identity of the metals under consideration. The addition of aluminum or magnesium nitrate is sometimes recommended. This is to aid in maintaining oxidizing conditions, speed the decomposition, and minimize interaction of analytes with the container wall.

Contamination during dry ashing can be a serious problem. Because the sample must be open to the air over an extended period, dust particles from the air can be easily trapped by the sample. Also scale from furnace walls and furnace heating elements can be a serious source of contamination. Dry ashing should therefore be done in as clean an area as possible. Blanks must always be run through the dry ashing procedure.

To give an idea of the serious nature of contamination of samples during dry ashing the following example from my laboratory is cited. Blood samples were to be analyzed for chromium, nickel, and cobalt. Expected values were in the range of 0 to 20 ppb. Because none of the above metals is very volatile, dry ashing was chosen. One gram samples of blood were placed in carefully cleaned silica crucibles, placed in an oven, dried, and then ashed at 550°C. Three blank crucibles were run with each batch of 15 samples. Table 4.3 gives the values obtained for the blanks with one set of samples.

Table 4.3 shows that nickel and chromium blank values (probably from the

Table 4.3 Contamination of Blood Samples (ppb)[a]

Element	B_1	B_2	B_3
Ni	6	1	2
Cr	2	8	1
Co	< 1	< 1	< 1

[a] Because of the procedure used these numbers also represent absolute values in nanograms.

furnace) are variable and sometimes relatively high (although the contamination was less than 10 ng absolute—a very tiny amount). However, because of the contamination it was impossible to run blood samples in this way. Thus a wet ashing scheme had to be adopted.

Organic matter commonly contains appreciate levels of substances that form an acid insoluble residue during dry ashing. Most common among these is silicon. The residue thus obtained can readily trap trace elements so that they will not be recovered during the subsequent acid leach step. It is crucially important not to discard residues from dry ashing until it is certain that no trace elements have been trapped. To put difficulty soluble residues into solution, hydrofluoric acid treatment or fusion with potassium persulfate or lithium metaborate is commonly used.

Organic materials vary as to the temperature required for complete ashing. Liquids such as petroleum and food oils generally need relatively high temperatures (i.e., over 700°C). Flour and products made from flour can often be ashed between 525 and 550°C. A good average temperature range for ashing organic matter would be 550–600°C. Ashing times also depend on the nature of the material and commonly range from 2 to 24 h.

4.1.2.2 Wet Ashing

A variety of acids and acid mixtures have been proposed for the decomposition of organic samples. In most instances it is essential to employ an oxidizing agent (may be an acid or other constituent mixed with the acid) to obtain complete decomposition. When the loss of an element by volatility is a possibility, the decomposition should be done under reflux or in a closed container.

Problems with losses of mercury during organic sample decomposition led to using a refluxing system to retain the mercury. It is obvious that the mercury can be lost at room temperature as the metal. Thus prevention of the mercury loss would appear to involve minimizing the presence of Hg^0 during decomposition. This is indeed true. However, Hg^{2+} salts such as $HgCl_2$ can also be lost at relatively low temperatures. Thus for the most accurate work many workers advise that samples for determination of mercury should be decomposed in a refluxing apparatus. Typical equipment consists of a round bottom flask joined to a refluxing condenser by a reservoir with a two-way stopcock. The reflux condenser is in turn connected to a splash head. All connections are made with ground glass joints. The round bottom flask is fitted with a thermometer. Decompositions when covalent hydride-forming elements are to be determined can also be done in such an apparatus.

When problems due to loss of volatile constituents are anticipated it is possible to use closed decomposition vessels. When liquids are placed into such vessels and heated a pressure builds up. It is possible to purchase commercially a variety of pressure decomposition vessels commonly called bombs. These usually consist of a Teflon insert or inner vessel contained in a metal outer vessel. The apparatus is then capable of being sealed. The metal casing prevents loss of sample through rupture of the Teflon reaction vessel.

Because of the expense involved with having individual metal/teflon bombs for each determination, on the advice of my colleague Dr. A. Brzezinska I use a household pressure cooker that contains seven heavy-walled Teflon closed-vessel decomposition containers. In this way the pressure built up inside the Teflon vessels can be compensated by pressure inside the pressure cooker.

4.1.3 Composition of Decomposition Vessels

It is commonly necessary to heat oxidizing acid mixtures to fumes of a higher boiling point acid such as sulfuric acid. For this purpose platinum evaporation dishes were usually employed until the 1970s. The expense of these dishes, coupled with the more recent availability of suitable plastics (particularly Teflon), has greatly decreased their use. At this time borosilicate glass and pure silicon dioxide glass beakers or Teflon dishes are commonly used for this purpose. Teflon has proven in my laboratory to be suitable even for use with perchloric acid. Size of the container is dictated by sample size, but a 100–250 mL vessel is very often employed.

4.1.4 Contamination From Acids

Reagent grade acids are available at reasonable cost from most major chemical supply houses. Metal impurities in these acids are at an acceptable level for many analyses. However, more and more researchers, particularly in the clinical and biological fields, are finding they are becoming limited by the blank values obtained from reagent grade chemicals. Therefore several chemical companies now provide a grade, particularly of the common acids, that is of higher purity and that often fulfils this need. Table 4.4 shows the lot analysis for selected metals reported in the 1982 J. T. Baker Canadian Laboratory Supplies Catalogue for nitric acid. The grades reported are Baker Analysed Reagent (meets ACS specifications), Baker Instra-Analysed Reagent (special grade for trace element analysis), and Ultrex Reagent (ultrahigh purity).

Table 4.4 Typical Lot Analyses for Selected Trace Metals[a]

Elements	Analysed	Instra-Analysed	Ultrex
As	5	5(As + Sb)	< 1
Cd	—	5	< 1
Cr	100	5	0.3
Co	—	5	< 0.1
Cu	50	5	0.07
Fe	200	20	0.5
Pb	—	5	0.5
Mn	—	5	0.05
Ni	50	5	0.2
Zn	—	10	< 1
Cost	$19.30 (2.5 L)	$19.35 (500 mL)	$55.10 (500 mL)

[1]Reported in J.T. Baker Catalogue 1982. Values are ppb in HNO_3.

From this it is obvious that analysts must carefully determine their reagent purity requirements before ordering chemicals for routine analysis of samples.

As pointed out earlier, another possibility exists if large volumes of very high purity and expensive reagents are required. Special subboiling stills can be constructed and used to prepare such reagents from reagent grade acid.

4.1.5 Decomposition Properties of Acids

4.1.5.1 Nitric Acid

Concentrated nitric acid contains 65–69% HNO_3. Concentrations above 69% are called fuming nitric acid. Concentrated nitric acid is employed for many purposes. However, nitric acid by itself is seldom adequate to decompose organic samples. A few substances such as urine (boiling nitric acid) and liver and serum (repeated treatments with fuming nitric acid) can be handled. Instead, nitric acid is commonly used with perchloric acid for oxidizing organic matter in most samples.

*4.1.5.2 Perchloric Acids**

Perchloric acid when hot and concentrated is an extremely strong oxidizing agent (perchloric acid must be both hot and concentrated for this purpose). However, very serious explosions can result if perchloric acid is used alone to oxidize

*WARNING: Anyone planning to work with $HClO_4$ should obtain and read data sheets (available from BDH Chemicals Ltd.) on precautions to be taken for safe work.

organic matter. Perchloric acid is available as the concentrated reagent as 70–72% $HClO_4$ (anything appreciably above this level, particularly approaching 100%, should be treated as dangerous). Work with perchloric acid should only be undertaken in specially designed fume hoods. Such hoods are usually constructed from high quality stainless steel and have safety glass in the adjustable hood opening. A water wash-down system must be installed to allow removal of condensates from the duct work (otherwise such condensates can cause explosions).

The analyst when made aware of the problems in working with perchloric acid may try to avoid its use. It is my opinion that some use of this acid is essential and hence any viable trace metal analysis laboratory must be equipped for such a purpose. In my laboratory we have worked almost every day for over 20 years with perchloric acid without any problems. The key to such success is to treat it with great respect, observing all safety precautions and using only an approved fume hood.

In my laboratory all decompositions with perchloric acid must be done with a mixture of nitric and perchloric acids. Perchloric acid is never used by itself and therefore an excess of nitric acid is used. A commonly employed mixture to start a decomposition is 3 to 5 parts of nitric acid to 1 part perchloric acid. Because perchloric acid is not oxidizing unless hot and concentrated, the nitric acid begins the oxidation process and thus at the critical stage near the end when only hot concentrated perchloric acid is present, all easily oxidizable substances have been decomposed.

One other precaution in using perchloric acid bears emphasis because it is not immediately obvious. When filtering solutions containing this acid, it is essential that the paper be very carefully rinsed to remove traces of perchloric acid if the filter paper is to be subsequently dry ashed, which is common practice in isolating a residue for further decomposition.

4.1.5.3 Sulfuric Acid

Concentrated sulfuric acid is 95–98% H_2SO_4. Its oxidizing properties do not appear until warm or hot. Because sulfuric acid is an excellent dehydrating agent, its use alone often results in a black residue of carbonaceous material that is difficult to treat. Thus sulfuric acid is usually used with an additional substance such as hydrogen peroxide or nitric acid to yield a cleaner decomposition mixture. The addition of such substances also has the beneficial effect of speeding up the otherwise slow reaction. Sulfuric acid precipitates lead and hence should not be used when lead is to be determined. Calcium sulfate will also be precipitated and this precipitate is known to occlude other metals.

Sulfuric acid has the highest boiling point of the commonly employed acids. This is a valuable property in many instances when it is desirable to remove traces of more volatile constituents of a decomposition mixture (e.g., Cl^- or F^-).

4.1.5.4 Aqua Regia

Aqua regia (1:3 HNO_3:HCl) is not very commonly employed for decomposing organic samples. However, in my view it is very often quite satisfactory. In many cases in biological or environmental analyses, such as trace metal analysis of soils and sediments by AAS or ICPAES, it is not important to obtain the complete removal of all organic matter. It is only necessary to release the metal. Aqua regia will often do this without the problems of using perchloric acid or sulfuric acid mixtures.

When this reagent is to be employed, I usually add nitric acid first and heat to destroy easily oxidizable material and any carbonates. The solution is then cooled and hydrochloric acid is added and the heating continued.

*4.1.5.5 Hydrofluoric Acid**

Hydrofluoric acid is 48–51% HF in the concentrated reagent. This acid has no oxidizing power but finds unique application in the decomposition of organic samples. Most organic substances contain silicon or silica in varying amounts. Some substances such as corn leaves contain these at quite high levels. During a decomposition, a residue containing silica is commonly formed. This residue readily traps metals of interest. Thus it is often important to treat the sample with an acid mixture such as perchloric, nitric, and hydrofluoric acids to ensure that all the analyte metal is obtained in solution.

Because hydrofluoric acid attacks silicon bonds it is essential that *nonglass* apparatus be used for its transfer and sample treatment. I recommend the use of Teflon dishes for decompositions in which hydrofluoric acid is a component of the acid mixture.

4.1.6 Fusions

Fusions, although very effective in decomposing organic samples, have the distinct disadvantage in trace element analysis, that they add greatly to the dissolved salt content of sample solutions and add contaminants. Fusions are very commonly used in inorganic sample analysis. I recommend that fusions not be used in preparing organic samples if another approach is satisfactory.

As an example of an important fusion application, hydroxide fusions are commonly employed when organic samples are prepared for determination of the

*WARNING: Hydrofluoric acid can cause severe and very painful burns. The analyst should obtain and read an acid data sheet (e.g., from J. T. Baker Ltd.) before working with this acid.

hydride elements. This is because for total hydride element analysis it is necessary to use a method that completely dissolves the sample, including siliceous components. Hydrofluoric acid cannot be used in this application because volatile compounds of some of the hydride elements are formed and these would be lost.

A variety of fusing agents are available. These are mixed, usually in large excess (up to ten times), with the sample. Some of the more common fusing materials that can be used with organic samples are sodium peroxide, sodium hydroxide, potassium pyrosulfate, potassium fluoride, and lithium metaborate. Of these, potassium pyrophosphate and particularly sodium peroxide produce oxidizing conditions during the fusion.

The fusion must be done in a heat-resistant and fusion-mixture-resistant vessel. No vessel is entirely resistant to attack by the fusing agent and this results in contamination. Fusion crucibles are commonly made of platinum or platinum alloys. However, platinum cannot be used with sodium peroxide or sodium hydroxide and in these cases a silver or nickel crucible should be chosen.

Fusing agents, like acids, contain impurities that result in trace element contamination. It is, however, more difficult to purify fusing agents than acids and thus the risk of contamination from their use is higher. Of the fusing materials listed above sodium peroxide is particularly difficult to obtain in an acceptable purity.

Subsequent to the fusing step the mixture is cooled to room temperature and then leached to bring the sample into solution. Leaching solutions generally consist of dilute acid in mixtures. Hydrochloric and nitric acids are the most commonly used acids for dissolving fusions.

4.2 COMPARISON OF ASHING METHODS

De Boer and Maessen (1) compared commonly used methods for the decomposition of biological samples. They employed for this purpose NBS SRM 1577 Bovine Liver. The elements determined were manganese, zinc, copper, iron, cadmium, and lead. Table 4.5 lists the methods tested.

The results obtained by ICPAES determination of some of the trace metals in NBS SRM 1577 Bovine Liver are summarized in Table 4.6.

All procedures gave acceptable results for manganese and zinc. Copper and iron could not be done in this type of sample using nitric acid extraction alone. Cadmium and lead were not determinable in TMAH solutions because of the low levels of these elements in Bovine Liver and the large dilution factor involved.

The H_2O_2/H_2SO_4 gave good results for all metals. Hoenig and DeBurger (2) also report excellent results for the determination of Cu, Zn, Fe, Mn, Ni, Cr,

Table 4.5 Methods[a]

Direct Approaches

Extraction with diluted HNO_3: 0.65 g sample mixed with 6.5 mL 2% HNO_3; 24 h contact time, room temperature. Centrifugation and filtration.

Solubilization in tetramethylammonium hydroxide (TMAH): 0.05 g sample mixed with 1.5 mL 10% aqueous TMAH solution. 2 h on waterbath at 65°C. Dilution to 5 mL.

Acid Digestion Methods

Digestion with HNO_3 and $HClO_4$ in microwave oven: 0.15 g sample, 5 mL 65% HNO_3, and 1 mL 70% $HClO_4$ in conical flask. Flask in microwave-heated oven (2450 MHz, 600 W). Heat until near dryness, 20 min. Dilution to 10 mL.

Digestion with H_2O_2 and H_2SO_4: 1 g sample mixed with 1.5 mL 96% H_2SO_4. Dropwise addition of 9–16 mL 50% H_2O_2 during 1 h heating up to 300°C. Dilution to 5 mL.

Digestion in Teflon-lined steel bombs: 0.25 g sample and 3 mL 65% HNO_3 in 23 mL capacity bomb. 2 h heating in stove at 140°C. Dilution to 5 mL.

Ashing Methods

Low temperature ashing: 1 g sample portions; oxygen pressure, 2 torr; net 27 MHz power, 70 W; exposure time, 20 h. Ash dissolved in 2% HCl. Final volume, 5 mL.

Muffle furnace ashing: 1 g sample during 24 h in temperature programmed muffle furnace. Final temperature, 520°C. Ash dissolved in 30% HCl. Dilution and filtration. Final volume, 5 mL.

[a] Reprinted with permission from (1), DeBoer and Maessen, *Spectrochim. Acta* **38B,** 739. Copyright 1983, Pergamon Press, Ltd.

Co, Cd, Pb, As, Sb, and Ti using the related $H_2O_2/H_2SO_4/HNO_3$ decomposition method followed by flame AAS. I, however, do not generally recommend H_2O_2 methods because of the very labor intensive nature of such procedures (H_2O_2 must be added dropwise frequently during the long digestion period).

4.3 WET ASHING PROCEDURES

Table 4.7 compares a number of methods for the decomposition of biological and related samples for analysis by AAS and ICPAES. Work in our laboratory indicates that closed-tube acid digestions (4, 9) are preferable to open vessel diges-

Table 4.6 ICPAES Results in NBS SRM 1577[a,b]

Sample treatment procedure	Elements Determined					
	Mn	Zn	Cu	Fe	Cd	Pb
Extraction with diluted HNO_3 at room temperature	10.6 ± 0.3	132 ± 3	62 ± 3	154 ± 6	0.30 ± 0.02	0.40 ± 0.13
Solubilization in TMAH	9.7 ± 0.3	131 ± 2	183 ± 5	256 ± 6	nd	nd
Digestion with HNO_3 and $HClO_4$ in microwave-heated oven	10.2 ± 0.2	131 ± 2	191 ± 2	264 ± 3	0.30 ± 0.06	nd
Digestion with H_2SO_4 and H_2O_2	10.3 ± 0.2	134 ± 3	188 ± 2	263 ± 3	0.30 ± 0.2	0.28 ± 0.10
Digestion with HNO_3 in Teflon-lined steel bomb	10.4 ± 0.2	133 ± 3	201 ± 3	268 ± 3	0.31 ± 0.02	nd
Low-temperature ashing	10.0 ± 0.3	131 ± 2	190 ± 3	260 ± 3	0.29 ± 0.02	0.37 ± 0.08
Muffle furnace ashing	10.0 ± 0.2	132 ± 3	183 ± 5	258 ± 6	0.31 ± 0.02	0.33 ± 0.03
NBS certified value	10.3 ± 1.0	130 ± 3	193 ± 10	268 ± 8	0.27 ± 0.04	0.33 ± 0.08

[a]Content ($\mu g/g$) and 95% confidence limits.

[b]Reprinted with permission from (1), DeBoer and Maessen, *Spectrochim. Acta* **38B,** 739. Copyright 1983, Pergamon Press, Ltd.

Table 4.7 Comparison of Sample Preparation Methods for AAS and ICPES[a]

Reference	Method	Sample Type	Comments
3	(a) $HF/HClO_4$ (b) Fusion $LiBO_2$	High in silicate	(a) Open vessel (b) High dissolved solids can be a problem.
4	$HNO_3/HClO_4/HF$	Marine sediments	Sealed Teflon vessel, then evaporated in Teflon vessels to dryness; centrifuge residue.
5	(a) $HNO_3/HClO_4$	Tissue serum	(a) Dichromate and vanadate added as catalysts and to indicate complete oxidation of organics.
	(b) Dry ash (485°C); dissolve in HNO_3	Organic samples	(b) Report no problems due to loss of volatile metals.
6	(a) $HNO_3/HClO_4$	Biological, agricultural	(a) In Teflon beaker evaporated to dryness
	(b) Dry ash (550°C); dissolve in $HCl/HNO_3/H_2SO_4$		(b) No loss of volatile metal reported.
	(c) HNO_3/H_2O_2		(c) Add H_2O_2 dropwise until clarified solution obtained.
7	$HNO_3/HClO_4/H_2SO_4$	Plant and animal tissue	In 100 mL Kjeldahl flask other acid mixture found to be inferior.
8	$HF/HNO_3/HClO_4$	Soils, air samples, biological tissues	In open beakers
9	$HNO_3/HClO_4$	Serum, blood	Teflon bomb

[a] From Brzezinska et al. (10).

tions, dry ashing, or fusions. This approach has three advantages: (1) contamination is minimized; (2) the decomposition is speeded up, and (3) the volume of acids needed is minimized. For plant tissue samples hydrofluoric acid is found to be necessary.

4.3.1 Teflon Closed-Tube Method for Sample Decomposition

Closely related, rapid, Teflon closed tube digestion methods are given that can be used to analyze a wide range of samples (10). The proposed procedures have been proven using standard reference samples. The elements covered are Ca, Mg, P, Al, Fe, Mn, Zn, Cu, Mo, Ni, Pb, Co, Cr, Cd, As, Se, Sb, and Ag. Samples analyzed include animal and plant tissue, algae, sediments, and particulate matter.

Equipment. A commercially available pressure cooker suitable for home cooking was employed. Teflon bombs were made at the University of Toronto from Teflon from Canplas Industries Ltd., Toronto, Canada.

Figure 4.1 is a photograph of the Teflon digestion vessels used. These are cleaned using a concentrated HNO_3 leach followed by $KMnO_4$ treatment, which in turn is followed by HCl cleaning to remove residual MnO_2. The vessels are rinsed with water in between steps in the cleaning process.

Reagents. Reagent grade (Baker Analysed) chemicals were found to be pure enough for the elements covered in the proposed procedures.

Procedure

1. Animal Tissue.

(a) Weigh 0.2–0.5 g of wet tissue into a Teflon vessel. Add 10 mL concentrated HNO_3, close the vessel, and keep at room temperature for about 1 h for

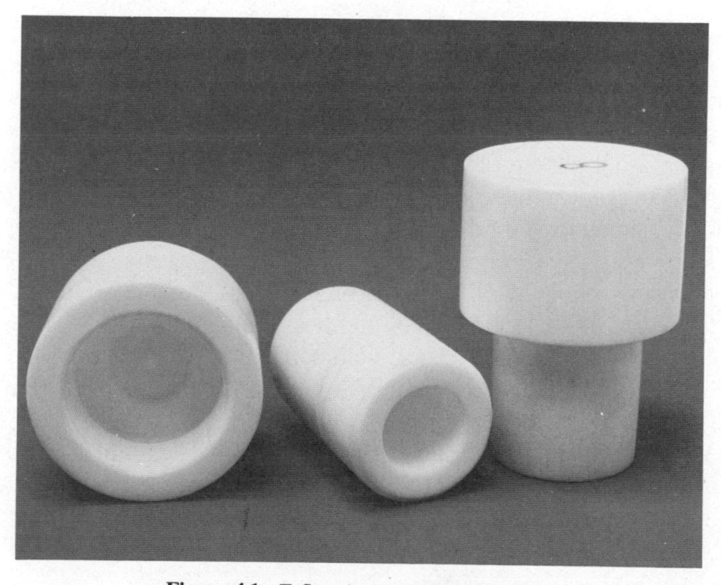

Figure 4.1 Teflon decomposition vessels.

initial decomposition. Heat the vessel in the oven at 100°C for about 2 h or in the pressure cooker (temperature about 120°C) for 45 min. After cooling, remove the top of the vessel and evaporate the sample on a hot plate to about 0.1 mL. Add 0.5 mL concentrated $HClO_4$ and evaporate the sample to near dryness. Dilute the residue in 10 mL 6% v/v mixture of 1:1 HNO_3 and $HClO_4$ and heat on a hot plate for about 10 min. After cooling transfer the digested sample to a 25 mL volumetric flask and adjust the volume with the same 6% acid solution. Store the samples in polyethylene bottles. Run 3 blanks with each series of samples.

(b) Weigh 0.25 g of dried or 0.3–0.4 g of wet tissue into a Teflon vessel. Add 1 mL of distilled water, 4 mL concentrated HNO_3, and 2 mL concentrated $HClO_4$. Close the vessel and allow to digest at room temperature for about 0.5 h. Heat the vessel at 140°C in the oven for 1 h. (This time may vary slightly depending on sample type.) Open the vessel and transfer the samples to 10 mL volumetric flasks and adjust to volume with distilled water. Store the samples in polyethylene bottles. Run 3 blanks with each series of samples.

Method (a) can be used for all the elements in the samples listed in Table 4.5. However, it is more time consuming than method (b) and contamination during the evaporation may be a problem. In our laboratory variable blanks were obtained for Zn, Pb, Cd, Ni, As, Se, and Sb. Method (b) is rapid and simpler and can be recommended for Ca, Mg, P, Al, Fe, Mn, Zn, Cu, and Mo.

2. Plant Tissue.

(a) Weigh about 0.5 g of dried or 0.8–1 g of wet plant tissue and place in a Teflon vessel. Add 4 mL of agua regia (1 mL concentrated HNO_3 and 3 mL concentrated HCl). Keep at room temperature in the closed vessel for about 0.5 h. Heat in the oven at 100°C for 2–3 h or in a pressure cooker for 1 h. Evaporate the digest on a hot plate to about 1 mL. Cool and add 3 mL of concentrated HNO_3 and 1 mL concentrated $HClO_4$. Heat again in the oven or in the pressure cooker for 0.5 h and evaporate the digest to white fumes. After cooling, add 1 mL concentrated HF and evaporate the sample to near dryness. (In case of high silica content the volume of HF must be increased.) Dilute using 10 mL of 6% v/v mixture of 1 : 1 HNO_3 and $HClO_4$ and heat on the hot plate for about 5–10 min. Transfer the solutions to 25 mL volumetric flasks and adjust the volume with the 6% acid solution. Store samples in polyethylene bottles. Run three blanks with each series of samples.

(b) Weigh 0.3–0.4 g of wet sample into a Teflon vessel. Add 4 mL concentrated HNO_3, 1 mL concentrated $HClO_4$, and 0.5 mL concentrated HF. Allow to react in the closed vessels at room temperature for about 0.5 h. Heat in the oven at 140°C for about 2–2.5 h or in the pressure cooker for 1.5 h. After cooling, transfer the sample to a 100 mL volumetric flask and adjust the volume with distilled water. Store the samples in polyethylene bottles. Run 3 blanks with each sample set. Prepare calibration standards to have the same acid composition as samples.

Note that HF is necessary for complete dissolution of plant material because of the presence of SiO_2.

Method (a) is much more time consuming than method (b) and is subject to contamination problems. Method (b) should be used when applicable. It is only when very low levels are encountered (e.g., Ni in these samples) that the greater dilution that is required in method (b) invalidates its use.

3. Sediments.

(a) Weigh 0.4–0.6 g of well-powdered sediment into a Teflon vessel. Add 3 mL concentrated HNO_3 and 1 mL concentrated $HClO_4$. Close the vessel and keep at room temperature for about 2 h. Heat in the oven at 100°C for 1 h or in a pressure cooker about 0.5 h. Evaporate the sample to fumes on a hot plate. Cool and add 10 mL HF. Close the tube and heat in the oven or in the pressure cooker as above. After cooling, evaporate the solution to near dryness. Dilute the residue with 10 mL of 6% 1 : 1 mixture of HNO_3 and $HClO_4$ and heat on hot plate for about 10 min. Filter the sample through a Whatman 540 filter into a 25 mL volumetric flask and adjust the volume with the same acid mixture. When the silica content is higher (e.g., sands), the volume of HF should be increased. Run 3 blanks with the samples by the same procedure.

(b) Weigh 0.1–0.2 g of sediment into a Teflon vessel. Add 2 mL concentrated HNO_3, 0.5 mL of concentrated $HClO_4$ and 4 mL HF. Keep the samples in closed vessels at room temperature for about 1 h. Heat the sample in the oven at 140°C for 2 h or in the pressure cooker for 1 h. Evaporate the sample to 2 mL on a hot plate at 250°C. After cooling, add 3 mL concentrated HNO_3. Place 1 g of boric acid into a 100 mL volumetric flask and then add about 50 mL of distilled water. Transfer the sample to the flask and shake the flask to dissolve the boric acid (1–2 h). Adjust the volume with distilled water. Run 3 blanks with the samples by the same procedure.

4.3.2 Pressure Method for Wet Digestion of Biological Materials*

This procedure (like the previous method), because it employs pressure, results in a more complete dissolution compared with conventional methods. It can be used on samples that are to be analyzed for mercury. Fish, bird, and plant tissues

*Procedure from Adrian (11).

have been successfully dissolved. Some samples containing lipids do not yield a clear solution by this method. Despite this, the cations calcium, magnesium, copper, iron, zinc, sodium, potassium, and lead are quantitatively extracted. Because of the low temperature and closed system, this decomposition method is well suited for the analysis of volatile metals. Technician time is also kept to a minimum.

Procedure. A 5 g sample is placed in a 2 or 4 oz Nalgene bottle with 1 mL $HClO_4$ and 2 mL HNO_3, sealed tightly, and allowed to stand to predigest overnight. Samples are then placed in hot running water for 2–3 h and then cooled. The caps are removed, 2–3 mL distilled water added, and samples are reheated in hot water in a fume hood to expel the excess acid. The radio of 1:2 $HClO_4:HNO_3$ seems to be the lower limit for good digestion.

4.3.3 Rapid Acid Dissolution of Plant Tissue*

This method was proposed and tested for the determination of cadmium. In my experience it is suitable for most metals in plant tissue. If zinc is to be done, the plastic wrap should be checked for absence of this element.

Reagents. Reagent grade chemicals are used. The acid mixture is made from a 2:1 ratio of concentrated HNO_3: 70% $HCLO_4$.

Deionized water was made by passing distilled water from the laboratory distribution system through a mixed cation–anion exchange resin column.

Procedure. Plant tissue was dried for 48 h in a forced-draft oven set at 70°C. The plant material was ground in a Wiley mill using a 40 mesh delivery tube. Some plant tissue required a 20 mesh delivery tube for homogeneity. Samples were stored in an appropriate vial or glass bottle.

Weigh 100 mg of plant material into 50 mL calibrated test tubes. With an automatic pipette, add 1 mL of acid mixture to the test tubes. Cover the test tubes with a small Pyrex funnel, transfer to the circular digestion rack (circular aluminum block with holes bored to hold test tubes), and place on a hot plate. Preheat at 60°C for 15 min or until the reaction subsides, then heat at 120°C until complete dissolution of the sample occurs, which takes about 75 min at 120°C. The entire process is complete in less than 2 h. Digestion should be conducted in a stainless steel $HClO_3$ fume hood to minimize the hazard associated with the powerful oxidizing capacity of $HClO_4$. After cooling, add deionized water to bring to the desired volume. Cover with Saran-wrapped rubber stoppers or appropriate polyethylene stoppers and mix thoroughly. Each sample

*Procedure from Ganje and Page (12).

goes through the entire dissolution process and is brought to volume in the original test tube in which the plant tissue was weighed and digested.

4.3.4 Wet Ashing for Organic Matter for Determination of Antimony*

The wet ashing of organic matter with the usual approach using nitric acid–perchloric acid mixtures results in a loss of antimony to the walls of glass vessels used for decomposition. A mixture of nitric acid/perchloric acid/sulfuric acid is recommended for avoiding this problem. The suitability of the procedure was ascertained using a tracer technique. Samples investigated were fat-free milk powder and wheat flour.

Equipment. Wet ashings can be made in 100 mL Pyrex conical flasks on a hot plate (380 × 180 mm; 1400 W) or in Teflon tubes (outer diameter, 23 mm; height, 170 mm; wall, 1 mm thick) in a temperature programmable aluminum heating block with holes (diameter, 27 mm; depth, 120 mm), type RNS2HR4 (Gebr. Leibisch, Bielefeld, GFR).

Reagents. The acids used where HNO_3 (65%), $HClO_4$ (60%), and H_2SO_4 (95–97%). The organic matter for the wet ashing experiments were fat-free milk powder and wheat flour. The dry ashing residues (700°C) were 7.5% and 0.42% of wet weight.

Procedure. Weigh 1 g of organic matter into glass or Teflon containers. Add 10 mL HNO_3, 5 mL $HClO_4$, and 5 mL H_2SO_4 (glass) or 10 mL HNO_3, 2 mL $HClO_4$, and 5 mL H_2SO_4 (Teflon). Place glass flasks or Teflon tubes in the appropriate heating apparatus. Use the following heating program:

Glass flasks—raise surface temperature of hot plate from 160 to 260°C over 1–2 h.

Teflon tubes—allow the temperature of block to rise from 20 to 150°C over 1 h. Hold the temperature at 150°C for 7 h. Then let the temperature rise from 150 to 150°C for 1 h. Finally, hold the temperature at 230°C for 2 h.

4.3.5 Soluene Method for Tissue Solubilization†

A quaternary ammonium hydroxide tissue solubilizer is used. This organic-based material has the distinct advantage that it enhances the analyte signal when

*Reprinted in part with permission from (13) Bajo and Suter, *Anal. Chem.* **54,** 50. Copyright 1982, American Chemical Society.

†Reprinted in part with permission from (14) Jackson, Mitchell and Schmachen, *Anal. Chem.*, **44,** 1004. Copyright 1972, American Chemical Society.

flame atomic absorption is used compared with that obtained in aqueous digests. The procedure has been tested for zinc, copper, iron, and manganese in unnamed tissue. The method of standard additions was used for the subsequent determinations.

Reagents. Reagent grade chemicals were used. Soluene 100 is available from Packard Scientific. The Soluene contains a 2% (w/v) solution of ammonium-1-pyrrolidene dithiocarbamate.

Procedure. Weigh the tissue sample into a 50 mL volumetric flask. Add 0.5–1.0 mL Soluene/100 mg of tissue. Stopper the flask and let stand at room temperature for 24 h. Heating to 60°C will speed solubilization. All tissues investigated gave a clear, homogeneous, and aspiratable solution suitable for atomic absorption. A three- to fourfold dilution of the preparation is made with Soluene. The samples were analyzed by the method of additions.

4.3.6 Alcoholic Solution of TMAH for Solubilizing Tissues*

Alcoholic solutions of tetramethylammonium hydroxide (TMAH) are used to decompose human adrenal, aorta, bladder, blood, bone, brain, cecum, fascia, hair, heart, jejunum, kidney, liver, lung, muscle, nails, nodes, pancreas, prostate, skin, spleen, stomach, teeth, testes, thyroid, and urine for flame or electrothermal atomic absorption analysis of cadmium, copper, lead, manganese, and zinc.

The method consists of solubilizing small quantities of tissues with alcoholic TMAH and using various dilutions of these solutions directly in the furnace or by aspiration into a flame. Unfortunately procedural details are sparse. The NBS Bovine Liver standard was analyzed and good recoveries were obtained for all the metals but copper, which gave low results by 20%. Blanks should be run, but are usually found to contain negligible quantities of analyte.

Procedure. Place 1 g of tissue, blood, or urine in a borosilicate liquid scintillation vial equipped with a plastic liner. Add 2 mL of 25% TMAH in alcohol (South Western Analytical Chemicals, Inc., Austin, TX) and heat the vial with shaking in a 70°C water bath for at least 2 h. Prolonged digestion time causes metal losses. Dilute the clear (usually amber) solutions to 10 mL with deionized water (1 : 10 dilution). Biological standards are used for instrument calibration.

*Procedure from Cross and Parkinson (15).

4.3.7 Alkaline Permanganate for the Dissolution of Biological Samples for Mercury Determination*

Sodium or potassium hydroxide solutions are very useful for the disintegration of biological samples. If potassium permanganate is added, the sample is effectively oxidized. In the following procedure, a potassium hydroxide/potassium permanganate solution is used to dissolve samples for mercury determination. The solution is then acidified with sulfuric acid and excess oxidant reduced by oxalic acid addition. The following substances have been successfully decomposed: human hair, fingernails, urine, fish products, blood, liver, rat fur, and plant tissue.

Reagents. All reagents were specially selected for their low mercury content: 1.5 *M* potassium hydroxide, 0.3 *M* potassium permanganate, 1.1 *M* oxalic acid, and 9 *M* sulfuric acid (analytical grade).

Procedure. Details of preparation of individual samples are summarized in Table 4.8. Gently warm weighed samples with potassium hydroxide solution until dissolution or homogenization is evident. Prepare blanks and samples simultaneously. After this initial treatment, add potassium permanganate and allow the samples to stand for up to 30 min, or until the oxidant had obviously been consumed. Add more oxidant until the permanganate color persists. Sam-

Table 4.8 Reagent Amounts Added

Sample	Weight[a] (g)	1.5 *M* KOH (mL)	0.3 *M* $KMnO_4$ (mL)	9 *M* H_2SO_4 (mL)
Human hair	0.01	1	1	2
Fingernail	0.01	1	1	2
Rat fur	0.01	1	1	2
Rat liver	0.1	1	2.5	2
Rat blood	0.5[b]	5	7	2.5
Urine	10[b]	4	14	3.5
Tinned fish	0.5	5	20	3
Dried fish	0.1	1	5	2
Vegetation	0.1	10	13	3.5

[a]Weight per determination. For most samples a larger quantity was homogenized with KOH and then subsampled into the reaction flasks.

[b]Volume taken (mL).

*Procedure from Chapman and Dales (16).

ples are then acidified and allowed to stand for several hours. If the permanganate is depleted, more is added. This process is repeated until the permanganate color persists for about 4 h. Add a 1 mL aliquot of oxalic acid to the blank. After mixing and standing, add a further 1 mL and repeat the process until a clear solution is obtained. The same quantity of reagent is added slowly to the other samples. As reduction with oxalic acid proceeds, carbon dioxide is produced. It is therefore necessary to ensure that the reaction flasks are kept only loosely stoppered to prevent them becoming pressurized.

4.4 DRY ASHING PROCEDURES

4.4.1 Dry Ashing Followed by Mineral Acid Decomposition*

The following procedure is commonly used for plant material including bark, wood, leaves, stems, seeds, fruit, roots, and highly organic soils. Dry ashing is done at 450°C. At this temperature mercury will be lost. Arsenic and the other covalent hydride-forming elements will most likely be lost in varying amounts. Other volatile metals such as lead, zinc, and cadmium are normally retained. When appreciable siliceous matter is present, for example, in leaves and stems of many plants, hydrofluoric acid must be included in the acid digestion mixture.

The main advantages of a dry ashing procedure is that large sample sizes can be used and perchloric acid is not needed. As emphasized elsewhere, perchloric acid requires the use of a special fume hood and has been known to cause violent explosions when handled improperly.

Procedure. The sample should be broken into small pieces. Although 30 mesh is best, up to 1 mm lengths are permissible. Place the desired sample weight into an appropriate size Pyrex beaker. Place in a large cool oven and begin heating at a very slow rate. The temperature should rise from room temperature to 450°C in 6–8 h. Heat at 450°C for 2 h or until ashing is complete. Then use the aqua regia method given below. If no HF is necessary, decomposition can be done in the original beaker. If HF is required, use Teflon dishes.

Nitric, hydrochloric, and hydrofluoric acids. Weigh a 0.2 g sample into a 100 mL beaker. Add 5 mL HNO_3 and 15 mL HCl. Place a watchglass over the beaker and digest at medium heat for 60 min. Evaporate to dryness. Add 5 mL HNO_3 and evaporate to dryness. Add 1 mL HNO_3 and warm. Add 1 mL of water and warm and filter into a 25 mL flask. Cool and dilute to volume. Make appropriate dilutions as required, maintaining the acid content at 1%. If the

*Procedure from Van Loon (17).

sample contains siliceous material, the following step should be inserted after addition of 1 mL HNO_3.

Wash the material from the beaker with a minimum of water into a 100 mL Teflon dish. Scrub beaker walls with a plastic stirring rod and rinse. Evaporate to dryness. Added 2 mL HF, 1 mL HNO_3, and evaporate to dryness. Add 1 mL HNO_3 and warm. Proceed as above.

4.4.2 Low Temperature Oxygen–Fluorine Ashing of Biological Samples*

Fluorine is used together with oxygen to greatly reduce the time required in the radio-frequency ashing of biological samples. Atomic fluorine is produced by a 2-step reaction resulting from the attack on PTFE crucibles that hold the samples by atomic oxygen. Loss of some elements may occur due to the formation of volatile fluorides. Arsenic, selenium, gold, and silver can be lost in an oxygen plasma alone. The elements for which this technique has been found to date to be satisfactory are tin, iron, lead, and chromium.

Equipment. Ashing instruments used were obtained from Tracerlab Ltd., Division of Electronics, Richmond, CA, Model LTA 600; Branson-International Plasma Corp. (IPC), Haykward, CA, Model IPC 4000/104B, and Nonotech (Thin Films) Ltd., Sedgley Park Trading Estate, Prestwich, Manchester, Model P100.

When in use each reactor must be equipped with a silicon or aluminum reactor chamber, silicone rubber gaskets and tube connectors, and a vacuum system using a halocarbon oil.

As a safety precaution all units should be fitted with a trap filled with disodiuim tetraborate crystals to neutralize fluorine emissions and should be vented to the external atmosphere. The manufacturers will supply data on radio-frequency shielding and the current leakage levels permitted.

A Kenwood Chef using a liquidizer attachment was used to prepare all foods to a slurry. The original liquidizer blades were replaced by titanium blades made in the laboratory.

A Mettler Model P1200 electronic balance was used for all weighing operations and was also used to measure (by mass) precise dilution volumes using the tare facility on the balance. The mass of ash remaining after oxidation was sufficiently small (0.02–0.06 g) to enable solutions to be prepared by simply adding fixed volumes of acid within the range 5–25 mL without correction.

Poly(tetrafluoroethylene) crucibles were machined from 15 mm thick PTFE sheet ("Fluon," Imperial Chemical Industries, Plastics Division, Welwyn, Garden City). The external dimensions were as follows: diameter 60 mm; depth 10

*Procedure from Chapman and Dales (16).

mm; and wall 3 mm. After cleaning in (1 + 1) HNO_3, they were placed in an oxygen plasma to etch and clean the surface for 15 min at 100 W. Nonvolatile impurities, which accumulated as a result of machining on the surface, were dissolved away by immersing the entire surface in a solution of HCl containing H_2O_2. Under the test conditions in a Branson–International plasma chemistry reactor, mass losses of 0.06 g/h at 100 W were typical, but they depended on the type of reactor and to some extent on the position of the PTFE crucible inside the reactor.

Reagents. All reagents were Aristar grade (BDH Chemicals Ltd., Poole, Dorset).

Hydrochloric Acid–Hydrogen Peroxide. A 40 mL aliquot of HCl (sp. gr. 1.17) was diluted with water and 2 mL H_2O_2 (30% m/v) were added, giving a total volume of 100 mL.

Nitric Acid. Nitric acid (40 mL) was diluted with water to 100 mL.

Procedure. Weigh 1.0–5.0 $\pm$ 0.02 g of a representative homogenized sample into a PTFE crucible of known mass using an electronic balance. Dry the sample for 2 h at 120°C. Use the reactor conditions recommended by the manufacturer for the reactor model, for example, for an IPC, Model 4000–104B: oxygen pressure 14–34 kN/m^2, oxygen flow rate 300 mL/min; vacuum 0.5 mm Hg; and radio-frequency power 100 W. Treat single 5.0 g samples for 4–6 h and increase the time proportionately for up to 3 $\times$ 5 g samples. Test the treated samples with 1–2 drops of distilled water to establish complete ashing, which will be evident by the absence of any black particles. If ashing is incomplete, return to the reactor for further treatment. To avoid ash loss by static electricity, allow the charge to dissipate with the radio-frequency power switched off before sample withdrawal from the reactor. Dissolve the sample ash in an acid or alkali suitable for the analytical technique to be used.

4.5 METHODS OF SEPARATION AND CONCENTRATION

A good review on separation and concentration methods for trace elements has been published by Bachmann (19). He classifies methods of separation and concentration as shown in Table 4.9. The most common approaches and those best suited to trace element analysis by the technique of analytical atomic spectrometry are solvent extraction and ion exchange chromatography.

Separations and concentration methods involve extra steps in an analysis and should be avoided if not essential. Problems due to contamination or losses are frequently encountered during separation and concentration steps.

Table 4.9 Classification of Concentration for Trace Elements According to Initial and Second Phase[a]

Initial Phase	Second Phase	Method of Concentration
Solid	Gas	Evaporation of trace elements of matrix elements: volatilization by reaction with reactive gas
Solid Liquid	Gas	Gas chromatography
Liquid	Liquid	Solvent extraction of trace elements or matrix elements
Liquid	Liquid	Extraction chromatography
Liquid	Solid	Sorption of trace elements or matrix
Liquid	Solid	Precipitation and coprecipitation of trace elements or matrix
Liquid	Solid	Electrolytic deposition of trace elements or matrix
Liquid/solid	Liquid/solid	Selective dissolution of trace elements or matrix

[a]Reprinted with permission from (19) Bachmann, *Critical Reviews Anal. Chem.* **12,** (1981). Copyright CRC Press, Inc., Bocas Ratan, FL.

The reason for undertaking a concentration step in a procedure is to bring the level of concentration of an analyte to a detectable level for the determinative technique chosen. Fortunately instrumentation is improving rapidly with new developments that are greatly improving detection limits. An excellent illustration of this is the history of furnace atomic absorption. Equipment until 1979 had poor temperature control coupled with relatively slow maximum heating rates. Thus it was difficult to reach the desired atomization temperature quickly and consistently. In addition, most workers atomized samples placed on the tube walls and encountered serious gas phase interference problems. The recent popularity of the L'vov platform is obviating most of these problems. As a result of these developments, detection limits in real samples for many elements by furnace atomic absorption have improved by up to one order of magnitude. Thus as time goes on the need for concentration steps should be continually decreasing.

Separations are sometimes necessary to remove the analyte from an interfering matrix. As in the case of concentrations the need for separations is continually decreasing with improvements in instrumentation. Sturgeon et al. (20) state that, because very large volumes of seawater cannot be subjected to solvent extraction, both from a theoretical and physical standpoint there is a limit to the concentration factors attainable by single-stage solvent extraction. However, in my laboratory, solvent extraction has been found to be the most satisfactory approach. In addition to yielding a good separation, solvent extraction has the advantage that the metals of interest are in an organic solvent. If the proper solvent is chosen, an enhanced signal compared with water solutions is obtained when flame atomic absorption is employed.

4.5.1 Solvent Extraction

Solvent extraction methods for concentrating trace metal ions in water, prior to atomic spectrometer analysis, abound. Unfortunately the majority of these have been developed without regard for important theoretical data available from such sources as Stary (21), Morrison and Freiser (22), and Zolotov (23). There have been a few critical studies of solvent extraction–atomic spectrometer procedures. As a result, available methods are seldom optimized with respect to pH range, buffer, ionic strength, stability, equilibration time, and so on. This means it is often impossible for the analyst to obtain good results on a routine basis. Proposed solvent extraction methods for the atomic absorption analysis of trace metals were examined in my laboratory (24). The following summarizes the important considerations that resulted. The solvent used to extract metal complexes must have a number of desirable characteristics. It must (1) extract the desired metal chelates; (2) be immiscible with the aqueous solution; (3) not tend to form emulsions; and (4) have good burning characteristics if a flame is used.

Work was carried out on a number of likely solvents. Benzene and xylene should not be used with a flame because of the turbulent and unstable nature of the flames. Decanol proved to have too pungent an odor. Chloroform, a solvent widely used in colorimetric work, evaporates too quickly, leaving the solid com plex behind. Ethyl acetate, methyl isobutyl ketone, isoamyl acetate, and *n*-butyl acetate were found to be good when used with flames. Of these, ethyl acetate and methyl isobutyl ketone gave the greatest enhancement as compared to the absorbance of the same quantity of metal in water. Ethyl acetate is too volatile for easy use.

The use of a buffer is mandatory in routine extraction work. This fact is not recognized by many workers. It is well known that the quantity of metal extracted is strongly dependent on the pH of the solution and that chelating agents will often alter the pH of the solution to which they are added.

The choice of the buffer is very important. It must be stable, have a high buffering capacity, and not participate in any reaction. A number of buffers that have been studied for solvent extraction–atomic absorption work are borate, phosphate, citrate, acetate, and formate. Solutions containing the formate buffer were found to be unstable and slowly decomposed (organic droplets appeared on container walls) after several days. An acetate buffer was also unfavorable because it would combine with any lead or silver in solution to form stable acetates that were not readily extracted. A citrate buffer, which was found to be stable and did not interfere with the extraction process, is best for most metals. This buffer, however, does contain considerable cadmium and iron. The buffer should be purified as well as possible of trace metal contaminants by an extraction wash using the chelating agent. In spite of this precaution, a blank must be run with each set of samples. If silver and lead are not to be analyzed, an acetate buffer is recommended because of lower cadmium and iron contamination.

Two of the main considerations affecting the choice of a chelating agent for most applicataions are that it should (1) extract the largest number of trace metals and (2) extract the metals equally well over some fairly wide range of pH of the solution. Many procedures available at present require pH adjustment to within one pH unit or less, which can result in serious errors in routine applications.

Table 4.10 lists the various chelating agents that show the elements extracted and organic solvent employed (18). This will be useful as a guide. However, if a flame atomizer is chosen, a burnable solvent as discussed above or back extraction into an aqueous phase must always be used. Experience with the ICP suggests that most solvents are applicable.

Table 4.10 Chelating Agents for Concentration of Trace Elements by Extraction[a]

Chemical Name	Elements That Are Extracted (More Than 50% are Extracted)	Organic Phase
Dithiocarbamate ammonium tetramethylendithiocarbamate	V, Cr, Fe, Cs, Ni, Cu, Zn, Ga, Ge, As, Nb, Mo, Tc, Ru, Rh, Pd, Ag, In, Sn, W, Re,	$CHCl_3$, CCl_4
Sodium diethyldithiocarbamate	Os, Ir, Pt, Au, Hg, U, Ti, V, Cr, Mn, Fe, Co, Ni, Cu, Zn, Ga, As, Se, Mo, Pd, Ag, Cd, In, Sn, Sb, Te, W, Au, Hg, Tl, Pb, Bi, U, Pu	$CHCl_3$, CCl_4
Sodium *N,N'*-phenylacetyl dithiocarbamate	V, Mn, Fe, Cs, Ni, Cu, Zn, As, Se, Mo, Re, Rh, Rd, Ag, Cd, Sb, Te, Os, Ir, Pt, Au, Hg, Tl, Pb, Bi	$CHCl_3$, CCl_4
Sodium *N,N'*-phtalyldithiocarbamate	Fe, Cr, Ni, Cu, Zn, As, Se, Mo, Rh, Pd, Ag, Cd, Sn, Sb, Te, Pt, Au, Hg, Tl, Pb, Bi	$CHCl_3$, CCl_4
Ammonium *o*-aminophenyldithiocarbamate	Cr, Mn, Fe, Co, Ni, Sn, Zn, As, Se, Mo, Pd, Ag, Cd, In, Sn, Te, Ir, Pt, Au, Hg, Tl, Pb, Bi	$CHCl_3$, CCl_4
Ammonium *m*-aminophenyldithiocarbamate	V, Cr, Fe, Co, Ni, Cu, Zn, Se, Mo, Pa, Cd, In, Sn, Ic, Ir, Pt, Au, Hg, Tl, Pb, Bt	$CHCl_3$, CCl_4

Table 4.10 (*Continued*)

Chemical Name	Elements That Are Extracted (More Than 50% are Extracted)	Organic Phase
Ammonium *p*-aminophenyl dithiocarbamate	V, Mn, Fe, Co, Ni, Cu, Zn, As, Se, Mo, Pd, Ag, Cd, In, Sn, Te, Ir, Pt, Au, Hg, Tl, Pb, Bi	$CHCL_3$, CCl_4
Ammonium anilinodithiocarbamate	V, Fe, Co, Ni, Cu, Zn, Mi, Pd, Ag, Ca, In, Sb, Te, W, Ir, Pt, Hg, Tl, Pb, Bi	$CHCI_3$, CCL_4
Diethylammonium-diethyldithiocarbamate	V, Cr, Mn, Fe, Co, Zn, Ga, Ge, As, Se, Mn, Pd, As, Cd, In, Sn, Sb, Te, W, Pt, Hg, Tl, Pb, Bi	$CHCI_3$, CCl_4
Diphenylthiocarbazone	Mn, Fe, Co, Ni, Cu, Zn, Ga, Pd, As, Cd, In, Sn, Te, Pt, Au, Hg, T1, Pb, Bi, Po,	$CHCI_3$, CCL_4
8-Hydroxiquinoline	Be, Mg, Al, Ca, Sc, Ti, V, Cr, Mn, Fe, Co, Ni, Cu, Zn, Ga, Sr, Y, Zr, Nb, Mo, Tc, Ru, Rh, Pd, Aq, Cd, In, Sn, Sb, Ba, La, Hf, W, Hg, Tl, Pb, Bi, Ce, Nd, Sm, Er, Th, Pa, U, Pu	$CHCI_3$, CCl_4 toluene
Acetylacetone	Be, Al, Sc, Ti, V, Cr, Mn, Fe, Co, Cu, Zn, Ga, Zr, Mo, Ru, Pd, In, Sn, Hf, Hg, Tl, Pb, Bi	Acetylacetone benzene, acetylacetone
Ammonium phenylnitroso-hydroxylamine	Ti, V, Fe, Co, Cu, Al, Ga, Nb, Mo, Pd, Sn, Sb, Pb, Bi, Co, Th, Pa, U	$CHCI_3$, MIBK ethyl acetate
Thenoyltrifluoracetone	Be, Ak, Ca, Sc, Cr, Mn, Fe, Ga, Ni, Cu, Sr, Y, Zr, Mo, Pd, In, Sn, Cs, La, Hf, W, Pt, Tl, Pb, Bi, Po, Ce, Eu, Th, Pa, U, Np, Pu	CCI_4
1-(2-Pyridylazo)-2-napthol	Sc, Ti, V, Mn, Fe, Co, Ni, Cu, Zn, Ga, Y, Zr, Rh, Pd, As, Cd, In, Sn, La, Ir, Pt, Hg, Pb, Bi, Co, En, Th, U	$CHCI_3$
n-Benzoyl-*n*-phenylhydroxylamine	Be, Sc, Ti, V, Cr, Mn, Fe, Co, Ni, Cu, Zn, Al, Ga, Ge, Y, Zr, Nb, Mo, In, Sn, Sb, La, Hf, Ta, W, Re, Hg, Tl, Pb, Bi, Co, Nd, Th, Pa, U, Pu	Benzene
Triphenylphosphine	Ag, Au	
1-Phenyl-3-methyl-4-benzoyl-pyrazolone-5	Cd, Co, Cr, Cu, Fe, Mn, Mo, Ni, Pb, Ti, V, Eu	$CHCI_3$ ethanol
Diphenylthiourea	Ag, Au, Pd, Pt, Ru, Rh, Ir	

[a]Reprinted with permission from (19), Bachmann, *Critical Reviews Anal. Chem.* **12**, 1(1981). Copyright CRC Press, Inc., Boca Raton, FL.

4.5.2 Ion Exchange

Ion exchange and chelating resins are used for two purposes in environmental and biological analysis: (1) preconcentration of the trace metals prior to the determinative step; and (2) separation of the trace metals from interfering concomitant substances. Chelating resins are particularly useful in trace metal water analysis because of the very high selectivity of polyvalent over monovalent ions.

Ion exchange resins consist of an insoluble polymer (commonly styrene) lattice with attached functional groups. The polymer matrix is porous to water and inorganic ions. Resins are cross-linked with divinylbenzene. The porosity of a resin is determined by the hydration of the matrix. Resins are formed into beads of various mesh sizes. Styrene polymer chains are cross-linked with divinylbenzene. Resins swell when hydrated, the amount of swelling being dependent on the degree of cross-linking.

Ion exchange resins are available with acidic or basic functional groups and may be either weakly or strongly acidic or basic depending on the nature of the group. Acidic resins exchange cations and are called cation exchange resins. Likewise, basic resins exchange anions and are termed anion exchange resins. For strong resins the order of selectivity favors polyvalent over monovalent ions. The ion exchange affinity is inversely proportional to the radius of the hydrated ion for ions of the same charge.

A composite affinity sequence of cations is

$$
\begin{aligned}
&Li^+ < Na^+ < NH_4^+ < K^+ < Rb^+ < Cs^+ < Ti^+ < Ag^+ < Mg^{2+} \\
&< Ca^{2+} < Sr^{2+} < Ba^{2+} \\
&< Fe^{2+} < Co^{2+} < Ni^{2+} < Cu^{2+} \\
&< Zn^{2+} < Al^{3+} < Sc^{3+}
\end{aligned}
$$

For anions the composite affinity sequence is

$$
\begin{aligned}
&F^- < Cl^- < Br^- \; CrO_4^{4-} < MoO_4^{2-} < PO_4^{3-} < AsO_4^{3-} \\
&< NO_3^- < I^- < SO_4^{2-}
\end{aligned}
$$

These series vary slightly depending on individual ion-exchange resins and differing conditions.

Chelating resins are also a styrene polymer cross-linked by divinylbenzene. A typical chelating group used in these resins is imminodiacetate. This reactive group chelates with trace metals giving a 5000 to 1 selectively of divalent to monovalent ions. Chelating resins therefore find useful application in trace metal analysis of seawater.

Table 4.11 lists some available ion exchange resins (25). It shows the functional group polymer support and pH range of use for each resin.

Table 4.11 Ion Exchange Resins

Classification	Functional Group	Polymeric Support	Useful pH Range	Trade Name	Source[c]
Strongly basic (strong anion exchange)	Tetraalkyl-ammonium hydroxide	S-DVB	0–14	ANGA-542 REXYN 201 OH	Baker Fisher
	$-CH_2N(CH_3)_3Cl$	S-DVB		Amberlite IRA 400 Dowex-1	R and H Dow
	$-CH_2N(CH_3)_3OH$	S-DVB		Amberlite IRA 400 Dowex–1	R and H
	Tetraalkyl-ammonium chloride	S-DVB		IONAC A-540 REXYN 201 Cl	MC/B Fisher
Moderately basic	$-N(CH_3)_2$	S-DVB	0–14	CGA-301	Baker
Weakly basic	$-HN_2$	S-DVB	0–9 0–7 0–12	Amberlite IRA 93 Amberlite IRA 45 ANGA-316	Baker
	$-NH_2$	A-DVB	0–7	Amberlite IRA 68	R and H
Strong acidic (strong cation exchanger	$-SO_3^-H^+$	S-DVB	0–14	Dowex-50 IONAC C-242	Dow MC/B
	$-SO_3^-NA^+$	S-DVB	0–14	Amberlite IR 120P Amberlite 200 Amberlite 252 CGC-241	R and H R and H R and H Baker
Weakly acidic	$-COO^-H^+$	S-DVB	0–12	CGC-270 Amberlite CG 50	Baker R and H
	$-COO^-H^+$	MA-DVB	5–14	Amberlite IRC 50	R and H

[a]From Braun (25)

[b]S-DVB is styrene–divinylbenzene; A-DVB is acrylate–divinylbenzene; MA-DVB is methacrylate–divinylbenzene.

[c]Baker is J. T. Baker Chemical Co.; Fisher is Fisher Scientific Co.; MC/B is Matheson, Coleman and Bell Manufacturing Chemists; Dow is Dow Chemical Co.; R and H is Rhom and Haas Co.

4.6 FIELD AND LABORATORY PRECONCENTRATION PROCEDURE FOR AS, SB, AND SE*

This procedure can be used in the field as a method for sample preservation. The hydride elements arsenic, antimony, and selenium are evolved from the sample and trapped on Whatman GF/C pads. The pads are saturated with 5% silver nitrate solution (traps selenium) or 5% mercuric chloride (traps arsenic and antimony). The pads can be analyzed at another time.

Procedure. Transfer 100 μL of each sample to a clean borosilicate glass bottle. Immediately add 2 mL 20% m/v stannous chloride in 20% v/v hydrochloric acid to the sample, followed by the addition of 2 mL 15% m/v aqueous potassium iodide. Allow the sample solution to stand for 5 min, during which time the hydride-forming elements are reduced to their optimum valence states, As (III), Sb (III), and Se (II).

Then transfer the sample to a modified gas wash bottle and insert a two-way head (Fig. 4.2). Connect a filter assembly containing two Whatman GF/C pads in line with the gas exit port (Fig. 4.3). Impregnate the pad nearest the incoming gases with either 100 μL of 5% aqueous silver nitrate (to convert seleane to selenium or selenium dioxide) or 100 μL of 5% aqueous mercuric chloride (to convert arisine or stibnine to mercuric orthoarsenate or mercuric antimonate). Add one of these solutions dropwise to the surface of the pad immediately prior to each extraction. The second pad acts as a buffer to prevent contact between the reaction solution and the filter holder. Pass argon through the reaction vessel at 100 mL/min. Generate the hydrides using either 2 mL of a zinc slurry (500 g of zinc powder in 400 mL water) for sample volumes greater than 50 mL or, if the initial sample volume is less than 50 mL, 2 mL 10% m/v sodium borhydride in 2% m/v sodium hydroxide solution.

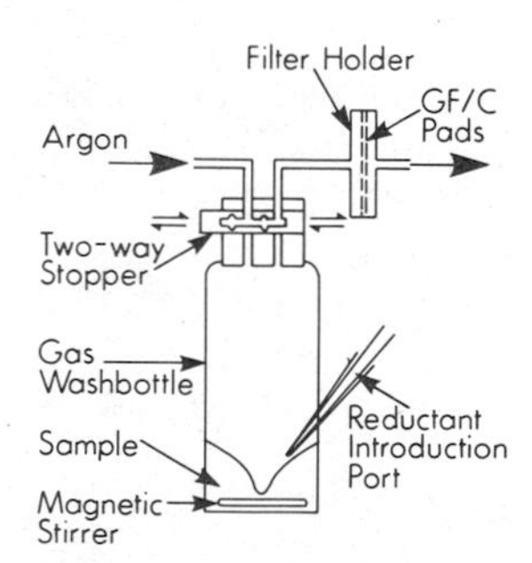

Figure 4.2 Field preconcentration apparatus for As, Sb, and Se (26).

*Reprinted with permission from (26), Watling and Watling, *Spectrochim. Acta.* **35B,** 451. Copyright 1980, Pergamon Press, Ltd.

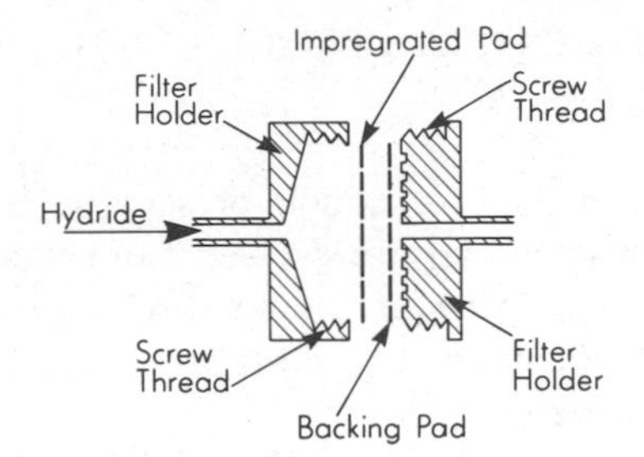

Figure 4.3 Filter assembly (26).

Back at the Laboratory. Reextract selenium and antimony from the filter pads using a mixture of 1 mL of nitric acid and 0.2 mL of perchloric acid; arsenic is extracted with 1 mL of sulfuric acid. Place each pad in a 10 mL beaker containing one of these solutions and heat until fumes of either perchloric acid or sulfuric acid are observed. Immerse the beaker and its contents in 50 mL of a 5:4:1 solution of distilled water, hydrochloric acid, and sulfuric acid, respectively, contained in a 100 mL beaker. Allow the mixture to react for 30 min, after which are added 2 mL of 20% m/v stannous chloride in 20% m/v hydrochloric acid followed by 2 mL of 15 m/v aqueous potassium iodide. After a further 5 min, transfer an aliquot of this acidified sample to a hydride generation apparatus.

REFERENCES

1. J. L. M. De Boer and F. J. M. J. Maessen, *Spectrochim. Acta* **38B,** 739 (1983.
2. M. Hoenig and R. DeBorger, *Spectrochim. Acta* **38B,** 873 (1983).
3. J. O. Burman and K. Bostrom, *Anal. Chem.* **51,** 516 (1979).
4. J. W. McLaren, S. S. Berman, V. J. Boyko, and D. S. Russell, *Anal. Chem.* **53,** 1802 (1981).
5. R. L. Dalquist and J. W. Knoll, *Appl. Spectros.* **32,** 1 (1978).
6. A. F. Ward, L. F. Maciello, L. Carrara, and V. J. Luciano, *Spectros. Lett.* **13,** 803 (1980).
7. J. W. Jones, S. G. Capor, and T. C. O'Haven, *Analyst* **107,** 353 (1982).
8. N. R. McQuaker, D. F. Brown, and P. D. Kluckner, *Anal. Chem.* **51,** 1082 (1979).
9. H. Uchida, Y. Nojiri, H. Haraguchi, and K. Fuwa, *Anal. Chim. Acta* **123,** 57 (1981).
10. A. Brzezinska, A. Balicki, and J. C. Van Loon, *Water, Air and Soil Pollution* **21,** 323 (1984).
11. W. I. Adrian, *At. Absorpt. Newslet.* **10,** 96 (1971).
12. T. J. Ganje and A. L. Page, *At. Absorpt. Newslet.* **13,** 131 (1974).
13. S. Bajo and U. Suter, *Anal. Chem.* **54,** 50 (1982).
14. A. J. Jackson, L. M. Mitchell, and H. J. Schmachen, *Anal. Chem.* **44,** 1004 (1972).
15. S. B. Cross and E. C. Parkinson, *At. Absorpt. Newslet.* **13,** 107 (1974).

16. J. F. Chapman and L. S. Dales, *Anal. Chim. Acta* **134,** 379 (1982).
17. J. C. Van Loon, *Internal Lab Method,* 1983.
18. E. V. Williams, *Analyst* **107,** 1006 (1982).
19. K. Bachmann, *Critical Reviews Anal. Chem.* **12,** 1 (1981).
20. R. E. Sturgeon, S. S. Berman, A. Desauliners, and D. S. Russell, *Talanta* **27,** 85 (1980).
21. J. Stary, *The Solvent Extraction of Metal Chelates*. Macmillan, New York, 1964.
22. G. H. Morrison and H. Freiser, *Solvent Extractions in Analytical Chemistry.* Wiley, New York, 1957.
23. Y. A. Zolotov, *Extraction of Chelate Compounds*. Ann Arbor-Humphrey Science Publications, MI, 1970.
24. J. D. Kinrade and J. C. Van Loon, *Anal. Chem.* **46,** 1894 (1974).
25. R. D. Braun, *Introduction to Chemical Analysis*. McGraw-Hill, New York, 1982.
26. R. J. Watling and H. R. Watling, *Spectrochim. Acta* **35B,** 451 (1980).

CHAPTER

5

BOTANICAL AND ZOOLOGICAL SAMPLES

5.1 INTRODUCTION

Biological samples consist of plant and animal material. Some biological material, for example, clinical and food samples, are of sufficient specialized interest to be treated separately in their own chapters. Thus in this chapter materials other than clinical and food samples is covered. There will, of course, be some redundancy in such a division of biological samples.

5.1.1 Plant Material

Plant analyses are commonly needed when plants are being used as foods for domestic animals. In this application the part of the plant that the animal eats is the sample used for analysis. For example, the whole plant should be analyzed when dealing with dairy cattle. Seeds are the sample for analysis when considering nutritional value to birds.

An area that is quickly growing in importance is the use of plant analysis as a guide to soil fertility for computing the requirements for fertilizer. The sample chosen would be a part of the plant that because of its stage of development would yield the desired information. In this application it is crucial that the analyses be done quickly because any delay may negate correcting measures.

In most cases, in both the above applications, the major elements would likely be of the greatest interest. However, the role of trace elements in nutrition and soil fertility is becoming increasingly important.

Plant analyses are also necessary for botanical and environmental purposes. In the case of the latter, the uptake of trace elements by plants can provide important information on environmental contamination. Plants are often an integral stage in the cycling of trace elements and as such are crucial to understanding such processes.

A recent very important application of trace element plant analyses is the use of plants in the search for minerals. Elevated levels of trace elements in plants can be used together with other geological and geochemical data in the search for minable ores. Some types of plants accumulate trace elements in soils to astonishingly high levels (>1000 ppm) and in extreme cases plant material has actually become the "ore" in a mining operation.

Uptake of selenium by plants provides a useful example of trace element accumulation in plants. In considering selenium uptake, plants can be divided into three groups.

1. Primary accumulators, selenium levels are greater than 1000 ppm. Selenium is present largely in organic and water soluble forms.
2. Secondary accumulators—selenium levels are seldom more than a few 100 ppm. Selenium is present mainly as selenate with only small amounts of organic forms.
3. Nonaccumulators—selenium levels are below 10 ppm. Inorganic forms of selenium predominate.

The toxicity to animals of organic forms such as $(CH_3)_2Se$ is only one five-hundredth of the toxicity of selenite and selenate. This underscores the importance of being able to determine the chemical form of the elements. (The analytical aspect of this subject, only recently receiving widespread attention, is treated separately in Chapter 11.)

5.1.1.1 Contamination

The plant sample must be carefully examined for contamination prior to analysis. Rinsing the tissue in distilled water may often be sufficient to remove contaminants. For more stubborn problems a gentle detergent wash or a rinse in dilute acid may be essential. When dilute acid is used, it is important that the procedure be evaluated to ensure that no trace elements actually present in the plant are removed. In a few instances it has been my experience to find contaminant particles actually embedded in the surface of plant material. In these cases no known procedure was effective in removing the offending material.

5.1.1.2 Drying and Sieving

Drying of plant material should be done as soon as possible. This prevents deterioration of the sample due to growth of mold and microorganisms or continued respiration. Samples can be carefully spread out in the sun for drying in field. If dried in an oven, samples should be heated at a temperature of less than 80°C. If mercury is to be determined, losses may occur if the sample is dried at elevated temperatures prior to analysis. With the advent of good commercial equipment, freeze drying of samples has become popular. This method of drying is very desirable.

Plant material should be ground up to pass a 30 to 40 mesh sieve. The grinding process is a serious potential source of contamination. Conventional mills used for this purpose contain metal parts that may release varying amounts of chromium, nickel, cobalt, molybdenum, copper, and iron. If the sample is small enough, grinding in a ball mill, constructed of high-purity alumina and containing alumina balls is preferable.

Now that the plant material is ready for analysis it must be dried again before weighing because moisture is readily reabsorbed by many plant tissues, especially during humid periods. Again, drying should be done at a temperature less than 80°C.

5.1.1.3 Ashing

As indicated earlier there is much controversy regarding the merits of dry versus wet ashing. In summary the most satisfactory approach is a treatment with nitric and perchloric acids (with or without sulfuric acid.) Mercury, however, may be lost during such a digestion.

Dry ashing must be done at a relatively low temperature. A temperature of between 450 and 500°C is usually chosen. However, loss of some metals (e.g., lead, cadmium, zinc) at this temperature may still occur. Arsenic, selenium, and mercury will, of course, be lost.

Plant tissue with the exception of seeds is often relatively high in siliceous material. Thus for most samples hydrofluoric acid should be included during the wet ashing steps. The residue resulting from dry ashing is frequently quite high in silica or silicates and must be treated with hydrofluoric acid to remove trapped trace metals. Acids used for decomposition of plants must usually be of very high purity because of the relatively low levels of trace metals in such material.

5.1.2 Animal Material

Animals fill important niches in ecosystems that are being studied for trace metal cycling. In recent times, for example, fish and shellfish have been widely analyzed in areas where heavy metal contamination has occurred. Unlike most plants, animals are mobile, which means that great care is necessary in understanding their movements and how this may affect the conclusions of trace metal studies.

Animal material covers a wide range of compositions. Sample types include muscle, fat, organs, body fluids, bones, and nails. Generally, however, the silicon content of most animal samples is lower than that of plant material. Thus there is much less frequently a need for hydrofluoric acid in decomposition mixtures.

As with plant material there is considerable controversy about suitable methods for sample preparation. Much of the discussion on plant samples also applies here.

Samples must be guarded against deterioration prior to analysis. For this reason samples should be kept in containers to prevent moisture loss and then stored cold or frozen. Freeze drying has become a popular method for preparing animal tissue samples.

As with plant material wet ashing using oxidizing acid mixtures is the generally preferred method for sample decomposition. Mixtures of nitric acid and perchloric acid with or without sulfuric acid are commonly employed. Some workers have used sulfuric acid with dropwise addition of hydrogen peroxide successfully for elements such as arsenic and selenium in both muscle and fat tissue. Dry ashing commonly results in losses of the more volatile trace metals and thus is seldom employed.

5.1.3 Trace Metal Content of Plant and Animal Materials

Tables 5.1 and 5.2 give selected trace metal contents of plant and animal samples, respectively (1).

5.2 PROCEDURES

A few ICPAES procedures have been developed for the simultaneous multielement determination of trace metals in biological samples. A selection is included here.

Procedures given in the following section involve wet digestion of the samples with an oxidizing acid mixture. Hydride generation is recommended for arsenic, bismuth, antimony, and selenium. In this case both AAS and ICPAES can be used for the determinative step.

5.2.1 Determination of 23 Elements in Agricultural and Biological Samples by ICPAES*

This method is for the determination of 23 elements (As, B, Ba, Bi, Ca, Cd, Co, Cr, Cu, Fe, K, Mg, Mn, Mo, Na, Ni, P, Pb, Sb, Se, Sr, Ti, and Zn) in agricultural samples. Accuracy was assessed using NBS standard materials. The NBS samples used were Bovine Liver, Orchard Leaves, Pine Needles, Rice Flour, Spinach, Tomato Leaves, and Wheat Flour. In addition, a Japan National Institute for Environmental Studies Pepperbush sample was also employed. Suitable agreement between the results and the certified values was obtained for the 23

*Procedure reprinted from reference 2 p. 803 by courtesy of Marcel Dekker, Inc.

Table 5.1 Selected Trace Metals in Dried Plant Materials in ppm

Element	Plankton[b]	Ferns[c]	Gymnosperms	Angiosperms[d]
Ag	0.25	0.23	0.07	0.06
Al	1000	—	65	550
As	—	—	—	0.2
Au	—	—	—	<0.00045
B	—	77	63	50
Bi	—	—	—	0.06
Cd	0.4	0.5	0.24	0.64
Co	5	0.8	0.2	0.48
Cr	3.5	0.8	0.16	0.23
Cu	200	15	15	14
Fe	3500	300	130	140
Hg	—	—	—	0.015
Li	—	—	—	0.1
Mn	75	250	330	630
Mo	1	0.8	0.13	0.9
Ni	36	1.5	1.8	2.7
Pb	5	2.3	1.8	2.7
Se	—	—	—	2.0
Sn	35	2.3	<0.24	<0.3
U	—	—	<0.35	0.038
V	5	0.13	0.69	1.6
Zn	2600	77	26	160

[a] From Bowen (1).

[b] Mainly diatoms.

[c] Does not include horsetails or club mosses.

[d] Woody species.

elements. The three decomposition procedures evaluated were dry ash, nitric acid–hydrogen peroxide, and nitric acid–perchloric acid. The dry ash procedure has problems due to losses of volatile elements. Results for the wet ashing methods were nearly comparable. The nitric acid–perchloric acid procedure is given here because it gives slightly better recoveries for some determined elements. Hydride generation was used for arsenic, bismuth, antimony, and selenium.

Equipment. The three plasma direct-reading spectrometers used are listed in Table 5.3. The operating conditions for the instruments used with a direct nebulization technique and a hydride generation cell are listed in Table 5.4. The analytical lines used to perform the determinations are listed in Table 5.5. In

Table 5.2 Selected Trace Metal Content of Dried Animal Tissue in ppm

Element	Mollusca	Crustacea	Pisces	Mammalia
Ag			11?	0.006
Al	50	15	10	<3
As	—	—	—	0.2
Au	—	—	0.0003	<0.009
B	20	15	20	<2
Co	2	0.8	0.5	0.3
Cr	—	—	0.2	<0.3
Cu	20	50	8	2.4
Fe	200	20	30	160
Mn	10	2?	0.8	0.2
Mo	2	0.6	1	<1
Ni	4	0.4	1	<1
Pb	0.7	0.3	0.5	4
Se	—	—	—	1.7
Sn	15?	0.2	3?	<0.16
U	—	—	—	0.023
Zn	200	200	80	160

[a] From Bowen (1).

general, for elements likely to be present over a fairly wide dynamic range in the samples, alternate lines were employed for at least one instrument. Background correction was used for all analytical lines in Table 5.5 except for the sensitive lines of major elements. In systems I and II, background was measured at a position of 0.04 nm to the high wavelength side of lines in the first order (0.02 nm for second order), which is equivalent to a wavelength shift of about 1.33 band passes. For system III, background was measured as the average of the two measurements made at both +0.04 nm and −0.04 nm in the first order (0.02 nm in the second order) from the spectral line used for the analytical measurements.

Figure 5.1 shows a schematic of the hydride generation cell used in this study. Sample solutions after stabilization, reduction to the correct valence state, and pH adjustment were pumped through one channel of a multichannel peristaltic pump. Sodium borohydride solution that was stabilized in sodium hydroxide was pumped through another channel of the same pump, and the two liquids merged in a recessed cup inside the polypropylene reaction cell. The hydrides and hydrogen formed by the chemical reaction in the small cup were then swept by an argon carrier flow into the plasma torch itself; the liquid in the cup overflowed and was expelled to waste.

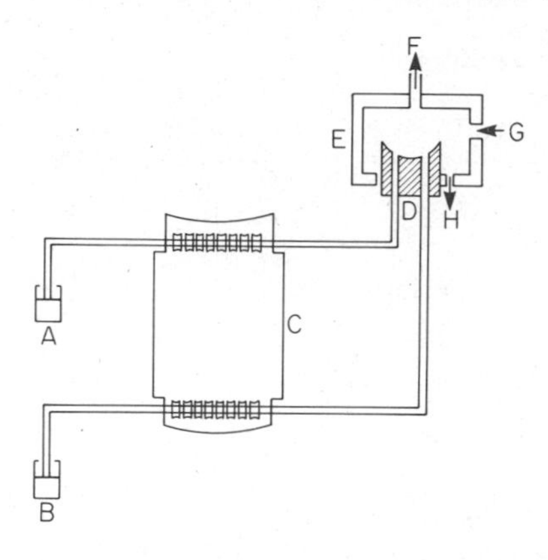

Figure 5.1 Hydride generation system (2).

Table 5.3 Instrumentation

	System I	System II	System III
Plasma direct reader	Model 96-976, Jarrell-Ash, Waltham, MA	Model 96-965, Jarrell-Ash, Waltham, MA	Model 96-1160, Jarrell-Ash, Waltham, MA
Mounting	Ebert		
Dispersion	1.6 nm/mm		
Spectral band pass	0.03		
Spectrometer	0.75 m	0.75 m	0.75m
Mounting	Paschen–Runge	Paschen–Runge	Paschen–Runge
Dispersion	0.55 nm/mm	0.55 nm/mm	0.55 nm/mm
Spectral band pass (1st order)	0.03 nm	0.03 nm	0.03 nm
Monochromator	96-978, 0.5 mete Jarrell-Ash	—	—
Computer	Model PDP-8E, DEC, Maynard, MA	Model PDP-8E, DEC, Maynard, MA	Model PDP-8E, DEC, Maynard, MA
Memory	8K	8K	124K
Terminal	LA36	LA36	LA36
Mass storage	RX02 disk	RX02 disk	RL01 disk
Storage memory	512K words	512K words	10M words

Reagents. All reagents were reagent grade. The borohydride reagent was prepared by dissolving 40 g sodium borohydride in powder or pellet form and 1 g sodium hydroxide in about 750 ml distilled water. After dissolution was complete, the solution was further diluted to 1 L.

Table 5.4 Operating Conditions

	Systems I and II	System III
	Direct Nebulization	
Gas flows		
Coolant or plasma	18 L/min	18L/min
Sample	0.7 L/min	0.7 L/min
Auxiliary	Off	Off
Nebulizer uptake rate	1.2 mL/min	1.2 mL/min
Peristaltic pump	Minipuls II	Minipuls II
Solution rate	1.3 mL/min	1.3 mL/min
Observation zone	18 mm above coil	18 mm above coil
RF power Forward	1.1 kW	1.1 kW
Reverse	<5 W	<5 W
Preexposure time	30 sec	30 sec
Integration time	14 sec	14 sec
Background		
high line wavelength	7 sec	7 sec
low line wavelength	0 sec	7 sec
Washout time	10 sec	10 sec
	Hydride Generation	
Gas flows		
Coolant or plasma	18 L/min	
Sample	0.8 L/min	
Auxiliary	Off	
Peristaltic pump	Minipuls II	
Sample pump rate	3.0 mL/min	
Borohydride pump rate	3.0 mL/min	
Observation zone	16 mm above coil	
RF power Forward	1.3 kW	
Reverse	<5 W	
Preexposure time	60 sec	
Integration time line	10 sec	
Washout time	30 sec	

Procedure. Transfer 1 g of oven-dried sample to a Teflon beaker; add 10 mL concentrated HNO_3 and 2.5 mL concentrated $HClO_4$. Bring the sample very slowly to boiling on a hot plate and heat to dryness. If sample blackening occurs during the fuming stage, add HNO_3 dropwise. Cool the sample, redissolve in 10 mL water, add 1 mL concentrated HCl, and bring to volume in a 25 mL volumetric flask. Analyze the solution against calibration curves established

Table 5.5 Analytical Lines Used (nm)

Element	System I	System II	System III
Ag	328.068	328.068	—[a]
Ai	308.215	308.215	308.215
As	193.696	193.696	193.696
B	249.773 × 2[b]	249.678 × 2	208.959
Ba	493.409	493.409	—
Be	—	234.861	234.861
Bi	223.061 × 2	223.061 × 2	223.061 × 2
Ca	317.933[c]	317.933[c]	315.887[c]
Cd	228.802 × 2	228.802 × 2	—
Co	228.616	228.616	228.616
Cr	267.716	205.552	205 × 552 × 2
Cu	324.754	324.754	219.958
Fe	259.940	259.940	238.204
K	776.490[c]	776.490[c]	404.721
Mg	279.553[c]	279.553[c]	383.231
Mn	257.610	257.610	257.610
Mo	202.030	202.030	281.615
Na	589.592[c]	588.995[c]	330.237
Ni	231.604 × 2	231.604 × 2	231.604 × 2
P	214.914 × 2[c]	214.914 × 2[c]	—
Pb	220.353	220.353	283.306
Sb	217.581	—	206.833
Se	196.026	—	196.026
Sn	189.980	283.999	—
Sr	421.552	421.552	421.552
Te	—	—	214.231 × 2
Ti	334.941	334.941	—
Tl	377.572	—	190.864 × 2
V	292.402	292.402	290.882
Zn	213.856	213.856	213.856

[a] Signifies element not present in that system.

[b] × 2 indicates line used in second order.

[c] Indicates no background correction employed.

using the standards listed in Table 5.6. Take a HNO_3–$HClO_4$ blank through the same sample preparation.

Hydride Generation. From the previous digestion, transfer 10 mL of the solution to a 25 mL volumetric flask. Add 1 mL of 10% ammonium iodide and 5 mL of concentrated HCl, and then bring the solution to volume. Analyze the samples using calibration curves established for the elements and the standards listed in Table 5.6 and a 10% HCl blank.

Calibrate the instrument using a blank and the standards listed in Table 5.6. When the sample is introduced into the plasma by nebulization, four replicate readings are taken for the standards. Use the average values to calculate the slope and intercept of a linear calibration. Then analyze using a 30-sec preexposure to permit the system to equilibrate before the integration period is started. A single determination is made on each sample and a 10-sec rinse is employed between samples. Thus the sample cycle time is either 61 sec for systems I and II or 68 secs for system III. The quality control standard is determined every tenth sample. The instrument is recalibrated if the drift-in elements at or above 1 mg/L are above 3% or above 10% for elements present at the 0.25 mg/L level.

A similar routine is used for the hydride generation technique. Perform instrument calibration using a blank and the standards listed in Table 5.7. Take four replicate exposures on each standard, and the average values are used to calculate the slope and intercept of the linear calibrations. A 60-sec preexposure is employed to permit system equilibrium, the extra time being required because of the extra dead volume in the pump tubing compared with nebulization. A 30-sec washout period is also used for the same reason. Duplicate determinations are made on each sample solution, resulting in a cycle time of 110 secs or,

Table 5.6 Calibration Standards for Nebulization Analysis

Standard #1	500 mg/L	Ca, K, P
	100 mg/L	Na, Mg
Standard #2	10 mg/L	Al, Ba, Cu, Fe, Mn, Pb, Sr, Zn
Standard #3	1 mg/L	Ag
Standard #4	1 mg/L	As, B, Be, Bi, Cd, Co, Cr, Mo, Ni, Sb, Se, Sn, Te, Tl, V
QC standard	100 mg/L	Ca, K, Mg, Na, P
	1 mg/L	Al, As, B, Ba, Cu, Fe, Mn, Pb, Sr, Zn
	0.25 mg/L	Cd, Co, Cr, Mo, Ni

Table 5.7 Calibration Standards for Hydride Generation

Standard #1	100 μg/L	As, Se, Te
Standard #2	100 μg/L	Bi, Sb, Sn
QC standard	20 μg/L	As, Bi, Sb, Se

allowing for a quality control standard every tenth sample, 29 samples each hour can be analyzed. Perform recalibration after a drift of greater than 10% in the quality control standard, which typicall occurs about every hour.

5.2.2 Determination of 19 Elements in Biological Materials and Soils by ICPAES*

This is an extensive study using an instrument capable of simultaneous multielement analysis. It contains valuable sections on interferences (spectral, ionization, and transport), solute vaporization interferences, and sample preparation methods. In addition, a good assessment of accuracy is given based on a comparison of results obtained with accepted values for a variety of standard reference materials. Results are also compared with those obtained with atomic absorption spectrometry. The elements determined were Na, K, P, Ca, Mg, Fe, Cu, Zn, Mn, Pb, Cd, Co, Cr, Ni, V, Ti, Al, Sr, and Ba. Best compromise instrumental conditions were employed.

Wet digestion was compared with three dry ashing procedures. The latter were as used by three independent laboratories (Illinois Natural History Survey; Department of Soil Sciences—University of Minnesota; and Hasler Research Center—Applied Research Laboratories). As has been stated earlier, there is potential for loss of the more volatile elements (in this study, lead, zinc, and cadmium) when dry ashing is employed. Table 5.8 is a compilation of the results obtained with NBS Orchard Leaves for the ashing procedures employed (see procedure for details).

There is a general trend to decreasing recovery with increasing ashing time. Also method C tends to give consistently low recoveries for iron, nickel, vanadium, and aluminum. In addition to the problem of possible losses, it is important to consider the upper practical limit on the amount of sample that can be obtained in a given volume of solution. With the wet ashing procedure the upper limit is about 1 g/50 mL. For dry ashing the upper limit is 1 g/5 or 10 mL of solution. Because for many biological samples the concentration of trace elements is very low it is important to have as little dilution as possible. Thus the dry ash would be preferred if applicable.

Spectral Interferences. It is important to emphasize that although background correction may be necessary at any concentration, generally the effect of background correction becomes more and more important as the detection limit is

*Procedure from Dalquist and Knoll (3).

Table 5.8 Comparison of Sample Preparation Methods on Orchard Leaves (NBS)[a]

	Method[b]			
Element	A	B	C	D
P	—	86.7	92.4	90.5
K	—	98.3	102	107
Ca	97.4	94.7	98.2	94.3
Mg	99.2	98.2	102	102
Fe	98.9	89.6	68.5	89.3
Cu	108	93.3	109	125
Mn	91.2	100	104	103
Zn	108	96	107	112
Pb	93	91	107	—
Ni	—	113	78	—
Cr	—	(94)	(39)	—
Co	—	(91)	(91)	—
V	—	(110)	(69)	—
Al	—	(133)	(58)	—
Sr	95	(97)	(99)	7
Ba	124	(106)	(108)	—

[a] Percentage recovery compared with assigned value. Numbers in parentheses are comparisons with results that are *not* certified.

[b] Methods: A—1 : 1 conc. HNO_3: conc. $HClO_4$;
B—500°C in Vycor 0, 14 h (HNO_3–HCl);
C—485°C in covered Vycor, 4–8 h (HCl);
D—200°C, 8 h; 485°C, 18 h in Pyrex tubes (HNO_3–HCl).

approached (of course, background may greatly affect the detection limit, that is, the detection limit is often much worse in solutions with high salt contents).

Wavelength scanning was used to evaluate interference problems. Corrections were made by establishing quantitative relationships between analyte and interferent. With biological samples problems could be encountered, particularly from the alkaline earth elements cadmium and manganese and from iron.

Transport (Nebulization) Effects

Using a concentric pneumatic nebulizer the effect of concentrations of various mineral acids was studied. For a 1.0 mm internal diameter tube the relative uptake rate compared with distilled, deionized water for 10% HCl, 35% HNO_3, 35% $HClO_4$ and 48% H_2SO_4 was 92, 73, 72, and 31%, respectively. If such a problem occurs, the authors recommend using an internal standard.

Solute Vaporization Interference. It is often difficult to delineate between solute vaporization interferences and transport effects. The authors cite the classic PO_4^{3-} effect on Ca^{2+} as a case in point. Even at very high PO_4^{3-} to Ca^{2+} ratios, when the signal depression was 60% there was no evidence that this problem was due to solute vaporization interference. By using an internal standardization method to correct for transport effects, no problem due to solute vaporization interference could be demonstrated even at PO_4^{3-} to Ca^{2+} ratios of up to 30,000.

Ionization Interference

Ionization effects caused by the presence of high and variable amounts of an easily ionizable element are a serious problem with classical atomizer–excitation sources. However, with the ICP these researchers found no appreciable ionization effect at the levels commonly encountered in biological samples.

Equipment. The inductively coupled plasma quantometer (ICPQ) used in this study was a standard commercially available instrument system consisting of a direct-reading polychromator model QA-137 and an ICP excitation source, with provision for pneumatic nebulization of liquid samples and direct introduction of the aerosol from the spray chamber into the plasma axial channel. Instrumental details are given in Table 5.9. Two-cavity interference filters with 1 nm bandwidth (full width at half maximum) were used for soil extract and dry-ashed plant tissue determinations of sodium and potassium (types J-15 and J-21, respectively; Dell Optics, North Bergen, NJ). Viewing height of the filter photometers was 50–60 mm above the radio-frequency coil. During the same work, 20 mm bandwidth filters (Acton Research Corp., Acton, MA) were installed in the optical system at the analytical wavelengths indicated in Table 5.9 for rejection of residual far scattered radiation.

A metal-free, all-plastic environment with prefiltered air holding the work space at positive pressure and enclosing class 100 vertical laminar flow fume hoods was used for sample preparation.

Fluorinated ethylene propylene or polyethylene (PE) used as a liquid sample or reagent containers were precleaned with 1% v/v HNO_3 and then "normalized" for 48 h minimum (typically for weeks to months) with the same acid reagent and concentration to be used for final determinations. Pyrex class A volumetric ware and silicon carbide boiling chips used for wet digestion were refluxed for 24 h minimum with concentrated $HClO_4$ prior to use.

Reagents. Distilled, deionized water used for cleaning and sample preparation was prepared by passing Arrowhead (Los Angeles, CA) distilled water succes-

Table 5.9 Apparatus and Operating Conditions

Inductively Coupled Plasma	Spectrometer
Generator: Air-cooled, 3 kW rating, continuous	*Mount:* 1.0 m Paschen–Runge, f30
Frequency: 27.1 MHz, crystal controlled, within ISM band, less than 2% ripple on rf envelope	*Slits:* 12 m entrance, 50 m
Power: 1600 ± 50 W forward power, <2 W forward power, <2 W reflected, <150 W coupling loss	*Gratings:* 1920 rulings/mm, interferometrically ruled quartz blank replica blazed at 270 mm; reciprocal linear dispersion, 0.48–0.52 nm/mm in first order; range, 180–460 nm; 1080 rulings/mm, interferometrically ruled quartz blank replica blazed at 600 nm; reciprocal linear dispersion: 0.92, 0.46, and 0.31 nm/mm in first, second, and third orders; range, 180–820 nm
Regulation: To forward power, $<1.0\%$	*Transfer optics:* A vertical 4 mm segment of the axial channel of the plasma focused on the primary slit with 0.7 demagnification using a 15 cm focal length fused quartz lens
Autostart: Automatic impedance match to coil for start, ignition, sample injection, and run at preset power level	*Observation height:* 17 mm above the rf coil (vertical aperture defines range of 15–19 mm)
Coil: Two-tern silver-plated copper 4.7 mm o.d. tube on 28.7 mm diameter form; water cooled	*Analytical wavelengths (nm):* $\overline{\text{Zn}}$ 202.5, $\overline{\text{P}}$ 213.6, Mn 257.6, Fe 259.9, Mg 279.5, Cu 324.8, Ca 393.4, Al 396.2, Sr 407.8, Ba 455.4, $\underline{\text{Na}}$ 588.9, $\underline{\text{K}}$ 766.5, $\overline{\text{Pb}}$ 220.3, $\overline{\text{Ni}}$ 231.6, $\overline{\text{Co}}$ 238.9, Cr 283.6, V 311.1, Cd 226.5 Elements underlined used 20 nm photomultiplier tube (see text).
Impedance matching network: Series-parallel capacitance	*Readout:* PDP 11/05, 16 K words, computer with DecWriter console and paper tape or floppy disk data and program (BASIC language integrated software systems); successively segmented third-order regression defining calibration curves; second-order regression defining calibration curves; second-order regression defining interference; corrections on a concentration basis.
Torch: Integral, quartz, 18 mm o.d., three concentric tube configuration	
Gas: Argon—welding grade coolant, 10.5 L/min; plasma, 1.5 L/min; aerosol carrier, 1.0 L/min	
Gas flow control: Three-stage regulation of pressure, capillary restrictors, interlocked	
Nebulizer: All glass, permanently aligned; coaxial pneumatic; computer-controlled desalter	
Spray chamber: Conical straight-through type, 30 mm p.d. by 90 mm; of "Scott" type, 100 cm^3	
Enclosure: rf gasketed to conform to FCC regulations, and safety interlocked	

sively through the Barnstead (Boston, MA) organic, layered bed, mixed bed, and cation exchange cartridges, and was finally filtered at 0.45 μm.

Inorganic compound solutions used for preparation of calibration reference solutions and as digestion catalysts were obtained from Alpha Inorganics (Ventron, Beverly, MA) or J. T. Baker (Phillipsburg, NJ). Concentrated HNO_3, HCl, and $HClO_4$ were twice distilled in Vycor; the HCL was stored in polyethylene, and the HNO_3 and $HClO_4$ used for wet digestions were stored in sealed Vycor ampuls (G. Frederick Smith Co., Columbus, OH).

One master solution was prepared containing matrix elements (Na, K, P, Ca, and Mg) and trace elements of interest. This multielement solution was then diluted to appropriate concentrations for use in calibrations. The synthetic reference solutions for calibration were prepared in the same acid matrix as the tissue samples, that is, 17% by weight $HClO_4$ for wet digestions and 1.25 M HCl for ashed tissue. Reference solutions for soil extracts were prepared by elemental additions to the blank extractant solutions.

Procedure. Wet Digestion (Method A). Lyophilized botanicals and animal soft tissue should be desiccated at 25°C for not less than 20 h in a diffusion-pumped vacuum system at not more than 0.05 Pa (2×10^{-6} torr). Lyophilized serum controls are reconstituted as directed by the manufacturer, and contact of the reconstituted sera with the rubber stopper of the vial must be avoided to minimize contamination.

The wet digestion procedure used is described by Feldman (4). Digestion is carried out in volumetric flasks, and samples are brought to volume without transfer. A 1:1 mixture of concentrated HNO_3 and $HClO_4$ is used in the proportions of 1 g tissue/25 mL or 2 mL sera/5 mL acid mixture. All tissue digestions are performed in 50 or 100 mL volumetric flasks and the sera digestions in 10 mL volumetric flasks. (Volumetric neck condenser extension as recommended by Feldman (4) were not employed because efficiency of the volumetric flask necks as condensers was improved by placement in the vertical laminar flow hood.)

Add 1 mg each of chromium and vanadium per gram of lyophilized tissue or per 2 mL serum to the digest as potassium dichromate and ammonium vanadate. These reagents serve both as catalysts and as indicators of completion of oxidation of organic compounds in the digest. When the digestion is complete, cool samples to room temperature, bring to volume, and run within the day or transfer to normalized polyethylene bottles and store at 8 to 10°C until use. Final $HClO_4$ concentration is 17% by weight.

Dry ashing (Method B). Ash lyophilized plant (2.0 g) and animal tissues (1.0 g) in Vycor for 14 h at 500°C. When the samples are cooled, add 5 mL concentrated HNO_3 and evaporate slowly to dryness on a hot plate. Return material to the muffle furnace for 30 min. After cooling, add 5 mL concentrated HCl, heat the mixture to dissolve the residue, and dilute to the final volume of 50 mL. Place solutions in polyethylene bottles.

Method C. Ash plant tissues at 485°C in covered Vycor crucibles in a temperature programmed and controlled muffle furnace for 4 to 8 h. Add 10 mL of 1.25 M Ultrex HCl to each sample in a polyethylene vial containing ash from 1.0 or 2.0 g of tissue. Shake the vials and allow to stand a minimum of 30 min. Pour the supernatant into acid-leached Pyrex tubes and centrifuge prior to aspiration.

Method D. Lyophilized plant tissues are desiccated as described previously, weighed to 1.000 ± 0.002 g, transferred to acid-leached 10 × 100 mm Pyrex tubes, and covered with similarly treated 13 × 100 mm tubes. Place the covered tubes in a drying oven for 8 h at 200°C, then transfer to a Lindberg furnace, brought to 485°C, and maintain for 18 h (time required to produce a uniform white or gray ash). Add 2 mL concentrated HNO_3 twice distilled in Vycor to each 1.0 g aliquot in the same tube used for ashing. Transfer to a 160°C heating block in a Class 100 vertical laminar flow fume hood and take to dryness. Return the tubes containing the ash residue to the 485°C oven for 20 min and cool. Add 10 mL 1.25 *M* Ultrex HCL. Stir the mixture in the tube while covered with parafilm and centrifuge before aspiration. (Use of the Pyrex tubes precludes accurate determination of sodium in tissue prepared in this manner. Additionally, over a period of three days the lead determinations are observed to increase with storage time, particularly at low concentrations.)

Soil Extraction. (Equilibrium extraction of exchangeable/extractable cations.) Add 30 ml of neutral 1.0 *M* ammonium acetate to 3 g of air-dried soil, shake for 30 min, and centrifuge. The supernatant is analyzed directly.

Extractable elements (Mn, Fe, Zn, Cu, Ni, and Pb) can also be obtained from 10 g of dry soil shaken for 2 h with 20 mL 0.01 *M* $CaCl_2$ and 0.1 *M* triethanolamine-buffered 0.005 *M* DTPA and adjusted to pH 7.3 with HCL. Reference solutions must be prepared in the blank extraction solutions described.

Determination. Bring samples to room temperature and centrifuge in Teflon (fluorinated ethylene propylene) or Pyrex test tubes that have been leached in 1% v/v HNO_3 for 2 months minimum before use and then have been rinsed copiously in distilled, deionized water. Centrifugation of the samples is employed to avoid either positive or negative contamination that is possible using filtration and to prevent particulate plugging of the nebulizer. Introduce solutions to the nebulizer 40 sec before integration of photomultiplier currents to allow the spray chamber to reach equilibrium. Rinse the nebulizer and spray chamber for 10 sec with the blank acid matrix and follow with a 30 sec deionized water rinse prior to introduction of the next sample.

Three separate and successive 10 sec integrations of photomultiplier currents are acquired for the elements of interest in each solution. The average measures of the three integrations are transformed to concentration units. Internal standardization is not used.

Data Corrections. Spectral interferences due to the influence of major concomitants and any potential spectral interferences based upon known unresolved spectral line overlaps are qualitatively identified by wavelength scans of the background structure near analyte wavelengths (above 0.08 nm).

Spectral interference correction coefficients are experimentally derived by relating the combined net measures of residual stray light and spectral overlaps to concentration of the interfering species using single-element solutions prepared in the solvent under consideration.

Identification of significant reagent impurities is made by wavelength scanning. Net line measures above off-peak background are used to determine impurity concentration either by comparison with single-element reference solutions and linear extrapolation or by standard additions. Nominal assigned concentrations of the multielement reference solutions for calibration are adjusted for intrinsic reagent impurities and residual scattered light due to concomitants when present at significant levels.

5.2.3 ICPAES and Hydride Generation AAS Method for 18 Elements in Plant and Animal Tissues*

The procedure employed is diagramed in Figure 5.2. The sample (plant or animal tissue) is digested in a three-acid mixture ($HNO_3/HClO_4/H_2SO_4$). Other acid combinations including $HNO_3/HClO_4$, and $HNO_3/HClO_4/HF$ were also evaluated, but these were found to be inferior. The ternary acid mixture with the addition of HF can be used when a siliceous residue is obtained. The digest is finally diluted to 100 mL.

Subsequently 95 mL is treated by a modification of the Chelex-100 procedure of Kingston et al. (6). Two separate fractions are taken: (1) containing Al, Co, Cd, Cu, Mo, Ni, Pb, V, and Zn for ICP analysis and (2) containing As, Se, and Sb for hydride atomic absorption analysis. The remaining 5 mL fraction is diluted to 25 mL. Then Ca, Fe, K, Mg, Mn, and P are determined directly by ICP.

The elements in brackets in Figure 5.2 can be determined under favorable conditions (as judged from results obtained on an appropriate standard reference

*Procedure from Jones et al. (5).

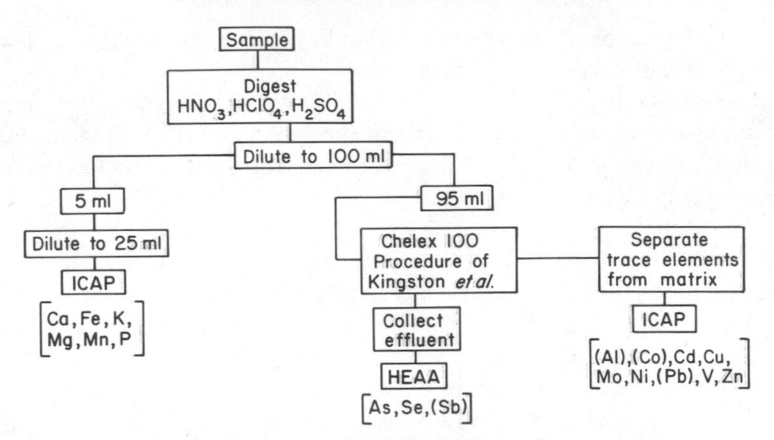

Figure 5.2 Procedural schematic diagram (5).

sample). For example, Al can be done, using the procedure, in samples not yielding siliceous residues. If these are obtained, HF must also be employed. The procedure was evaluated using NBS standard reference samples. For some elements there is no NBS certification for a given standard. In these cases values obtained from literature sources are employed. The following are comments on the behavior of the individual elements.

Aluminum. Losses of Al were obtained when using the ternary acid and Chelex procedure. This element must usually be done with the addition of HF to the ternary acid mixture.

Cadmium and Nickel. Good agreement with literature values was obtained for Ni in spinach and pine needles. Also good agreement with literature values was obtained for Cd in tomato leaves. Nickel seemed to be associated with the silicate phases, so HF may be necessary.

MOLYBDENUM. In the case of brewer's yeast, the addition of HF gave higher results. Generally, however, Mo did not seem to be trapped in siliceous residues.

LEAD. In general, Pb results were poor by the proposed procedure. Lead values seemed, in most cases, to be affected by the siliceous residue. The elution with ammonium acetate caused variable recovery of Pb. However, a good Pb value for pine needles was obtained using the Chelex procedure without ammonium acetate elution.

CHROMIUM AND MANGANESE. These two elements behaved poorly by the chelation procedure. Chromium was removed from the elements covered by the proposed procedure. Manganese could frequently be detected and analyzed in the fraction not subjected to the Chelex procedure.

IRON. Recoveries of soluble iron for most SRMs using the proposed ternary acid/Chelex procedure were greater than 90%. However, in most cases, Fe could be measured directly in the fraction not subjected to the Chelex procedure.

VANADIUM. Good results for V depend on the use of the proposed ternary acid mixture. Strangely, other mixtures including those with HF give lower results in most cases (only brewer's yeast with ternary acid plus HF mixture gave higher results).

ZINC. In some samples such as pine needles and brewer's yeast, Zn values were slightly low without an HF treatment. In many instances Zn could be detected and determined without Chelex separation.

COBALT. The ternary acid mixture plus HF gave good Co results in tomato leaves and in spinach. The authors, however, state that more verification is needed.

ARSENIC, SELENIUM, AND ANTIMONY. Little difference in the results were obtained for HF both present and absent with As and Se. With As, the occurrence of refractory organic As compounds made it essential to use the ternary acid mixture. There were not enough SRMs with suitable values of Sb to assess its behavior in the proposed procedure.

Table 5.10 gives the minimum analyte quantity that is necessary for reliable determination of a given element.

Table 5.10 Approximate Minimum Analyte Masses Required for Accurate Analysis by Proposed Multi Element Scheme

Element	Mass/μg	Step[a]	Element	Mass/μg	Step[a]
Al[b]	2.0	ICAP-C	Mn	2.5	ICAP–NC
As	0.20	HEAA	Mo	0.90	ICAP-C
Ca	2.5	ICAP-NC	Ni	0.15	ICAP-C
Cd	0.15	ICAP-C	P	250	ICAP-NC
Co	0.30	ICAP-C	Pb	1.5	ICAP-C
Cu	0.15	ICAP-C	Se	0.20	HEAA
Fe	5.0	ICAP-NC	Sb	0.20	HEAA
K	250	ICAP-NC	V	0.15	ICAP-C
Mg	5.0	ICAP-Nc	Zn	0.15	ICAP-C

[a]C = Chelexed fraction; NC = non-Chelexed fraction.

[b]Aluminum appears to be amenable to the scheme, provided that glassware contamination can be avoided.

Equipment. The ICAP measurements were performed with a Jarrell-Ash Model 975 Plasma Atom Comp. 0.76-m direct-reading spectrometer operated at 1.1 kW source power at 27.12 MHz. Potassium measurements were performed on an attached Jarrell-Ash 0.5-m scanning monochromator (n + 1). A limited number of cobalt measurements were obtained on a second 0.75-m Jarrell-Ash plasma spectrometer. Analytical lines and standardization concentrations used for all measurements are included in Table 5.11. Emission measurements were performed at 16 mm above the induction coil. The argon coolant plasma flow was 18 L/min. The Jarrell-Ash cross-flow nebulizer system was

Table 5.11 ICAP[a] and Hydride AAS Analytical Lines and Measurement Conditions

Element	Line (nm)	Standardization Concentration (μg/mL)	Background Correction
Cd	226.5	1.00	Yes
Al	308.2	10.0	No
Co[b]	228.6	1.00	Yes
Fe	259.9	10.0	No
Cr	357.9	1.00	Yes
Mo	203.8	1.00	Yes
Ni	231.6	1.00	Yes
V	292.4	1.00	Yes
Ca	393.3	5.00	No
Cu	324.7	5.00	Yes
Mn	257.6	5.00	Yes
Pb	220.4	5.00	Yes
Zn	206.2	5.00	Yes
Mg	279.5	25.0	No
P	214.9	100	No
Ca	370.6	200	No
K[c]	766.5	200	No
As[d]	193.7	—	Yes
Se[d]	196.0	—	Yes
Sb	217.6	—	Yes

[a]ICAP measurements were made at 16 mm above the coil.

[b]Cobalt measurements were performed at 228.6 nm using an independent ICAP polychromator system.

[c]Potassuim measurements were obtained on an n + 1 0.5-m scanning monochromator.

[d]Determined by HEAA using calibration graphs prepared from 25 to 500 ng of each element.

slightly modified by bubbling the nebulizer argon (0.5 L/min) through a fritted 20 × 5 cm cylindrical column of deionized water located at the gas inlet of the nebulizer. The resulting water-saturated nebulizer argon reduced crusting of salts at the cross-flow nebulizer needle orifices. Solutions were delivered to the nebulizer by a peristaltic pump (Gilson Minipuls 2) at 1.1 mL/min. Wavelength modulation (spectrum-shifter) background correction was used for some elements, as indicated in Table 5.11, to compensate for minor baseline drift during the analyses.

Hydride evolution AAS measurements were performed on a Perkin-Elmer Model 403 atomic absorption spectrometer using a semiautomatic hydride generator as described in Fiorino et al. (7).

Borosilicate volumetric ware was first handwashed with hot tap water and rinsed with distilled water, then soaked in 20% v/v HNO_3 for 48 h, and rinsed with distilled, deionized water before use. Borosilicate 100 mL Kjeldahl digestion flasks were cleaned by boiling them in a mixture of concentrated HNO_3, $HClO_4$, and H_2SO_4 and rinsing with copious amounts of deonized water before use.

Resin columns consisted of 200 × 8 mm polypropylene columns with a polyethylene frit resin support (Kontes Co., No. K-420160). Sample and reagent reservoirs were 125-mL FEP Teflon separating funnels (Nalgene 4301–0125). The drain stems of the separating funnels were removed and replaced with polypropylene micropipette tips cut to fit securely into the Telfon drain stopcock of the funnel. Reservoirs and columns were cleaned before the resin was loaded by soaking in warm 20% HNO_3 for 2–3 h, followed by rinsing with copious volumes of distilled, deionized water.

Reagents Concentrated nitric and sulfuric acids: distilled by subboiling procedure from quartz.

Concentrated perchloric acid; redistilled from Vycor (G. F. Smith Chemical Co., Columbus, OH).

Concentrated hydrofluoric acid: analytical reagent grade.

Concentrated ammonia solution: analytical reagent grade.

Chelex 100: iminodiacetate chelating ion exchange resin, 200–400 mesh, Na^+ form (Bio-Rad Laboratories, Richmond, CA).

Metal standard solutions, 1000 ppm: commercial or prepared from high-purity metals or metal salts.

Water: distilled and deionized (Milli-Q, Millipore Corp.).

Ammonium acetate solution: 1 *M*. Prepared from analytical reagent grade crystals. As prepared, the solution was contaminated with unacceptable levels of copper, zinc, and lead; it was cleaned before use by adjusting the pH to 5.3 with HNO_3 or ammonia solution and passing the solution through columns of Chelex-100 resin (NH_4^+ form). Although the resulting reagent was clean enough for this study, a better preparation is to mix high-purity acetic acid and ammonia solution.

Sodium tetrahydroborate (III): high-purity pellets (Ventron Corp. Beverley, MA).

Sodium iodide crystals: analytical reagent grade.

Concentrated hydrochloric acid: analytical reagent grade.

Procedure. Weigh 1–3 g of dry sample into a 100 mL Kjeldahl flask. Add 25 mL conc. HNO_3, 5 mL conc. $HClO_4$, and 2 mL conc. H_2SO_4. Heat until the vigorous reaction of $HClO_4$ has ceased. Then raise the heater temperature and react until all the $HClO_4$ has been expelled and dense white SO_3 fumes are given off. Cool and dilute to 100 mL with distilled, deionized water. If HF is needed, filter the sample, obtaining the residue on a polyethylene frit (prewashed with HNO_3). Dissolve the residue by adding 0.5–0.9 mL hot 47% HF directly to the material in the frits. Return this solution directly to the original solution and dilute to 100 mL as before.

The separation conditions used by Kingston et al. (6) were followed as closely as possible except for the column reservoir system, which was inconvenient for this application. Instead, the simpler gravity flow system described under apparatus was used.

Prepare resin columns by pipetting 10.0 mL of a magnetically stirred slurry of 30 g of resin in the Na^+ form plus 150 mL of water in to the column. Treat the approximately 2 g of water-saturated resin successively with two 15 mL portions of water, two 15 mL portions of 15% v/v HNO_3, two 15 mL volumes of water to rinse excess of acid from the resin, 10 mL of 15% v/v ammonia solution to convert the resin into the functional NH_4^+ form, and finally, two 15 mL volumes of water to remove excess of ammonia. A 2 cm layer of water should be left above the resin bed.

Divide the 100 mL of diluted digest solution into two fractions. Transfer a 5 mL volume of this solution into a 25 mL calibrated flask and dilute to volume with 20% v/v $HClO_4$ (the non-Chelexed fraction). Transfer the remaining 95 mL into the Teflon separating funnel. Adjust the pH to 5.3 $\pm$ 0.2 with concentrated ammonia solution followed by dilute ammonia solution and dilute HNO_3 as required. Add a 1 mL volume of 1 *M* ammonium acetate solution (pH 5.3) to buffer the solution. (At this pH some NBS reference materials developed a slight flocculent white precipitate, which did not interfere with the separation other than to slow the flow rate of sample through the resin.) Pass the sample solution through the gravity-flow column by repetitive manual filling from the separating funnel. Collect the effluent in a Teflon bottle, cap and store for the determination of As, Se, and Sb (the hydride fraction).

Wash the column with 40 mL 1 *M* ammonium acetate solution (pH 5.3) to remove sequestered alkali and alkaline earth metals. Rinse the excess of ammonium acetate from the resin with 10 mL of water. Next add 10 mL of 15% v/v HNO_3 to the column to strip the sequestered trace elements from the resin. Collect the HNO_3 eluent in calibrated 25 mL stoppered, graduated cylinders. Rinse residual HNO_3 from the resin with a single 4 mL volume of water, which is collected in the cylinders. Dilute the eluents to 15 mL with water.

(The shrinking and swelling of the Chelex-100 resin with changes in sorbed ions do not drastically affect the separation time (usually 4–8 h), although different eluent flow rates are observed for different cationic forms of the resin.)

Inductively Coupled Plasma Procedure. Determine concentrations of all solutions under the conditions in Table 5.11. Perform two-point calibrations using a standard blank and the standard concentrations shown in Table 5.11. Base calibrations on the average of two sequential 21 sec exposures of multielement standards, that is, 14 sec on line and 7 sec on background. Measure all sample intensities by averaging two back-to-back 21 sec integration periods. Correct these intensity integrations for interelement interferences and/or background fluctuations as required.

Measure each non-Chelexed fraction and the corresponding Chelexed fraction three times: first, against one calibration using multielement standards in 20% v/v $HClO_4$ or 10% v/v HNO_3 for non-Chelexed and Chelexed fractions, respectively, and then by two subsequent calibrations and measurements, for a total of six plasma measurements of each digested sample. These data provide sufficient information to establish an estimate of the plasma contribution to the overall analytical variability. Measure the non-Chelexed fractions directly by plasma against multielement standards in 20% v/v $HClO_4$. The purpose of the relatively concentrated $HClO_4$ is to provide a sufficiently viscous solution matrix such that minor differences in residual acid concentrations between sample digests would be "swamped" by $HClO_4$ and therefore would not contribute to variability in aerosol transport to the plasma.

Hydride AAS Procedure. Collect the initial column effluents of the neutralized sample solution in Teflon bottles (treated with 48 mL concentrated HCl) and dilute to 160 mL with water. Transfer aliquots of this solution ranging from 1 to 20 mL (depending on the analyte concentration) into test tube reaction cells and dilute to 20 mL as necessary with 30% v/v HCl. Prepare calibration standards covering the range 25–500 ng of each analyte in 30% HCl. Treat samples for arsenic measurement with sodium iodide prereductant to convert the pentavalent species into trivalent states. Generate hydrides of arsenic and selenium by adding a basic 4% m/v solution of sodium tetrahydroborate (III) from a semiautomatic hydride generator. Pass analyte gases directly into a hydrogen (nitrogen-diluted) flame for measurement of the transient absorption signals.

5.2.4 Determination of Pb, Cd, Ni, and Cr in Biological Samples by Furnace AAS*

The procedure has been used for the analysis of blood, urine, liver, oyster tissue, tuna, spinach leaves, pine needles, and tomato leaves. A matrix modifier of ammonium phosphate is used to allow increased ashing temperatures to be employed. The sample is placed on a L'vov-type platform. No significant matrix interferences were found. The procedure was tested on NBS standard reference samples and the results were in acceptable agreement with the certified values.

Equipment. All analytical work was performed using a Perkin-Elmer Model 603 atomic absorption spectrophotometer with the HGA-2200 graphite furnace and temperature-ramp accessory. Background correction was done on all measurements. The maximum power mode of heating was used to obtain a faster temperature increase during the atomization stage of the heating cycle. A Perkin-Elmer Model 056 strip-chart recorder provided readout for peak-height measurements.

Sample introduction to the furnace was done via a Perkin-Elmer AS-1 autosampler. Analysis was based on duplicate 10 μL sample injections. Argon was used for the sheath and purge gas. The internal gas flow of 300 mL/min was reduced to 20 mL/min during atomization. Temperature calibration was checked with an Ircon Model 2-30C30 recording optical pyrometer.

Pyrolytically coated graphite tubes were used. All tubes were tantalized using the method of Zatka (9) to extend tube life. Platforms were made by cutting the two ends of a tantalized graphite tube into eight (four from each end) 7 $\times$ 5 mm grooved, curved sections. A stainless steel milling blade was used to make two perpendicular cuts parallel to the tube axis into the ends of the tube. A narrow piece of masking tape was wrapped around the ends, and a third cut was made perpendicular to the tube axis and 7 mm in from the end, giving four platforms.

The platform was placed in a graphite tube, which was then positioned in the furnace head. The right window was temporarily removed and the platform centered within the graphite tube directly beneath the sample port using a metal rod. Adjustment of the injector tip was made to ensure that it did not come into contact with the platform surface.

Use of the platform required an additional 10 sec cool-down period to return to ambient temperatures for the next sample. This was accomplished by making a minor electrical adjustment on the HGA-2200 power supply–controller board (Perkin-Elmer Part No. 0290–9014). Instead of the autosampler being triggered by the end of the atomization stage, the trigger was taken from the recorder

*Procedure from Hinderberg et al. (8).

interrupt signal. Users who wish to make a similar modification should consult with their service engineer for details on how to proceed.

Reagents. All standards were made in 1% v/v HNO_3 by appropriate dilution of 1000-μg/mL solutions from Fisher Scientific Co. A 5% solution of ammonium phosphate ($NH_4H_2PO_4$) was prepared from the reagent grade compound. Impurities were removed by complexing with ammonium pyrrolidine carbodithioate (APCD) followed by extraction into methyl isobutyl ketone (MIBK). The modifier solution was diluted 1 + 5 by the sample to give 0.8% $NH_4H_2PO_4$ in the injected solution. High-purity NHO_3 was prepared by subboiling distillation from an all-quartz still.

Procedure. Blood and urine samples are digested in order to obtain homogeneous solutions. Add 2 mL of purified HNO_3 to 1 g of whole blood in a 2 oz polyethylene bottle, heat in a hot water bath at 100°C for 1 h, and dilute to 20 mL with deionized water. This gives a 1:20 dilution of the original sample in 10% HNO_3.

Urine samples are prepared by a Kjeldahl NHO_3 digestion procedure. Place 25 mL urine aliquots in 100 mL Kjeldahl flasks with 10 mL HNO_3 and heat. After reduction of the volume to 5 mL, dilute the digested samples to 25 mL with deionized water, giving about 20% HNO_3 solutions.

Liver and vegetation samples are also prepared by Kjeldahl HNO_3 digestion. Digest samples of 0.25 to 0.50 g with 15 mL HNO_3. Dilute the digested samples (approximately 5 mL final volume) to 50 mL with deionized water, giving about 10% HNO_3 solutions. Run the samples in the furnace using conditions in Table 5.12.

5.2.5 Determination of Mo in Botanical Samples by Furnace AAS*

The samples of plant material are dissolved using nitric acid alone (no perchloric acid is employed). Recovery studies showed a 94% recovery of added molybdate. I question the validity of such an addition approach because the molybdate is not taken up by the sample. Recovery of molybdenum from NBS Orchard Leaves was excellent (0.39 μg/g certified value). Presumably it would be possible to use perchloric acid if desired, which would involve only a slight modification of the procedure.

Equipment. A Varian-Techtron Model AA6 atomic absorption spectrophotometer and Model 90 carbon rod atomizer were used for the flameless determination of molybdenum. A spectral band pass of 0.5 nm was used to isolate

*Procedure from Neuman and Munshower (10).

Table 5.12 Furnace Temperature Settings

Element	Wavelength (nm)	Spectral Band Pass (nm)	Char Temperature (°C)[a]	Atomization Temperature (°C)[b]
Pb	283.3	0.7	950	2100
Cd	228.8	0.7	750	2100
Cr	357.9	0.7	900	2600
Ni	232.0	0.2	900	2600

[a]All char settings provide for a 15-sec temperature ramp with a hold time of 15 sec.

[b]Atomization temperatures were by maximum power mode with an auto-burnout (2700°C for 5 sec) added for Pb and Cd. Drying was at 100°C (15 sec ramp, 25 sec hold) from the tube wall and 200°C from the platform.

the wavelength of 313·3 nm and peak-height absorbance signals were displayed on a Varian A-25 recorder set at 2 mV range. The carbon rod atomizer parameters (temperature, time) for the dry, ash, and atomize cycles were 90°C/30 sec, 700°C/30 sec, and 2750°C/3 sec, respectively. Glassware used in the determination was rinsed before use with dilute HNO_3.

Reagent. Standards containing 0, 10, 20, 30, 40, and 50 ng Mo/mL prepared from an ammonium molybdate solution containing 1000 mg Mo/L were treated identically as the samples.

Procedure. Grind dried plant tissues with a stainless steel Willey mill to pass a 40 mesh screen and place a 2.000 g sample in a 250 mL Erlenmeyer flask. After addition of 20 mL 7 *M* HNO_3 (Baker Ultrex), heat the mixture to a gentle boil for 20 min, cool, and filter into a 100 mL volumetric flask using Whatman No. 42 paper. Dilute the solution to volume with distilled, deionized water. Pipette 5 μL aliquots onto the carbon rod atomizer for molybdenum determination. Analyze the sample using the above heating cycle.

5.2.6 Determination of As in Botanical Samples by Hydride Generation and ICPAES*

Hydride generation is done in a disposable plastic syringe of 20 mL capacity. A reaction time of 20 sec is employed. During this period the syringe and contents are gently shaken. Samples that have been analyzed by the procedure include orchard and tomato leaves and kale and cabbage samples. Signal integration was employed to obviate the need for close replication of the timing of the

*Procedure from Pickford (11).

hydride insertion step. Standards were prepared by addition of arsenic solutions to arsenic-free cabbage that was then taken through the procedure. Results obtained on standard reference samples agree satisfactorily with the certified values.

Equipment. A Plasmatherm 2500A ICP source was used and was operated at 1.5 kW with a coolant flow rate of 13 L/min and a sample injector flow rate of 1 L/min. The observation height was 16 mm above the load coil. Emission signals were measured with a Spex 1704 1 m monochromator at a band pass of 0.3 nm. The arsenic 228.8 nm line was used throughout. Analog output from the Spex DPC-2 photon counting unit was displayed on a chart recorder, using a 10 sec photometer integration time. This displayed emission signals as one or more steps, which were summed for the total intensity.

Reagents. For preparing standards, known amounts of arsenic (as a 1000 μg/mL solution) in the range 1–10 μg were added to 1 g portions of a dried cabbaged sample that, from previous measurements using instrumental neutron activation analysis and gamma activation analysis, were known to contain insignificant amounts of arsenic. This material, with the arsenic additions, was then carried through the wet ashing step. A filtered 2% sodium tetrahydroborate (III) solution containing 1% of sodium hydroxide was used for the hydride generation. This was prepared daily.

Procedure. Weigh 1 g vegetation samples, previously oven dried at 80°C, into cleaned 50 mL calibrated flasks, add 10 mL concentrated HNO_3 and 1 mL of concentrated $HClO_4$ (both should be Aristar grade reagents). After dispersion of the solid material, heat the flasks to 50°C over a period of 20 min until frothing has ceased, and then heat over a period of 2 h until refluxing of $HClO_4$ occurs. At this point a small amount of colorless slurry remains. Dilution to 50 ml with 25% HCl produces a clear or slightly turbid solution.

After peaking the monochromator on the chosen line, using a 100 μg/ml solution of arsenic and the conventional pneumatic nebulizer, activate the chart drive, and run samples at discrete time intervals.

Draw a 4 mL sample of the acid solution into a disposable syringe. Everett 20 mL disposable syringes with No. 1 needles were used. This type of syringe was found particularly convenient because it has a silicone sealing ring and a ridged plunger that acts as a reproducible stop. Dip the needle into the alkaline sodium tetrahydroborate (III) solution and draw the plunger back rapidly to the stop. Invert the syringe and hold the needle tip with a paper tissue. Gently shake. Immediately insert the needle into the open end of the silicone nebulizer tube of the ICP and depress the plunger steadily so as to inject the 20 mL of hydrogen and arsine into the nebulizer chamber in about 5–10 sec. Provided signal integration is used, the timing of this step is not critical. The arsenic signal is then seen as successive steps on the chart recorder, which are summed for the total intensity. One sample per minute is found to be a practical sample throughout.

5.2.7 Determination of Hg in Biological and Environmental Samples*

The following is a procedure involving wet digestion of organic samples followed by cold vapor atomic absorption spectrometry that is applicable to 0.5 g sample sizes. Digestion is done in Folin digestion tubes held in an aluminum block. Addition of V_2O_5 acts as an oxidizing agent and the digestion proceeds first with HNO_3. Following this, H_2SO_4 is added and the digestion is continued. Samples that are known to be handled by this procedure include grass, leaves, fish tissue, liver, sewage sludges, and soils. A detection limit of 0.01 μg/g of Hg was obtained. Accuracy was assessed using NBS standard orchard leaves and bovine liver.

Equipment. A weighing device consisting of the barrel of a 50 mL disposable plastic syringe for holding a Folin digestion tube and hung in the balance was employed. Samples were digested in Folin digestion tubes calibrated at 25 and 50 mL. The tubes were heated in an aluminum hot block (26.5 × 19.0 × 7.5 cm) with 24 holes 2.54 cm in diameter and 5.6 cm deep. The block was heated on a small hot plate capable of generating and maintaining a temperature of 160°C.

Mercury measurements were made using a Fisher Scientific Hg-3 mercury meter equipped with a Hewlett-Packard recorder (Model 7101B-24). A 250 mL bottle fitted with an aspirator was used to generate mercury vapor for measurement.

Reagents. Stannous chloride (10%) was prepared by diluting a 40% stannous chloride solution (in concentrated HCl) with distilled water. Both solutions should be stored in dark bottles.

Mercury standard solution (1000 ppm) was prepared by dissolving 1.353 g of reagent grade $HgCl_2$ in 1 L of distilled, deionized water. A working standard solution (10 ppm) was prepared from the stock by diluting 10 mL of the stock standard solution of 1 L to which 10 mL concentrated HNO_3 and 13 mL 0.25 *N* $K_2Cr_2O_7$ solution had been added. This solution was stable for at least 3 months. Nitric acid and $K_2Cr_2O_7$ may contain some mercury and the working standard solution should be checked against a similar dilution of the stock solution that does not contain any preservatives in order to obtain its true concentration.

Procedure. Weigh 0.5 g of sample into a Folin digestion tube using the special weighing device. Add 80 to 100 mg reagent grade V_2O_5 followed by 10.0 mL

*Reprinted in part with permission from (12) Knecthel and Frazer, *Anal. Chem.* **51,** 315. Copyright 1979, American Chemical Society.

of concentrated HNO_3. After the foaming subsides, heat the tube and contents in a hot block (at 160°C) for 5 min. Remove the tube and cool before adding 15.0 mL concentrated H_2SO_4. Replace the tube in the hot block for 15 min, then remove and allow to cool. A blank digestion containing all the reagents must also be done along with samples.

Transfer the digested sample cautiously to a graduated cylinder and make up to a total volume of 100 mL with deionized water. Transfer the whole sample or a suitable aliquot diluted to 100 mL to the Hg reduction flask. After addition of 10 mL of 10% stannous chloride, place the aspirator into the Hg reduction flask as rapidly as possible, and volatilize the Hg into the absorption cell using compressed air (1800 mL/min).

5.2.8 Determination of Sn in Marine Organisms by Hydride Generation and AAS*

Samples that were analyzed by the procedure include algae, mollusc tissue and shell, visceral mass and gills, and fish muscle. The sample is dissolved in a 2-step method, first with HNO_3 and then with a mixture of HNO_3, H_2SO_4, and $HClO_4$. Tin is then coprecipitated as the hydroxide using lanthanum and the precipitate is dissolved in H_2SO_4. Stannane is produced by reaction with sodium tetrahydroborate (III). Interfering elements include (μg/mL): Ag(I) $>$ 10; As(V) $>$ 5; Cr(VI) $>$ 100; Co(II) $>$ 10; Cu(II) $>$ 5; Fe(III) $>$ 100; Hg(II) $>$ 50; Mo $>$ 50; Ni(II) $>$ 1; Sb(V) $>$ 10; Se(VI) $>$ 10; and Te $>$ 50. The detection limit is 0.9 ng/mL.

Equipment. A Varian-Techtron Model 1200 atomic absorption spectrometer fitted with a silica T-tube mounted 1 cm above a stoichiometric air–acetylene flame was used. The T-tube was 14 cm long, 0.8 cm i.d. and had a 0.3 cm i.d. side arm. The side arm was shielded by a silica sheaf with an annulus to prevent premature decomposition of stannane. The hydride generation system was a modification of the system described by Thompson and Thomerson (14). For convenience a burette was added to measure the sodium tetrahydroborate (III) solution and a stopcock was fitted at the bottom for the removal of waste solution. A mixture of 1% v/v air in nitrogen was used to flush stannane from the hydride generation system through a silicone rubber tube to the T-tube furnace.

Procedure. Samples were stored frozen to minimize possible changes by bacterial action. Samples of macroalgae were washed with distilled water to remove salts, then freeze dried and ground to pass a 300 μm sieve. Fish and molluscs were separated into their component tissues. Samples of each tissue were homogenized and subsamples were used.

*Procedure from Maher (13).

Weigh biological material (<2 g dry) into a 100 mL Erlenmeyer flask. Add 10 mL concentrated HNO_3 and place a glass bulb on top. Allow the mixture to stand for at least 12 h and then heat until the evolution of brown fumes ceases. After cooling, add 10 mL HNO_3–H_2SO_4–$HClO_4$ mixture (10 + 2 + 3, v/v) and continue heating until dense fumes of SO_3 appear. Transfer the sample digest to a 30 mL polypropylene centrifuge tube with 10 ml of distilled water. Add 1 ml of 5% La(III) solution and three drops of phenolphthalein solution with stirring. Add 25% ammonia solution until a faint pink color appears. Separate the La precipitate containing Sn by centrifugation (discard the supernatant liquid) and redissolve in 10 mL 0.35 M H_2SO_4.

Measure a 2 mL portion of 2% sodium tetrahydroborate solution into the hydride generation chamber. Adjust the inert gas flow rate to 2 L/min and pass the gas until a stable baseline is established. Inject the sample solution (1.0 mL) as quickly as possible into the sodium tetrahydroborate solution with a plastic syringe. Measure the tin absorbance at 286.33 mn. Turn off the recorder when the signal returns to the previously established baseline.

The apparatus adsorbs small quantities of stannane; thus before sample solutions were processed, a concentrated tin standard is injected through the system to saturate any stannane adsorption sites. Blanks for the entire procedure are typically less than 3 ng/ml and derived mainly from the sodium tetrahydroborate(III) solution.

REFERENCES

1. H. M. Bowen, *Trace Elements in Biochemistry.* Academic Press, London, 1966.
2. A. F. Ward, L. F. Marciello, L. Carrara, and V. J. Luciano, *Spectrosc. Lett.* **13,** 803 (1980).
3. R. L. Dalquist and J. W. Knoll, *Appl. Spectrosc.* **33,** 1 (1978).
4. C. Feldman, *Anal. Chem.* **46,** 1606 (1974).
5. J. W. Jones, S. G. Capar, and T. C. O'Haver, *Analyst* **107,** 353 (1982).
6. H. M. Kingston, I. L. Barnes, T. C. Rains, and M. A. Champ, *Anal. Chem.* **50,** 2064 (1978).
7. J. A. Fiorino, J. W. Jones, and S. G. Carper, *Anal. Chem.* **48,** 120 (1976).
8. E. J. Hinderberg, M. L. Kaiser, and S. R. Koirtyohann. *At. Spectrosc.* **2,** 1 (1981).
9. V. J. Zatka, *Anal. Chem.* **50,** 538 (1978).
10. D. R. Neuman and F. F. Munshower, *Anal. Chim. Acta* **123,** 325 (1981).
11. C. J. Pickford, *Analyst* **106,** 464 (1981).
12. J. R. Knechtel and J. L. Frazer, *Anal. Chem.* **51,** 315 (1979).
13. W. Maher, *Anal. Chim. Acta* **138,** 365 (1982).
14. K. C. Thompson and D. R. Thomerson, *Analyst* **99,** 595 (1974).

CHAPTER

6

FOOD SAMPLES

6.1 INTRODUCTION

The recognition of a need for determining metals in foods has existed for at least 160 years. This resulted, for example, in England, from the 1860 Parliamentary passage of the "Act for Preventing the Adulteration of Articles of Food and Drink." It is, however, only relatively recently (i.e., after the mid-1930s) that methodology and equipment existed for reliable trace analysis work. Prior to the advent of analytical atomic absorption spectrometry in 1955 trace element analysis was done by solution absorption spectrophotometry, emission spectrometry, and electroanalytical chemical techniques. The importance of atomic absorption spectrometry to trace element analysis has been emphasized repeatedly throughout this book.

6.1.1 Essential Trace Elements

With the exception of a few trace metals in foods there are still a large number of problems being encountered in the determination of these elements in foods. The elements that at present are known to be essential in human nutrition are Fe, I, Cu, Mn, Zn, Co, Se, Cr, Mo, Sn, F, V, As, Ni, and Si. Table 6.1 compiled by Wolf (1) lists the essential trace elements with the date of the discovery of a requirement, whether a recommended dietary allowance (RDA) has been established, whether subadequate intake has been identified, and whether fortification recommendations have been made.

It is interesting that except for Fe, I, Cu, Mn, Zn, and Co the essentiality of the trace elements listed in the table was discovered in the second half of the twentieth century. The uses of trace metal analytical data in nutrition are as follows:

1. Establishing that a trace metal is or is not essential.
2. Determining the amount of metal required for proper nutrition.
3. Identification of inadequate or excessive levels of trace metals in diets.
4. Prevention of nonoptimal intakes of trace metals.

To determine a metal's essentiality and the level at which it is needed it is important to identify the biological role the element plays. Once a biological role

Table 6.1 Essential Trace Elements[a]

Element	Discovery of Requirement	Recommended Diet Allowance	Subadequate Intakes Identified	Fortification
Fe	17th century	Yes	Yes	Yes
I	1850	Yes	Yes	Yes
Cu	1928	Pending	?	More data needed
Mn	1931	No	No	No
Zn	1934	Yes	Yes	Yes
Co	1935	No (B_{12})	No (B_{12} yes)	No
Mo	1953	Pending	No	No
Se	1957	Pending	Yes	More data needed
Cr	1960	Pending	Yes	More data needed
Sn	1970	No	Experimental	Unknown
V	1971	No	Experimental	Unknown
F	1972	Pending	Experimental	Unknown
Si	1972	No	Experimental	Unknown
Ni	1973	No	Experimental	Unknown
As	1975	No	Experimental	Unknown

[a]Reprinted with permission from (1), Wolf, *Anal. Chem.* **50,** 190A. Copyright 1978 American Chemical Society

has been established, a biological parameter affected by the trace metal can then be chosen for monitoring that will allow establishment of the optimum level of the trace element being studied. The classical example of this would be the monitoring of blood hemoglobin levels as a way of determining the effect of dietary iron. In this way the recommended dietary level of the nutrient metal can be established. From Table 6.1 it can be seen that only for iron, iodine, and zinc have recommended dietary allowances been established. The recommended dietary allowance for copper is pending.

If dietary deficiencies are established, then recommendations for fortification of foods can be made. Only for iron, iodine, and zinc have such recommendations been forthcoming. One of the best known of these would be the fortification of salt with iodine to help in the prevention of goiter. In the case of selenium, although there is no supplementation of human foods, the feeding of selenium to animals is widely practiced.

6.1.2 Status of Trace Metal Food Analysis

At the present time there is a tremendous need for trace metal food compositional data. This is particularly important for the trace metals iron, zinc, copper, chromium, and selenium. These data will only be useful if they are obtained by analytical methods that have been properly and critically evaluated. Basic parameters that are important in establishing a method for analysis of foods are precision, accuracy, sensitivity, and suitability for quality control and ability to automate. Freedom from interference is crucial because of the wide range of matrices that may be encountered. Variations in food composition that are geographical or due to processing and change in seasons must be discernable; thus the precision of the method must be good enough to detect significant variations of this type. Generally precision must be better than 5%. Sensitivity can be a serious problem, for the method must be sensitive enough to determine background levels of the nutrients. For this purpose the method should have a detection limit that is at least 5 times lower than these background levels.

Accuracy of methods must be thoroughly tested. This can be done using standard reference samples, spiking, and recovery studies, and through interlaboratory comparisons (round robins). Performance of the method in routine use must continually be assessed by using internal laboratory standards and standard reference materials.

Simultaneous multielement methodology is important to cut down on time required for analysis. Thus ICPAES is potentially an attractive approach to analysis of trace metal nutrients and toxicants. Unfortunately in many cases the detection limits by this technique are not adequate to determine background levels of these substances.

Also important in speeding up analyses and cutting down on manipulative errors is automation. In this regard the relatively new technique of flow injection analysis is important. This approach not only results in rapid analyses but also requires very small amounts of equipment, reagents, and samples and is thus very economical.

Nutrient metals for which good methodology exists are iron, copper, and zinc. However, improvements can usually still be made regarding automation of such procedures. For the other elements existing methodology is still deficient.

Methodology for determining the different species of the metals present in food does not exist for most applications. It is the form of the element and not its total amount that determines its bioactivity. It is crucially important that speciation methodology be developed as soon as possible. Even more difficult will be the development of standard reference samples for trace metal speciation.

6.1.3 Food Sample Preparation

Sample preparation, sampling, and sample preservation are important in food analysis. Contamination and losses during such steps can be a serious source of error.

Washing may or may not be needed in food analysis, depending on the form in which the food exists. For example, market garden vegetables may be contaminated with soils. Such material can usually be cleaned by gently washing with a mild detergent followed by a thorough rinsing with water.

Homogenization of a sample can usually be done using a blender. Care must be exercised not to contaminate the food through contact with the metal parts of the blender, although soft foods do not present a great problem. However, some types of plant material could pick up significant contamination from this process. The sample is then commonly sieved to 30 or 40 mesh at this stage. Sieving can also be a source of contamination.

Results from a food analysis should be expressable on a wet weight basis. Thus it is important that no loss of moisture occur until the sample is weighed. Subsequently the sample can be dried either conventionally or by freeze drying. Freeze drying is favored by many workers today.

A good review on determining metals in foods has been published by Crosby (2). This publication gives a good coverage of the literature up to 1977.

6.2 PROCEDURES

There is an important need in the analysis of foods for methods that are simultaneous multielement. In this regard ICPAES methods are given for 14 commonly determined trace metals in foods and for the hydride-forming elements. Atomic absorption methods are given for a variety of trace metals in several combinations. The determination of lead in foods is particularly important in recent times and methods are given for its determination in a large variety of foodstuffs.

6.2.1 Determination of 14 Metals in Foods by ICPAES*

The sample is decomposed using a pressure dissolution technique. This involves treatment for 30 min under pressure in a linear polyethylene bottle using 6 *M* HCl. Samples that have been analyzed include lettuce, peanuts, potatoes, soybeans, wheat, corn, spinach, and beans. The elements determined are Be, Ca, Cd, Cr, Cu, Fe, K, Mg, Mn, Mo, N, P, Pb, and Zn.

Standardization is carried out using solutions that are 6 *N* in HCl to match the samples. The low standard is 6 *N* HCl and the higher standards are NBS reference agricultural standards in 6 *N* HCl into which appropriate aliquots of standard solutions are added. The procedure was evaluated using NBS Wheat and NBS Spinach samples. Satisfactory agreement with accepted values was obtained.

*Reprinted in part with permission from (3), Kuennen, Wolnicki and Friocke, *Anal. Chem.* **54,** 2146. Copyright 1982, American Chemical Society.

Equipment. The ICAP–polychromator used was a Jarrell-Ash Model 1140 Atom Comp (Jarrell-Ash Division, Fisher Scientific Co., Waltham, MA) capable of analyzing 16 elements simultaneously. The analytical wavelengths used with corresponding detection limits are given in Table 6.2. The data acquisition and readout system consist of a PDP 11/04 digital computer (Digital Equipment Corp., Maynard, MA).

The operating parameters are 0.9 kW forward radio-frequency power, 15.5 mm (above the load coil) observation height, 1.5 mL/min sample introduction rate (pumped), 23 L/min (air calibrated) argon coolant flow, and 5 sec signal integration time.

A nonadjustable cross-flow pneumatic nebulizer (No. 003444, Jarrell-Ash Division, Fisher Scientific Co., Waltham, MA) with sample argon flow of 0.68 standard liters per min (mass flow controller, Model FC-260, Tylan Corp., Torrance, CA) was used in this study. The sample argon flow was humidified by passing it through a glass frit positioned in a segmented Teflon column (Savillex Corp., Minnetonka, MN) filled three quarters fill with 18 MΩ /cm water prior to entry into the nebulizer. A two-piece glass impinger (Sherritt, Gordon Mines, Ltd., Alberta, Canada) was inserted into the spray chamber to increase aerosol residence time. Solution uptake was controlled with a peristaltic pump (Gilson, Minipuls II, Gilson Medical Electronics, Middleton, WI). A rotary shaker was used that was purchased from Fermentation Design, Allentown, PA, with 36-

Table 6.2 Wavelengths

Element	Ionization State	Wavelength, (nm)	Order	Detection limits, (ng/mL)	Background Correction[a]
Be	II	313.04	1	0.2	Y
Ca	II	317.93	1	5	N
Cd	I	228.80	2	1	Y
Cr	II	267.72	1	2	Y
Cu	I	324.75	1	2	Y
Fe	II	259.94	1	2	Y
K	II	766.49	1	100	N
Mg	II	383.20	1	20	N
Mn	II	257.61	1	0.5	Y
Mo	II	202.03	1	3	Y
Ni	II	231.60	1	6	Y
P	I	214.91	2	30	N
Pb	II	220.35	1	30	Y
Zn	I	213.86	1	1	Y

[a]Background measurement at +0.03 nm from the analytical wavelength; N + no, Y + yes.

sample capacity. The shaker was located inside a forced air oven (Hotpack, Philadelphia, PA).

All stages of sample preparation and analysis were carried out in a clean air environment. Prior to digestion or sample analysis, all glassware and plasticware were scrupulously cleaned by soaking overnight with 30% HNO_3, rinsed three times with distilled deionized water, and placed in a clean air environment until dry. Millipore Teflon filters, Nalgene Buchner funnels, and linear polyethylene bottles, which are reused after the sample dissolution, were placed into a warm Alconox solution to remove organic residues prior to the acid cleaning procedure. Thirty percent HNO_3 was passed through the Teflon filters, followed by rinsing with distilled deionized water. The filters were then allowed to dry prior to use.

Reagents. Distilled, deionized water (18 MΩ/cm), redistilled $HClO_4$, Instra-Analyzed; HN_3, Instra-Analyzed; H_2SO_4, Ultrex (J. T. Baker Chemical Co., Phillipsburg, NJ); doubly distilled $HClO_4$, 70% ACS reagent (G. F. Smith Co., Columbus, OH).

An ICP calibration standard containing 14 elements in 20% v/v H_2SO_4 was prepared from 1000 μg/mL (Be, Cd, Cr, Cu, Fe, Mn, Mo, Ni, Pb, and Zn) and 10,000μg/mL (Ca, K, Mg, and P) stock standards (Spex Industries, Metuchen, NJ) at final concentrations of 1 and 100 μg/mL, respectively. A working standard solution containing 100 μg/mL of Be, Cd, Cr, Cu, Fe, Mn, Mo, Ni, Pb, and Zn in 2% HNO_3 and stock standard containing 10,000 μg/mL of Ca, K, Mg, and P were used for making standard additions.

Procedure. Weigh 1 g dry weight portions of sample composite and transfer into preheated (80°C) 60 mL, wide-mouthed, linear polyethylene bottles containing 25 mL of 6 *M* HCl. Tightly seal the bottles, agitate, and place on a rotary shaker operated at 200 rpm for 30 min. Place the shaker in the forced air oven maintained at 80°C. Cool sample solutions to room temperature before filtration under vacuum through a Millipore teflon filter (Mitex, 47 mm, 10 μm pore size) housed in a Nalgene Buchner funnel (42.5 mm). Collect filtrates in 50 mL Erlenmeyer flasks and analyze by ICP–AES. The vacuum manifold shown in Fig. 6.1 is constructed of acrylic Plexiglass that is resistant to 6 *M* HC1, and therefore minimizes contamination. A solvent (mL) to sample (g) ratio of 25 is used to minimize organic matrix interferences. Determinations by ICP should be made within 24 h because after this time precipitates begin to form in the sample solutions. The ICP polychomator is calibrated by using two-point standardization. Six molar HCl is used as the low standard. The sample matrix standard used as the high standard consists of standard solutions added to aliquots of HCl dissolution of NBS standard reference materials and of laboratory reference materials (Table 6.3). A reference material corresponding to the crop

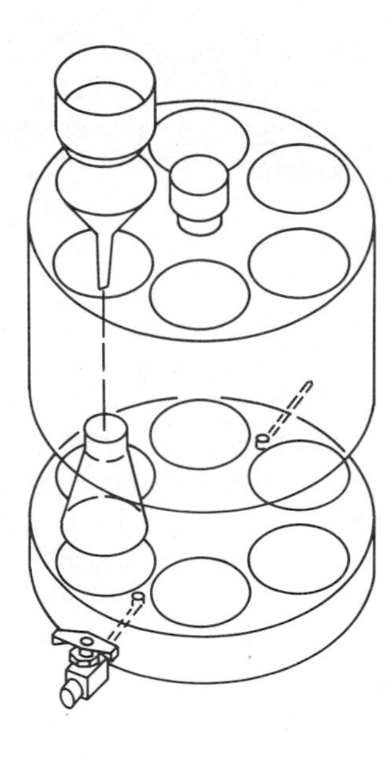

Figure 6.1 Vacuum manifold (3).

Table 6.3 Real Sample Matrix Calibration Technique

	Standard Contains:
Low standard no. 1 (14 elements)	6 *M* HCl blank
High standard No. 2 (Be, Cd, Cr, Cu, Fe, (Mn, Mo, Ni, Pb, Zn)	5 mL aliquot of real sample matrix solution (e.g., NBS Wheat) plus 5 μg Be, Cd, Cr, Cu, Fe, Mn, Mo, Ni, Pb, Zn
High standard No. 3 (Ca, Mg, K, P)	5 mL aliquot of real sample matrix solution (e.g., NBS Wheat) plus 500 μg Ca, Mg, K, P

Example

Element	NBS Wheat Certified Values (μg/g)	HCl Dissolution 1 g (dry wt.) /25 mL + Std. Added (μg/mL)	High Standard Computer Value (μg/mL)
Cu	2.00–0.3	0.08 + 1	1.08
Ca	190–10	7.6 + 100	107.6

being analyzed (e.g., NBS Wheat for wheat samples, laboratory reference sweet corn for sweet corn samples), is taken through the HCl pressure dissolution procedure with every set of samples. The sample matrix standard should be prepared (i.e., standards added) just prior to standardization.

6.2.2 Determination of Hydride-Forming Elements in Foods by ICPAES*

Simultaneous multielement detection of antimony, arsenic, bismuth, germanium, selenium, and tin separated by hydride generation and injection into an ICP forms the basis for this method. The gas–liquid separator U-tube is interfaced to the base of the ICP torch using a 165 ft coil. This coil is to minimize instabilities with the plasma when the sample argon flow must be interrupted. The presence of carbon dioxide in the gas phase causes a spectral interference with arsenic, germanium, and tin. It was not possible to eliminate carbon dioxide from the system with absorbants because these had a deleterious effect on some of the hydrides. Background correction should be employed. The detection limits for As, Bi, Ge, Sb, Se, and Sn are 0.02, 0.3, 0.6, 0.08, 0.1 and 0.8 (ng/mL), respectively. The RSDs vary from 2.5 to 8.9%. Several NBS standard reference materials—Rice flour, Wheat flour, Spinach, and Orchard Leaves—were analyzed as a test of accuracy. Results were for the most part in good agreement with the accepted values.

Equipment. The ICAP polychromator used in this work was a Model 1160 Atom Comp (Jarrell-Ash Division, Fisher Scientific, Co., Waltham, MA) capable of simultaneous analysis of 34 elements and equipped for three-point dynamic background correction at selectable wavelengths. The analytical wavelengths and operating parameters used in this study are listed in Tables 6.4 and 6.5. The hydride generation condensation apparatus is shown in Fig. 6.2. A simple three-way valve (I) is inserted prior to the reaction tube–carrier gas switching valve (II). A $\frac{1}{4}$ in. o.d. packed section of corrugated Teflon tubing is used as a condensation tube. The condensation tube (CT) is connected through a three-way, on–off valve (III) (Omnifit, Cedarhurst, NY) to a 2 in. column ($\frac{1}{4}$ in. o.d. PFA tubing) packed with an inert support material (Fluoroport T, 60 mesh) and 165 ft of open 0.04 in. i.d. PFA tubing. The end of open tubing is inserted through a rubber stopper at the base of the plasma torch to within 1 in. of the tip of the sample inlet tube.

Reagents. Distilled, deionized water with a metered resistance of 18 MΩ/cm and ultrahigh purity commercial acids were used to prepare all reagents, standards, samples and, where possible, the sample preparation was carried out in

*Reprinted in part with permission from (4), Kahn, Wolnick, and Fricke, *Anal. Chem.*, **54**, 1048. Copyright 1982, American Chemical Society.

Table 6.4 Analytical Wavelengths

Element	Wavelength (nm)	Background Correction[a]	Order
As	193.70	±	1
Bi	223.06	−	2
Ge	199.82	±	1
Sb	217.58	−	1
Se	196.03	−	1
Sn	189.98	−	2

[a] +, Background measurement at +0.03 nm (first order) −, background measurement at −0.03 nm (first order); ±, background measurement at ±0.03 nm (first order).

a class 100 clean air nonmetallic environment. The acid solution used for the hydride reaction with standards and samples was 15% v/v concentrated HCl with 10% concentrated H_2SO_4. Sodium borohydride, 4.0% w/v (98% powder, Alfa-Ventron, Danvers, MA) was dissolved in 10% w/v NaOH (ACS reagent) and filtered with suction through a medium porosity scintered glass filter to obtain a clear solution. Desiccants used in the generation–condensation system were $Mg(ClO_4)_2$ (ACS reagent) and $CaCl_2$ (anhydrous, 4 mesh, ACS reagent). The digestion acid was 3:2:1 v/v/v mixture of HNO_3–$HClO_4$–H_2SO_4 with 0.67 g/L NH_4VO_3 (ACS reagent) added as an oxidative catalyst. Standards were prepared by a serial dilution of 1000 μg/mL emission standards (Spex Industries, Metuchen, NJ).

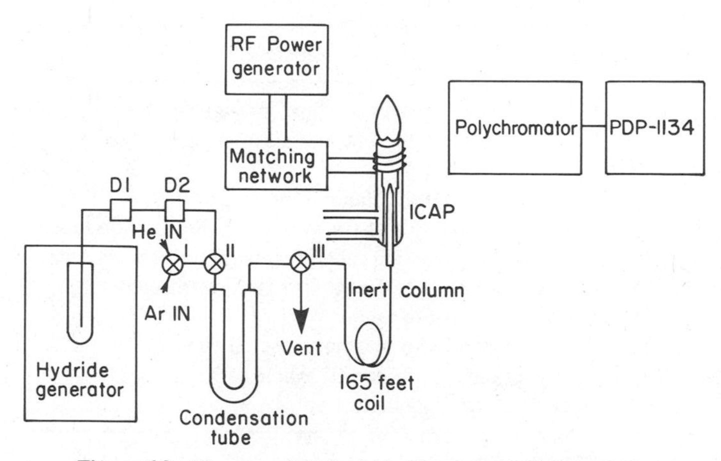

Figure 6.2 Diagram of the hydride Generation ICP System (4).

Table 6.5 Operating Parameters

ICAP Polychromator	
Forward Power (kW)	0.9
Reflected power (W)	<5
Obsvn. height (mm)	14
Argon coolant (L/min)	18
Auxiliary (L/min)	1
Carrier (L/min)	0.25
Exposure cycle (s) (8 cycles summed)	1, line; 1, high background; 1, low background; 1 line
Hydride Generation Reaction	
Acid mixture (10 mL)	15% v/v HCl–10% v/v H_2SO_4
$NaBH_4$ conc.	4% w/v in 10% NaOH
$HaBH_4$ delivery	1 mL/sec for 8 sec
pause (sec)	4
H_2O rinse	1 mL/sec for 4 sec

Procedure. Dissolve 2 g samples in 30 mL of digestion acid by heating to SO_3 fumes. Dilute to 50 mL in 15% HCl–10% H_2SO_4 v/v.

Tables 6.5 and 6.6 summarize the plasma operating parameters and the hydride evolution procedures, respectively. Use the steps outlined in these tables.

Table 6.6 Detection Sequence

Time (sec)	Operation[a]
0	I and II set for He flow through the system; II open to vent, closed to plasma.
0	
20	Place CT in liquid argon; start timer; II set so that CT is open to the reaction tube.
30	Place reaction tube on generator; start reaction.
55	Switch II to reestablish He flow through CT.
75	Switch I to argon.
85	Open III to plasma; close III to vent.
90	Transfer CT to 60° water bath.
100	Begin data collection.

[a]CT = condensation tube.

Table 6.7 Detection Limits to Be Expected from Proposed Method

Metal	Detection limits (μg/g)[a]	Normal Levels Found in Fish (μg/g)
Cadmium	0.02	0.02–0.05
Nickel	0.04	0.05–0.19
Lead	0.08	0.1–0.2
Chromium	0.09	
Copper	0.02	0.57–1.3
Zinc	0.2	0.9–1.5

[a]Twice the standard deviation at the lowest detectable level.

6.2.3 Determination of Cr, Cu, Zn, Cd, Ni and Pb in Fish by Flame AAS*

Samples of fish are digested in a heated aluminum block using a mixture of nitric and sulfuric acids at 150°C. Up to 5 g samples can be handled. Direct flame atomization of Zinc, copper, and chromium using the resulting solutions is employed. For cadmium, nickel, and lead a prior solvent extraction using ammonium tetramethylene dithiocarbamate and methyl isobutyl ketone is used to concentrate these elements for flame atomic absorption determination.

This method is reproduced here in keeping with my advice to use flame atomic absorption when possible. Even though an extra step, that is, solvent extraction, is required compared with furnace atomization, the flame method is probably more rapid than a furnace approach. This is because of the lengthy thermal program required for each element. The procedure was checked for accuracy using NBS Orchard Leaves and Bovine Liver and International Atomic Energy fish flesh. Results were in good agreement with certified values. Detection limits are given in Table 6.7.

Equipment. The aluminum hot block is constructed as shown in Fig. 6.3. Calibrated digestion test tubes are used in the block to digest the samples. The block is placed on a hot plate capable of heating the block to a constant temperature of 150°C. The temperature is monitored with a thermometer suspended in mineral oil placed in a tube in the block.

The aluminum block used in this study is a modification of that used by

*Procedure from Agemanian et al. (5).

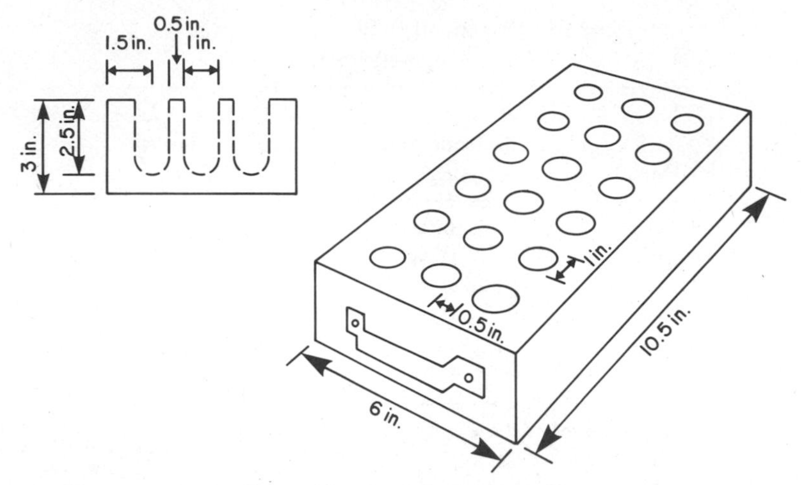

Figure 6.3 Aluminum hot block (5).

Bishop et al. (6). As large samples and large volumes of acid are used, it is essential to have as much of the digestion tube in the hot block as possible. The depth of the holes was therefore changed from 1.5 to 2.5 in. This change allows the sample–acid mixture to be heated more uniformly.

All the metals were analyzed using a Perkin-Elmer Model 603 atomic absorption spectrometer equipped with a triple-slot burner and deuterium arc background corrector. An acid-resistant nebulizer was used for the direct analysis of chromium, copper, and zinc from the acid extracts. A conventional nebulizer was used for the analyzes of cadmium, nickel, and lead after chelation and solvent extraction with the ADPC–MIBK system. Deuterium arc background correction was used for the direct flame analyses.

Reagents. High-purity certified analytical reagents were used throughout. The acids required are 16 *N* HNO_3 and 36 *N* H_2SO_4. A 1% w/v solution of ammonium pyrrolidinedithiocarbamate is prepared. Methyl isobutyl ketone and 30% H_2O_2 are also required.

Procedure. Weigh about 5 g of fish into a calibrated digestion tube. Add 5 mL 16 *N* HNO_3 and then 5 mL 36 *N* H_2SO_4 to the sample. Allow the reaction to proceed, taking care not to allow any overflow of the sample. When the reaction slows, place the digestion tubes in the hot block, heated to 60°C, for 30 min. Remove the tubes from the hot block and allow to cool for 5 min and then add another 10 mL HNO_3. Return the tubes to the hot block and increase the temperature, in steps, to 120°C (the contents of the tubes should be boiling) until the liquid is about level with the top of the block. Increase the temperature to 150°C. Remove the tubes when the samples go black, allow to cool for 5 min, and add 1 mL additions of H_2O_2 until the samples are clear. Remove the tubes and make up to 50 mL with deionized water when cool.

Zinc, Cu, and Cr are analyzed directly by flame atomic absorption spectrometry. Standards must be treated in the same way as samples and carried through the whole procedure. It is very important that the standards contain the same amount of acid as the samples, especially H_2SO_4, for the final solution contains 10% of this acid and its effect on the viscosity of the solution results in a significant suppression of sensitivity. This is strictly a physical effect and results in suppressions of sensitivity of 15, 10, and 25% for Cu, Zn, and Cr, respectively.

Nickel, Pb, and Cd are concentrated by chelation–solvent extractions as follows. A volume of the sample digest (40 mL) is made up to 100 mL; 5 mL of the APDC solution and 5 mL of MIBK are added to the sample. The mixture is shaken vigorously for 5 min. Nickel, Pb, and Cd are analyzed in the MIBK phase by flame atomic absorption spectrometry. Standards containing 1–100 μg/L of each metal and 4% v/v H_2SO_4 are also run.

Use the instrument manufacturers recommend for operating conditions.

6.2.4 Determination of As and Se in Foods by Hydride Generation ASS*

A dry ashing procedure with $Mg(NO_3)_2$–MgO is employed. Ten-gram samples are treated in this way. An aliquot of sample containing no more than 6.0 mg/mL of magnesium is used for analysis. Amounts of magnesium above this level interfered. Prior to hydride generation a prereduction step of As(V) to As(III) and Se(IV) to Se(II) was accomplished using potassium iodide and boiling in 4 *N* HCL, respectively. The procedure was verified using NBS SRM 1577 Bovine Liver, NBS SRM 1568 Rice Flour and NBS SRM 1571 Orchard Leaves. Detection limits (3 × standard deviation of blank) are 5 ppb for both arsenic and selenium using a 10 g sample.

Equipment. Varian Techtron AA 120 spectrometer, Varian Model 64 vapor generation accessory, and Vycor tube furnace were used.

Reagents

(a) Magnesium oxide: Ash analytical grade $Mg(NO_3)_2 \cdot 6H_2O$ overnight in 500°C muffle furnace and grind to fine powder. This MgO will give a lower blank for As and Se compared with commercial MgO.

(b) Ashing aid: Dissolve 80g $Mg(NO_3)_2 \cdot 6H_2O$ in 200 mL water. Add 8 g MgO and shake well before use.

*Procedure from Tam and Lacroix (7).

(c) Potassium iodide solution, 30% w/v: Prepare in water and store in refrigerator.

(d) Sodium borohydride solution 3% w/v: Dissolve 3 g $NaBH_4$ (Fisher Scientific Co.) in 100 mL 0.5% NaOH solution and filter through Millipore paper (type DA, 0.65% μm). Store in refrigerator.

(e) Arsenic standard stock solution, 100 μg/ml: Dissolve 1.320 g As_2O_3 in minimum volume of 20% w/v NaOH solution and neutralize with HCl. Dilute to 1 L with water.

(f) Arsenic standard working solutions: (1) 10.0 μg/mL. Pipette 1.00 mL As standard stock solution into 100 mL volumetric flask, add 5 mL 6 *N* HCl, and dilute to volume. (2) 100 μg/L: Pipette 1.00 mL of the 10.0 μg/mL solution into a 100 mL volumetric flask and dilute to volume. Prepare fresh daily.

(g) Selenium standard stock solution, 1000 μg/ml: Dissolve 1.000 g Se in HNO_3 and dilute to 1 L with water.

(h) Selenium standard working solutions, 10.0 μg/mL and 100 μg/L: Prepare as described in (f).

Procedure. Add 10 mL ashing aid to sample (1–10 g) in 150 mL beaker and mix well. Cover beaker with watchglass and dry contents in 110°C oven. Transfer beaker to cold muffle furnace and slowly raise temperature of furnace to 500°C. Ash overnight according to the method of Tam and Conacher (8). Cool sample to room temperature. Add 5 mL water to wet ash and dissolve in 36 mL 6 *N* HCl. Heat solution on hot plate almost to boiling and maintain 20–30 min to reduce selenate to selenite. Transfer to 50 mL volumetric flask and dilute to volume with 6 *N* HCl. Prepare duplicate blanks by carrying 10 mL ashing aid solution through the procedures.

Pipette duplicate aliquots containing <100 ng As or Se into hydride generation tubes. Add 6 *N* HCl to bring volume to 20 mL. To determine selenium, inject 3 mL 3% $NaBH_4$ solution through a syringe to generate hydrogen selenide, and measure peak height of signal. To determine As, add 1 mL 30% KI solution and let mixture stand 15–20 min before addition of $NaBH_4$. (Arsenate is reduced to arsenite by KI).

Pipette standard solution containing 10, 20, 40, 60, 80, and 100 ng As or Se into 20 mL 6 *N* HCl solution. Treat standards in same manner as samples. Calculate concentration of As or Se in sample using standard calibration curve.

6.2.5 Determination of Pb and Cd in Foods by Furnace AAS*

Food samples (liver, leafy vegetables, skim milk, beef tissue, wheat flour, and beef kidney) are dissolved using HNO_3–$HClO_4$. Lead and cadmium complexes

*Reprinted in part with permission from (9), Dabeka, *Anal. Chem.*, **51,** 902. Copyright 1979, American Chemical Society.

formed with ammonium pyrrolidine dithiocarbamate are extracted at pH 1.4–1.8 into methyl isobutyl ketone. The metals are then stripped from the solvent using 3% HNO_3 and 8% H_2O_2.

Samples up to 15 g can be handled by the procedure with about 1% coefficient of variation. Because of the extraction step, interferences are minimal. Standards need not be taken through the extraction procedure. Both acid-soluble and acid-insoluble lead can be determined. Detection limits (not defined) of 20 and 1 ng for lead and cadmium, respectively, were obtained.

The procedure was tested on NBS Standard Orchard Leaves, Spinach Leaves, Tomato Leaves, Bovine Liver, and Pine Needles. Good agreement with certified values was obtained.

Equipment. A Perkin-Elmer Model 403 atomic absorption spectrophotometer equipped with HGA-2000 graphite furnace and simultaneous deuterium arc background correction was used for all determinations. Readings were taken on a fast-response (0.3 sec FS) strip-chart recorder. Argon was used as the purge gas in the automatic stop-flow mode. Graphite tubes were not pyrrolytically coated. Wavelengths for lead and cadmium were 283 and 229 nm, respectively. Lead and cadmium electrodeless discharge lamps (EDL) were used for all determinations.

Peak-height measurements were made because of their availability on most instrumentation and because of their precision when properly applied.

Reagents. The NBS certified lead nitrate and high-purity iron and copper were used, whereas other chemicals were reagent grade.

Concentrated acids were purified by subboiling distillation in quartz and polypropylene stills. The MIBK was stored over 0.5% HNO_3 to remove lead. Concentrated ammonia solution was prepared by saturating cooled deionized water with 99% ammonia, and filtered through a membrane filter (0.3 μm porosity). Solutions of 10% ammonium dihydrogen phosphate and 10% ammonium hydrogen citrate were purified by extraction with chloroform and dithizone. Excess chloroform was removed in vacuo. A solution of 1% APDC was prepared fresh daily and filtered through a membrane filter.

Polypropylene bottles were cleaned up by immersing them for 48 h in 5% HNO_3 saturated with MIBK. The MIBK penetrates the plastic and provides a better interface for leaching impurities. Other labware was cleaned as described elsewhere.

High and erratic lead contamination was encountered from a commercial repipette with a Teflon-coated plunger and from the methyl violet indicator. (A powdered form of the latter contained 600 μg/g lead.) Hydrogen peroxide contributed significant but constant amounts of lead and cadmium.

Analyses and purification of reagents were not performed in a special clean room, and results reflect those attainable, with care, in a "normal" laboratory.

Procedure. Weigh (1–15 g dry weight) samples into 250 mL Vycor beakers covered with nonribbed watchglasses and dry overnight at 120°C if necessary. At least three blanks are carried through the entire procedure. Add concentrated HNO_3 according to the formula

$$\text{mL } HNO_3 = 10 + (5 \times \text{dry weight of sample in grams})$$

Digest the samples at room temperature for 2–12 h. For large samples or for those that froth excessively, use cold 50% HNO_3. Boil the solutions until the volume decreases to about one half the original volume of HNO_3 added, and add concentrated $HClO_4$ according to the formula

$$\text{mL } HClO_4 = 20 + (20 \times \text{weight of fat or oil present in grams}) + (7 \times \text{dry weight of sample in grams})$$

Heat and maintain boiling until only $HClO_4$ fumes remain (usually 0.5–1 h). If the evolution of gases becomes excessive or if sections of froth above the solution or the solution itself begin to darken, 2 mL portions of concentrated HNO_3 are added. Boil solutions at a high temperature to a volume of 4–8 mL. The comparatively large volumes of $HClO_4$ used in the digestion step are for safety and to obtain consistent blanks, because digestions usually proceeded to completion without the need for addition of HNO_3 to pacify the reaction. Wash the watchglasses and sides of beakers with about 40 mL of water, and heat the solutions to about 90°C and transfer to 125 mL separatory funnels. Rinse the beakers with 10 to 15 mL of water and heat and transfer as before.

A stripping solution containing 3% HNO_3 and 8% H_2O_2, and a modifiction solution of 0.33% ammonium dihydrogen phosphate and 0.096% ammonia are prepared.

Adjust the acidity of each digest with concentrated ammonia solution in the presence of a drop of methyl violet (0.1% in water) until a blue-green hue appears. The color of the indicator is stable only for several minutes and rapid adjustment is necessary. Samples containing large amounts of metals (>3 g kidney or liver) interfere with the indicator and are adjusted to pH 1.6 using a pH meter.

After cooling, extract and drain the solution in each separatory funnel in turn as follows: Add 5 mL 1% APDC solution with swirling for 10 sec. Immediately after, add 70 mL MIBK, stopper the funnel, and shake manually for 60 sec. Remove the stopper, separate the phases (20 sec), and drain the aqueous phase, leaving about 1 mL. Five minutes from addition of APDC, drainage of aqueous phase is complete. Pipette stripping solution (5 mL when Pb < 0.2 g; 10 mL when 0.2 g $<$ Pb $<$ 0.4 μg, and 20 mL when Pb $>$ 0.4 μg) into the funnel. Stopper the latter and rinse the point with water.

Shake the funnels gently for 5 min, invert on their stands to avoid seepage through the drainage spouts, let stand overnight, shake another 5 min, and return to an upright position. After 1 h, remove the stopper of each, drain the aqueous phase into a clean beaker, discarding the first 1–2 mL. Pipette 2.00 mL into a clean 25 mL polypropylene bottle containing 6.00 mL of modification solution. (When the volume of stripping solution permits, these amounts are increased by 2.5 times to reduce pipette errors.)

Use a dilution solution containing 0.75% HNO_3, 0.25% ammonium dihydrogen phosphate, 0.072% ammonia, 2% H_2O_2, and 0.43% MIBK for diluting samples and preparing standard Pb and Cd solutions. Prepare standards fresh monthly.

Pipette aliquots (20 μL) of samples and standard into the graphite furnace, dry (125–150°C), ash (750°C for 40 sec), and atomize (2300°C for 15 sec). Absorbances of samples are bracketed with those of standards. Measure blanks for both sets of solutions and include in the final calculation. When absorbances of samples exceed those of 60 ng/mL Pb or 6 ng/mL Cd, standard solutions are diluted to appropriate amounts with dilution solution.

Condition graphite tubes initially by running through atomization cycles for 15 aliquots of dilution solution and use for as many as 900 cycles. Clean out carbon deposits every 100–200 cycles. Gradual erosion of surface grapite during lengthy use of tubes causes 20–30% increases in actual temperatures for the same current settings, and adjustment of the latter is made every 300–500 firings. (More modern furnaces with temperature control obviate this problem. Also, with a more modern furnace a 5-step program would be better using ramp to ash and a 20 sec cooling step between atomize and final burnout step at 2700°C.*)

Determination of Acid-Insoluble Lead. Samples are digested according to the digestion procedure to $HClO_4$ fumes, diluted to about 150 mL with water, and heated to near boiling. Blanks containing lead standards equivalent to the acid-soluble Pb content of the samples are simultaneously carried through the entire procedure. Filter the hot digest through membrane filters (cellulose acetate–nitrate, 0.3 μm porosity, 25 mm diameter), prewashed with 5% HNO_3. Wash the precipitates thoroughly with 5% HNO_3 and then water, and transfer with the filter to clean 50 mL Teflon-FEP beakers. Add 5 mL concentrated HF and 2 mL HNO_3 and heat the solutions until the filters dissolve. Add 7 mL concentrated $HClO_4$ and evaporate the solutions to $HClO_4$ fumes. If any precipitate is visible, 7 mL HF is again added and the solution evaporated as before. Dilute the solutions with water, transfer to a separatory funnel, and analyze according to the general procedure.

6.2.6 Determination of Zn and Cu in Legumes by Flame AAS*

The following procedure was used in an interlaboratory comparison in which atomic absorption and X-ray fluorescence were employed. National Bureau of Standards (NBS) Orchard Leaves and Bovine Liver were analyzed with each set and the results used to calibrate the unknown samples.

Wet ashing is employed. Concentrated sulfuric acid is added first, followed by heating at a temperature less than 50°C. Hydrogen peroxide is then added drop by drop until the solution becomes clear.

Equipment. A Perkin-Elmer 503 atomic absorption spectrometer equipped with a 3-slot burner was employed. Test tubes, 15 × 150 mm, were heated in a Multitemp Block #2093 (Labine Instrument Co., Melrose Park, IL).

Reagents. The concentrated H_2SO_4 was Ultrex brand (J. T. Baker).

Procedure. Standard, water-soaked beans are prepared by adding 150 g of whole dry beans to 450 mL of distilled water, allowing the beans to imbibe water and hydrate at 21°C for 24 h. Half of the drained beans or about 150 g of the raw, hydrated produce is slurried with an equal volume of distilled water in a Waring Blender. The slurry is lyophilized and equilibrated in ambient air at about 50% relative humidity for 48 h to minimize changes in moisture content during subsequent handling.

Quick-cooking, salt-soaked beans are prepared from whole, raw dry beans as shown in Table 6.8. Hydration in the salt solution of large lima beans was facilitated by Hydravac treatment. Steam blanching is employed to accelerate hydration of pink beans. No special treatment is employed for blackeye and garbanzo beans, which hydrate rapidly at ambient temperature. As indicated in Table 6.8, beans are soaked in 3% w v of the appropriate salt solution for the specified hydration period, drained, and rinsed lightly. A one-half portion of about 150 g of each hydrated bean sample is slurried with an equal volume of water, and lyophilized and equilibrated in ambient air.

Weight triplicate samples of 0.5–1.0 g into glass test tubes (15 × 150 mm). Add 1 mL of concentrated H_2SO_4 water and 1.0 mL metal-free distilled water. Heat the samples gently at less than 50°C for 1 h. Carefully add 1 mL of 50% H_2O_2 until evolution of the bubbles ceased. Additional aliquots of H_2O_2 are added until the solution becomes clear. Filter the cooled solution into 15 mL plastic graduated tubes through Whatman 40 medium filter paper prewashed with dilute HCl. Wash the precipitate with water and make the solution up to 10.0 mL. If excessive precipitate is noted, concentrated HNO_3 is substituted for

*Reprinted from *Journal of Food Science*, 1979, **44**, 1711–1713. Copyright © by Institute of Food Technologists.

Table 6.8 Processing Conditions for Quick-Cooking Beans

Processing Variable	Bean Variety			
	Large Lima	Blackeye	Garbanzo	Pink
Bean weight (g)	150	150	150	150
Hydravac time (min) (5 min on–off)	50	—	—	—
Steam blanch (min)	—	—	—	3
Salt solution				
Volume (ml)	450	450	450	450
Sodium chloride (%)	1.0	2.0	2.0	2.0
Sodium tripolyphosphate (%)	0.5	1.0	0.0	0.1
Sodium bicarbonate (%)	0.45	0.45	0.375	0.45
Sodium carbonate (%)	0.15	0.15	0.125	0.15
Soaking time (h)	6	24	24	24
Subjective cooking time for salt-soaked beans (min)	8	10	15	11
Subjective cooking time for water-soaked beans (min)	45	25	120	50

the H_2SO_4 in the digestion procedure. Dilute aliquots of the filtered solution by a factor of three or four for analysis by flame atomic absorption spectrometry. Use standard operating conditions (Zn, 213.9 nm, Cu, 324.7 nm). Triplicate samples of NBS SRM Bovine Liver and Orchard Leaves are run with each set of unknown samples and used for calibration.

6.2.7 Determination of Pb in Vegetables by Flame AAS*

Lead is determined in garden vegetables by this procedure. Dry ashing at 560°C was found to be permissible. A pretreatment with HNO_3-H_2SO_4 was used to eliminate cadmium loss. The ash is dissolved in a mixture of HNO_3. Extraction of the lead with dithizone into $CHCl_3$ was used to check that the results were satisfactory. Detection limit (twice the standard deviation of the noise near the detection limit) is about 2 μg/g.

Equipment. A Perkin-Elmer Model 603 atomic spectrometer equipped with a deuterium arc background corrector and PRS-10 printer sequencer, electrodeless discharge lamps (EDL) for lead and cadmium, single-slot burner, and air–acetylene flame were used. Pyrex tall-form beakers of 200 mL with watchglass cov-

*Procedure from Preer et al. (11).

ers were employed. These were discarded when they became discolored or etched.

Reagents. Unless otherwise stated, water means deionized (type II) water.

Nitric acid, sulfuric acid, hydrochloric acid, ammonium hydroxide (ACS reagent grade).

Chloroform, hydroxylamine hydrochloride, phenol red, diphenylthiocarbazone (certified ACS grade).

Sodium acetate (certified grade).

Standard solutions, 1000 μg/mL solutions of lead and cadmium (certified atomic absorption standard).

Standard reference materials (National Bureau of Standards SRM 1570 Spinach Leaves; SRM 1571 Orchard Leaves; SRM 1573 Tomato Leaves; SRM 1575 Pine Needle.

Procedure. Take care to protect sample from contact with any object not known to be free from Pb and from prolonged exposure to air. Rinse with HNO_3 all glassware that is to come in contact with sample, and rinse well with water. Wash fresh vegetable sample thoroughly in low-Pb tap water (<10 μg/L) and blot dry on clean paper towel. Peel root vegetables and chop roots and garden fruits on clean surface with stainless steel knife. Dry samples to constant weight in forced-air oven at 100°C for leaves or 80°C for roots and fruits. Grind dry samples in Wiley mill fitted with a 40 mesh screen and store in polyethylene vials. Dry powdered sample to constant weight before analysis. Set up blanks and carry through procedure, omitting muffle furnace step. Weigh 1.6 g sample into ashing vessel and add 4 mL HNO_3 and 2 mL (1 + 1) H_2SO_4. Cover and heat to dryness on hot plate, manipulating vessel if necessary to ensure uniform charring. Place in cold muffle furnace and set to attain equilibrium temperature of 560°C. Remove after 16 h. Add 2 mL HCl and 2 mL HNO_3 and heat. Wash down the vessel with 2 mL water, transfer to 10 mL volumetric flask, rinse, and dilute to volume.

Let the electrodeless discharge lamp and deuterium arc background corrector stabilize (30 min–1 hr) and position the EDL to maximize signal at detector. Using standard solution of the element, adjust aspiration rate, burner positioning, and gas pressure and flow rate to give maximum absorbance reading. Set integration time to 3 sec. Aspirate water to set the zero between samples. Aspirate standards to set S1 value (end of linear range) and S2 value (2.5 × S1). Print out concentration readings. Check standard after every 5 samples; reset if reading error exceeds 1% of nominal value, and remeasure previous set if error exceeds 5%.

6.2.8 Determination of Pb in Foods by Furnace AAS*

A perchloric, nitric, and hydrofluoric acid digestion mixture is employed. A 1% $(NH_4)_2HPO_4$ solution is used for matrix modification. With this substance an ashing temperature of 800°C can be employed. Atomization is done from a L'vov platform and pyrolitically coated tubes are used. A cation to analyte ratio of 40,000:1 can be tolerated without interferance. The procedure was verified using NBS SRM 1566 Oyster Tissue and NBS SRM 1570 Spinach Leaves.

Equipment. A double-beam atomic absorption spectrophotometer (Perkin-Elmer Model 5000 was used. The electrothermal atomization was carried out in a heated furnace (Perkin-Elmer HGA-500) with pyrolytically coated graphite tubes. An automatic pipetting system (Perkin-Elmer AS-40) was used to deposit a 20 μL sample on a thin, pyrolytically coated graphite platform (L'vov platform). This platform is constructed as described by Hinderberger et al. (13). A 5 × 8 mm curved section is cut from the grooved end of a pyrolytically coated tube. The tube can be either a used or an unused tube. The sides of the chip or platform are filed so that the platform will fit into the furnace, and the platform is inserted into the graphite tube with a pair of tweezers. The platform is centered directly under the sample port by using a metal probe. The window is reinstalled and the injection tip is adjusted to ensure that it does not come in contact with the platform. The platform should not be resting on the bottom of the graphite cell but should have a 1 mm air gap. If the platform rests on the bottom of the graphic tube, a longer cooling-down period will be required after atomization.

Reagents.

(a) Lead stock solution, 10,000 μg/mL: Dissolve 1.000 g NBS SRM 49c in 10 mL HNO_3 and dilute to 100 mL with water. From this standard stock solution, prepare standard solutions of 1000 and 100 μg Pb/mL in 1% HNO_3 weekly. Prepare daily working standards in 1% HNO_3 from the 100 μg Pb/mL standard solution.

(b) Purified ammonuum phosphate, 10% w/v: Prepare by dissolving 20 g $(NH_4)_2HPO_4$ in 100 mL water. Transfer the solution to a 250 mL separatory funnel. Add 20 mL freshly prepared 2% solution of ammonium pyrrolidinedithocarbamate (APDC) and mix. Add 40 mL methyl isobutyl ketone (MIBK) and equilibrate solutions 2 min by shaking. Drain the aqueous phase into a 200 mL volumetric flask and dilute to volume.

(c) Acids: All acids were prepared at NBS by subboiling distillation.

*Procedure from Rains et al. (12).

Procedure. Transfer a 1–1.5 g sample to a weighing bottle and dry 24 h at ambient temperature in vacuum oven equipped with liquid nitrogen trap. Transfer a 1 g test portion of dried material to a Teflon beaker. Add 10 mL HNO_3 and 1 mL HF, and cover beaker with a Teflon lid. Digest overnight at 90°C on a hot plate. Remove the lid and add 5 mL $HClO_4$. Evaporate the solution to ~0.5 mL. Add 1 mL HNO_3 plus 10 mL water, and warm the solution to dissolve solids. Transfer the solution to a volumetric flask and dilute to calibrated volume. Subsequently transfer the solution to conventional polyethylene bottle for storage until analysis by AAS.

Starting with 100 μg/mL standard solution, prepare series of working standard lead solutions of 0.00, 0.02, 0.04, 0.06, 0.08, and 0.1 μg/mL that contain 1% HNO_3 and 1% $(NH_4)_2HPO_4$. Transfer duplicate test portions of unknown samples to volumetric flasks. (*Note:* Select an aliquot that will provide a final concentration of 0.02–0.06 μg/Pb/mL.) To one volumetric flask add a lead spike of 0.02 μg/mL. Add a volume of 10% $(NH_4)_2HPO_4$ to give a final concentration of 1% $(NH_4)_2HPO_4$ and dilute to volume. Start the instrument as described in the instruction manual. Turn on the lead hollow-cathode lamp and adjust the lamp current to recommended instrument manufacturer's value. Set AAS instrument controls as follows: wavelength, 283.3 nm; slit width, 0.7 nm; peak height; integration time, 5 s; scale expansion, 2.0. Let hollow-cathode and deuterium arc lamps warm up 15 min and then balance the two beams. Remove quartz windows to furnace atomization cells and clean. Insert L'vov platform as described previously and reinstall quartz windows. Adjust pipetting arm of automatic pipette to proper height.

Set furnace controls as follows: Dry for 40 sec at 130°C with 20 sec linear increase; ash for 40 sec at 800°C with 20 sec linear increase; atomize for 6 sec at 2400°C. Set automatic pipette control to 20 μL and start instrumental sequence. Establish calibration curve and repeat calibration curve until absorbances are within ±2%. Determine net absorbances for unknowns.

Calculation. Calculate concentration of unknown sample as follows: Set up calibration curve, using least squares fit on hand calculator. (*Note:* If hand calculator or computer is not available, plot on rectilinear graph paper the absorbance corrected for background, as ordinate, and concentration of standards, expressed in grams of analyte/milliliter, as abscissa. Calibration curve should pass through origin.) Determine concentration of unknowns from calibration curve. Check recovery using method of standard addition. If recovery is incomplete, correct concentration by the equation

$$C = x \cdot \frac{s}{y - x}$$

where x is the micrograms of analyte per milliliter in unknown as determined from calibration curve; s the amount of standard added in micrograms per milliliter in final volume; y the micrograms of analyte per milliliter found in unknown with standard added as determined from calibration curve; C the concentration in micrograms per milliliter.

Calculate analyte concentration of original sample as follows: Let W = weight of solid sample in grams; V_s = volume of liquid in which W is dissolved in milliliters; V = test portion of liquid sample or aliquot of V_s in milliliters, and Z = dilution factor:

$$Z = \frac{V_1 V_2 \cdot \cdot \cdot V_n V_f}{a_1 a_2 \cdot \cdot \cdot a_n}$$

where V_1 is the volume to which V is first diluted; a_1 the aliquot of V_1; V_2 the volume to which a_1 is diluted; a_2 the aliquot of V_2, and so on, to V_n and a_n (all in milliliters) V_1 the final volume on which measurement is made in milliliters; C is the analyte concentration of V_f in micrograms per milliliter; then the analyte in the original sample, (in micrograms per gram) is

$$\frac{CZV_s}{VW}$$

6.2.9 Determination of Pb in Milk by Flame AAS*

This procedure was used in a collaborative study involving five laboratories, three companies producing canned milks, and two government agencies. Results in this study were also obtained by anodic stripping volametry. Both the atomic absorption and anodic stripping volametry methods were found to be suitable for the lead determination. The atomic absorption method was adopted as "official first action."

A dry ashing step was employed in a stepwise fashion with 500° the highest temperature used. Lead was extracted with 1-pyrrolidinecarbodithioate into butyl acetate. Flame atomic absorption using the 283.3 nm line was employed. Background correction was apparently not necessary because of the extraction step.

Equipment. Atomic absorption spectrophotometer: Spectrometer equipped with 10 cm single-slot burner head and digital concentration readout attachment (DCR) or 10 mV strip-chart recorder, and capable of operating as follows (Per-

*Procedure from Fiorino et al. (14).

kin-Elmer Model 403 or equivalent): wavelength, 283.3 nm; flame, air–C_2H_2; range, 0.1–1.0 ppm based on sample.

Ashing vessels: Approximately 100 mL, flat-botted platinum crucible or dish, Vycor or quartz tall-form beaker dish (Corning Glass Works, NO. 13180 or equivalent). Discard Vycor vessels when inner surfaces become etched.

Centrifuge: Capable of holding 15 mL conical tubes and centrifuge at 2000 rpm.

Furnace: With controlling pyrometer to cover the range 250–600°C with variation of 10°C.

Reagents. Nitric acid is 1 *N*. Add 128 mL redistilled HNO_3 to 800 mL distilled or deionized water and dilute to 2 L. Redistilled HNO_3 (G. Frederich Smith Chemical Co.) may be rediluted and used without redistillation.

Butyl acetate: Special grade water saturated.

Ammonium 1-pyrrolidinecarbodithioate (APDC), 2%: Dissolve 2.00 g APDC in 100 mL distilled or deionized H_2O. Remove insoluble free acid and other impurities normally present by 2–3 extractions with 10 mL portions of butyl acetate.

Lead standard solutions: (1) Stock solution, 1 mg/mL Pb in 1 *N* HNO_3. Dissolve 1.5985 g dried $Pb(NO_3)_2$ crystals in about 500 mL 1 *N* HNO_3 in a 1 L volumetric flask and dilute to volume with 1 *N* HNO_3. (2) Intermediate stock solution, 5.0 μg/mL Pb. Pipette 5 mL stock solution into a 1 L volumetric flask, add 1 mL HNO_3, and dilute to volume with H_2O. (Solution is stable for several months if stored in polyethylene bottle.) (3) Working solution: Pipette 20, 10, 5, and 2 mL intermediate solution into separate 100 mL volumetric flasks and dilute to volume with 1 *N* HNO_3 (1.0, 0.50, 0.25 and 0.10 μg Pb/mL, respectively). Pipette 10 and 5 mL solution containing 0.50 μg Pb/mL into separate 100 mL volumetric flasks, and dilute to volume with 1 *N* HNO_3 (0.05 and 0.025 μg Pb/mL, respectively).

Citric acid solution, 10%: Weigh 10.0 g Pb-free citric acid into a 100 mL volumetric flask, dissolve in water, and dilute to volume. Stopper flask and shake thoroughly. If necessary, remove Pb impurity by extraction.

Bromocresol green, 0.1%: pH range, 3.8 (yellow) to 5.4 (blue). Transfer 0.100 g bromocresol green, Na salt, to 100 mL volumetric flask, and dilute to volume with H_2O. Use 1 drop/10 mL of analytical solution.

Procedure. Clean all glassware thoroughly in (1 + 1) HNO_3. Weigh about 25 g (to nearest 0.1 g) sample into an ashing vessel. Dry sample overnight in 120° forced-draft oven before ashing. (Sample must be absolutely dry to prevent flowing or spattering in furnace.) Place sample in the furnace set at 250°. Slowly (50° increments) raise temperature to 350° and hold at this temperature until smoking ceases. Gradually increase temperature to 500° (~75° increments).

(Sample must not ignite.) Ash 16 h overnight at 500°. Remove sample from furnace and let cool. Ash should be white and essentially carbon free. If ash still contains excess carbon particles (i.e., ash is gray rather than white), proceed as follows: Wet with minimum amount of H_2O followed by dropwise addition of HNO_3 (0.5–3 mL). Dry on a hot plate. Transfer to furnace at 250°, slowly increase temperature to 500°, and continue heating 1–2 h. Repeat HNO_3 treatment and ashing if necessary to obtain carbon-free residue. (*Note:* Local overheating or deflagration may result if sample still contains much intermingled carbon and especially if much potassium is present in ash.)

Dissolve residue in 5 mL 1 *N* HNO_3, warming on a steambath or hot plate for 2–3 min to aid solution. Filter, if necessary, by decantation through S & S 589 black paper into a 50 mL volumetric flask. Repeat with two 5 mL portions 1 *N* HNO_3 filter, and add washings to original filtrate. Dilute to volume with 1 *N* HNO_3.

Prepare duplicate reagent blanks for standards and samples, including any additional water and HNO_3 if used for ashing. (*Note:* Do not ash HNO_3 in furnace because Pb contaminant will be lost.) Dry HNO_3 in ashing vessel on stream bath or hot plate, and proceed as above.

Pipette 20 mL each of working solution, reagent blank for the standard (if different from that used for samples), sample solution, and appropriate reagent blank(s) for samples into separate 60 mL separators. Treat each solution as follows: Add 4 mL citric acid solution and 2–3 drops bromocresol green indicator. (Color of solution should be yellow.) Adjust pH to 5.4, using stock NH_4OH initially and then 1 + 4 in vicinity of color change (first permanent appearance of light blue). Add 4 mL APDC solution, stopper, and shake 30–60 sec. Pipette in 5 mL butyl acetate. Stopper the separator and shake vigorously 30–60 sec. Let stand until layers separate cleanly; drain and discard lower aqueous phase. If emulsion forms or solvent layer is cloudy, drain solvent layer into 15 mL centrifuge tube, cover with aluminum foil or parafilm, and centrifuge ~1 min at 2000 rpm.

Set instrument to previously determined optimum conditions for organic solvent aspiration (3–5 mL/min), using 283.3 nm Pb line and air–C_2H_2 flame adjusted for maximum Pb absorption. Flame will be somewhat fuel lean. Optimum position in flame for maximum absorption should be just above burner top. If using a recorder, DCR, and so on, adjust to manufacturer's specifications. Depending upon signal-to-noise ratio, scale expansion up to 10× may be used. Check zero point while aspirating H_2O-saturated butyl acetate. Aspirate sample and standard solutions flushing with H_2O and then butyl acetate between measurements. Record absorbance of each solution.

Prepare standard curve by plotting absorbance corrected for blank of each standard against the concentration of that standard in micrograms Pb per milliliter of butyl acetate. Concentration of standard in butyl acetate is four times that in the aqueous standard. Determine Pb concentration from the standard curve using absorbance corrected for the sample reagent blank, if used.

$$\text{ppm Pb} = \mu\text{g Pb/mL from curve/2 g/mL}$$

REFERENCES

1. W. R. Wolf, *Anal. Chem.* **50,** 190A (1978).
2. N. T. Crosby, *Analyst*, **102,** 225 (1977).
3. R. W. Kuennen, K. A. Wolnick, and F. L. Fricke, *Anal. Chem.*, **54,** 2146 (1982).
4. M. H. Kahn, K. A. Wolnik, and F. L. Fricke, *Anal. Chem.* **54,** 1048 (1982).
5. H. Agemanian, P. D. Sturtevant, and K. D. Austen, *Analyst* **105,** 125 (1980).
6. J. N. Bishop, L. A. Taylor, and P. L. Diosady, *High Temperature Acid Digestion for Determination of Hg in Environmental Samples*, Ontario Ministry of the Environment Lab Services Branch, 1975.
7. G. K. H. Tam and G. Lacroix, *J. Assoc. Off. Anal. Chem.* **65,** 647 (1982).
8. G. K. H. Tam and H. B. S. Conacher, *Environ. Sci. Health*, **B12,** 213 (1977).
9. R. W. Dabeka, *Anal. Chem.* **51,** 902 (1979).
10. L. B. Rockland, W. R. Wolf, D. M. Hahn, and R. Young, *J. Food Sci.* **44,** 1711 (1979).
11. J. R. Preer, B. R. Stephens, and C. Wilson-Bland, *J. Assoc. Off. Anal. Chem.* **65,** 1010 (1982)
12. T. C. Rains, T. A. Rush, and T. A. Butler, *J. Assoc. Off. Anal. Chem.* **65,** 994 (1982).
13. J. Hinderberger, E. J. Kaiser, and S. R. Koirtyohann, *At. Spectrosc.* **2,** 1 (1981).
14. A. Fiorino, R. A. Moffitt, A. L. Woodson, R. J. Gajan, G. E. Huskey, and R. G. Scholz, *J. Assoc. Off. Anal. Chem.* **56,** 1246 (1973).

CHAPTER

7

CLINICAL SAMPLES

7.1 INTRODUCTION

7.1.1 Types of Clinical Samples

Clinical samples consist of body fluids and a wide range of tissue material. Small amounts of a number of metals are required for the proper functioning of the human body. The trace metals that are known to be essential are Fe, Cu, Mn, Zn, Co, Mo, Se, Cr(III), Sn, V, Si, Ni, and As. It is likely that others will be found to be essential as analytical methodology improves and detection limits become better.

Other metals at present either show no observable effect, are stimulants, are therapeutic, or have been found to be toxic. The worst metals of the toxic category are Hg, Pb, Cd, Tl, Be, and Cr(VIII). Some metals in the essential category are toxic at levels only slightly higher. These include Se, As, Ni, and V. Exposure to combinations of metals may result in synergistic or antagonistic interactions.

One of the major problems in assessing the mode of action of a metal with a biological system is the lack of information on the forms of metals in tissue material. For example, $(CH_3)_2Hg$ is readily produced from inorganic mercury by bacteria. This substance is lipid soluble and hence enters the body easily. Thus mercury in this form is much more bioactive than inorganic mercury species. It is important therefore to be able to distinguish between the forms of mercury in undertaking a trace metal study. At present there are very few methods that allow the determination of the forms of metals in complex samples. This is a high priority in trace metal analytical chemistry research.

Body fluids constitute the largest fraction of clinical samples. The most common fluids for analysis are urine, whole blood, and blood plasma and serum. Urine consists of a large variety of inorganic and organic constituents. Some of these may precipitate on the cooling of urine. Precipitated material may consist of uric acid, oxalate salts, phosphates, and urates. The pH of urine is in the range 4.5–8.2 and averages near 6. The precipitation of solid material depends on the pH to some extent. At room temperatures urine may undergo changes on standing as a result of bacterial action. The most common problem is a conversion of urea to $(NH_4)_2CO_3$ resulting in a loss of NH_3 and hence a pH change.

Blood contains cells floating in plasma. Serum results when blood is allowed to clot and the fibrinogen is removed. On standing, fatty material may separate out of plasma and serum samples. Bacterial action during storage at room temperature causes compositional changes in these samples. If the distribution of a metal among the various solid and liquid phases of blood must be known, then the exchange of metals among the components must be prevented.

7.1.2 Sampling and Sample Storage

These topics are actually beyond the scope of this book. However, in the case of clinical samples some guidelines must be given because of the dynamic nature of such smaples and the ease with which contamination can occur.

Urine should be kept refrigerated at 4 °C or be frozen. The sample should be kept in a tightly closed clean container. Acidification has been used in the past to allow storage at room temperature. However, this can result in precipitation of uric acid, which results in analytical problems. Material such as chloroform can be employed as a fungicide or bactericide.

Blood should be taken with a Teflon intravenous catheter and not with a stainless steel needle. An anticoagulent (e.g., heparin) should be present in the container. Centrifugation is used to separate plasma and the plasma should be removed from contact with the solids immediately. To obtain serum the blood is clotted for about 1 h and the solid materials separated. Storage at 4 °C in a refrigerator or freezing is used for serum and plasma samples. Freeze drying is another approach that is employed for blood samples.

Containers used for storage of all body fluids for trace metal analysis should be acid washed. High density plastics are recommended. Containers should be capable of being tightly capped.

Contamination of body fluids during collection can be a serious problem. The best procedure to prevent this problem is for the donor to shower thoroughly prior to the donation. If this is not possible, the area from which the sample is taken must be very carefully washed. No metal objects should be employed in taking the sample.

7.1.3 Preparation of Clinical Samples for Trace Analysis

Wet ashing mixtures of HNO_3, $HClO_4$, and H_2SO_4 have found wide application in clinical sample analysis. This approach is relatively rapid when compared with dry ashing that commonly takes up to 24 h. If dry ashing is employed, temperatures greater than 500 °C should be avoided. Even at such a temperature, losses of mercury, lead, zinc, cadmium, arsenic, selenium, and tellurium should be anticipated. With lead, zinc, and cadmium, losses may not be significant. Sulfuric acid with dropwise addition of hydrogen peroxide has been used to wet ash tissue samples. This technique is laborious and its effectiveness is open to question.

Body fluids have been analyzed directly by dilution with water. Generally this technique is not recommended. However, dilution of sample with surfactant solution such as Triton-X 100 has been successfully used for blood and serum samples for elements such as lead, nickel, and chromium. When problems occur, destruction of organic matter in body fluid samples using wet ashing with oxidizing acid mixtures can be recommended. Again, combinations of H_2SO_4, HNO_3, and $HClO_4$ can be employed.

When arsenic, selenium, tellurium, or mercury are to be determined, special procedures are necessary. Oxidizing conditions must be maintained throughout the decomposition interval. Thus an oxidizing agent such as potassium persulfate or potassium permanganate is often added in addition to oxidizing acids. With care, mixtures of HNO_3 and $HClO_4$ can be employed without loss of the hydride-forming elements. The samples, however, must be watched carefully for appearance of reducing conditions during the decomposition period.

Because of the very low levels of some of the trace elements in clinical samples, preconcentration may be necessary. When this is the case, the technique of solvent extraction is most often employed. Of course, preconcentration steps are time consuming and subject to problems of loss and contamination and should be avoided unless absolutely essential. With the advent of better designs in furnace AAS, detection limits are improving and it is to be hoped that there will be little need for preconcentration steps in the near future.

Until recently most determinations of the hydride-forming elements in clinical samples were done by hydride generation procedures. Now furnace AAS is becoming popular for the determination of these elements. Success with this method is dependent upon adding an element such as nickel or silver (depending on the element) to the sample to prevent loss of analyte at relatively high ashing temperatures. Prior to the use of such reagents ashing was restricted to temperatures below 500°C and was very ineffective in removing interfering matrix constituents.

The importance of proper sample preparation is brought out nicely in a paper by Bailey and Kilroe-Smith (1). These authors tried seven different sample preparations for the determination of lead in blood as follows:

1. 0.5 mL of blood + 4.5 mL of double-distilled water.
2. 0.5 mL of blood + 4.5 mL of 0.01 *M* HCL.
3. 0.5 mL of blood + 4.5 mL of 2% Triton-X.
4. 0.5 mL of blood + 4.5 mL of Unisol; the mixture was allowed to stand overnight at room temperature until the solution was completely clear and then 3.5 mL of double-distilled water were added.

5. Procedure (4) with Unisol was repeated, but after a clear solution had been obtained 3.5 mL of 0.01 *M* HCL was added.
6. One volume of blood was extracted for 1 h at room temperature with one volume of a mixed perchloric acid–trichloroacetic acid preparation; after digestion, the mixture was diluted with 8 volumes of 0.01 *M* HCL, resulting in a final dilution of the blood to 10 volumes.
7. One volume of heparinized blood was placed in an acid-washed test tube fitted with a "Finn-Cap." An equal volume of 70% $HClO_4$ was added, the sample was mixed well on a vibration mixer, and one volume of 30% H_2O_2 was then added; the sample, after mixing again, was placed in a water bath at 70–80°C for 30–60 min with occasional agitation until the blood had completely dissolved. The mixture was then diluted with 7 volumes of double-distilled water to give a 10-fold dilution of the blood.

In their discussions these authors (1) point out that, among the sources of error in this determination, loss of lead during ashing (above 500°C) and contamination are the two most important. Of the seven methods investigated, method (6) was found to be superior mainly because of low background signal. Method (7) was found to give equally good results, but is more time consuming.

Use of a surfactant such as Triton-X 100 as a diluent for blood samples has become popular as a treatment prior to furnace atomic absorption analysis. This causes hemolysis of the erythrocytes. The approach has been shown by Fernandez and Hilligoss (2) to give excellent results for the determination lead in blood. In this procedure dihydrogen phosphate is also added as a matrix modifier.

7.1.4 Analytical Problems

The levels of the trace metals in body fluids are generally in the low ppm to ppb range. Thus electrothermal atomic absorption spectrometry is most often the determinative method of choice. Neutron activation analysis may be employed for many elements, but the lack of availability of this approach to most workers militates against its use in this application. Detection limits using conventional nebulization approaches with inductively coupled plasma emission spectrometry are too poor for most elements and make their direct determination (without preconcentration) impossible.

Boone et al. (3) report the results of an interlaboratory comparison study on the determination of lead in blood. There were 113 participants and their results were compiled and compared with each other and those obtained by the National Bureau of Standards using isotopic dilution, and mass spectrometry. Bovine blood samples were used from animals that had been fed lead nitrate. There were 12 samples covering the concentration range of 130 to 1020 μg/L. Table 7.1 summarizes the average accuracy (against NBS value) and precision of the methods.

Table 7.1 Summary of Lead Method Performance[a]

Sample	NBS Value μg/dL	Delves Cup, AAS		Extraction, AAS		Graphite Furnace, AAS		Carbon Rod, AAS		Tantalum Strip, AAS		Anodic Voltametry Stripping	
		B	SD	B	SD	B	SD	B	SD	B	SD	B	SD
A	12.4	+ 2.2	5.12	+3.3	11.5	+ 2.8	5.02	+10.8	16.7	+ 5.4	7.12	+ 4.1	4.79
B	13.0	+ 4.7	12.4	+6.9	17.7	+ 5.3	13.5	+ 0.8	6.50	+ 2.7	4.23	+ 2.2	2.43
C	15.9	+ 4.1	5.80	+2.9	8.65	+ 3.6	13.3	− 1.4	6.38	− 0.2	1.41	+ 4.0	9.95
D	22.2	+ 3.0	5.54	+1.6	6.19	+ 1.3	7.05	+ 4.2	13.7	+ 7.0	10.2	− 0.5	3.04
E	27.3	+ 1.6	6.36	+0.5	7.87	+ 1.5	6.62	+ 5.0	10.0	+ 8.4	13.6	− 2.5	5.21
F	53.8	− 3.9	13.0	−1.1	11.6	− 2.1	11.4	− 9.2	10.3	− 5.7	0.96	− 1.9	5.19
G	68.4	− 5.0	10.8	−1.5	13.4	− 6.6	19.2	− 8.0	10.9	− 7.3	2.44	− 5.7	11.9
H	81.6	− 7.5	8.15	−4.0	17.8	− 6.0	12.1	−15.7	23.1	− 6.4	12.8	− 6.1	6.77
I	81.6	− 5.0	10.7	−11.2	21.8	− 9.2	16.7	− 3.1	32.2	− 0.6	9.7	− 4.2	7.77
J	95.0	−11.5	11.7	−6.7	23.8	−11.3	20.9	−11.5	23.4	−14.3	24.2	−15.9	8.70
K	102.0	−13.9	17.6	−2.4	16.9	−11.4	19.9	−18.6	30.9	− 6.1	18.2	−10.7	8.22
L	102.0	− 9.1	12.1	−12.8	30.3	−14.9	19.2	− 0.1	32.6	− 8.2	28.1	− 3.5	9.85
Average	56.3	6.0	9.79	4.56	15.6	6.3	13.7	7.4	18.1	6.0	11.1	5.1	6.99

[a]Average method accuracy (bias versus NBS) and precision (standard deviation in μg/dL. From Boone et al. (3).
B = bias; SD = standard deviation.

As can be seen, all the methods used had a positive bias at the low concentration, and a negative bias at the high concentration region. Also the values obtained by most methods were in agreement with the NBS method only in the region of 40 μg/dL. Thus the authors conclude that "the average of all methods' results does not provide a reliable estimate of the actual Pb content of samples for blood and lead except in the restricted analytical region around 40 μg/dL."

The determination of trace metals in body fluids is fraught with many difficulties from which high or low results may be obtained. Contamination, as in most trace element analysis, can be a severe problem. This may result, as is common, from impurities in reagents or from dust in the laboratory. Particularly serious contamination has been noted from the addition of reagents such as Triton-X and trichloroacetic acid commonly used during sample preparation. Thus a blank is essential in all analytical work of this type. In addition, to minimize contamination it is desirable that the method employed involve the least number of reagent additions and manipulations that are consistent with good metal recoveries. Therefore there is some impetus to develop methods that work directly on the sample without the requirement for sample pretreatment. In general, however, reliable methodology of this type is difficult to find. Contamination can be a serious problem during sample acquisition, and storage and precautions to minimize such problems are essential. Zinc and iron are particular problems in this regard.

Low values commonly result from losses due to spillage, spitting, and so on, during sample preparations. However, an experienced analyst will minimize such losses. Any techniques that require ashing of the sample may yield low results for the most volatile metals (e.g., cadmium, lead, zinc). Thus care in using electrothermal atomic absorption methods is essential by keeping the ashing temperatures below the levels at which these metals are lost. Low results may also be obtained for these metals because the metal is lost during the ramp to atomization but prior to reaching the desired atomization temperature. A good example is loss of lead as lead chloride. Thus it is desirable to use as fast a ramp to final atomization temperature as possible. The L'vov platform is useful in preventing this type of loss. Conversion of lead to the nitrate by matrix modification also helps overcome this problem.

Of the metals commonly determined in body fluids zinc and iron are usually high enough to be analyzed by flame atomic absorption. If flame atomic absorption or inductively coupled plasma emission must be used for trace elements in clinical samples at very low levels, solvent extraction using a chelating agent can be employed. However, because of the problems from contamination, as discussed above, the time-consuming extraction procedures cannot be as highly recommended as furnace atomic absorption methods.

In the late 1960s and early 1970s cuvette-in-flame methods were pioneered by Delves (4) and others for the more volatile metals. These methods came into favor especially in clinical laboratories. Although they were an excellent contribution in their time, most workers now generally agree that graphite furnace atomic absorption is superior. This is because of:

1. increased sensitivity;
2. applicability to a wider range of elements;
3. conditions of drying and ashing that had to be very closely controlled and be repeatable, which was difficult with cuvettes;
4. positioning of the cuvette in the flame during atomization being crucial and the repeatability of exact placement essential; and
5. interference problems being severe.

The need for standard reference body fluid samples must be reemphasized here. The NBS and a few other groups have produced standard bloods for a few elements. Much more work is needed in this regard.

7.1.5 Levels of Trace Metals in Clinical Material

It is possible with present knowledge to give approximate levels of essential and nonessential trace metals in the human body. These are given in Tables 7.2 and 7.3, (5) respectively. The areas where the metals are stored are also indicated.

Table 7.2 Essential Trace Element Content of the Human Body and Blood[a]

Element	Body		Blood Total	Plasma Total	Repository
	(mg)	(μg/g)	(mg)	(mg)	
Iron	4200	60	2500	3.6	70.5% in hemoglobin
Zinc	2300	33	34	5.6	65.2% in muscle
Copper	72	1.0	5.6	3.5	34.7% in muscle
Selenium	13	0.2	1.1	—	38.3% in muscle
Manganese	12	0.2	0.14	0.025	43.4% in bone
Molybdenum	9.3	0.1	0.083	—	19% in liver
Chromium	1.7	0.02	0.14	0.074	37% in skin
Colbalt	1.5	0.02	0.0017	0.0014	18.6% in bone (marrow)
Nickel	10	0.1	0.16	0.09	18% in skin
Vanadium	<18	0.3	0.088	0.031	$>90\%$ in fat

[a]Data of Tipton (6) and Bowen (7). From Schroeder and Nason (5).

Table 7.3 Abnormal Trace Elements of Interest in the Human Body and Blood[a]

Element	Body (mg)	Body (μg/g)	Blood Total (mg)	Plasma Total (mg)	Repository
Lead	120	1.7	1.4	0.14	91.6% in bone
Aluminum	61	0.9	1.9	1.3	19.7% in lung, 34.5% in bone
Cadmium	50	0.7	0.036	—	27.8% in kidney and liver
Boron	<48	0.7	0.52	—	Essential for plants
Tin	<17	0.2	0.68	0.10	25% in fat and skin
Mercury	133	0.2	0.026	0.009	69.2% in fat and muscle
Titanium	9	0.1	0.14	0.12	49.1% in lung and lymph
Gold	<10	0.1	0.00021	—	52% in bone
Antimony	?7.9	0.1	2.024	0.16	25% in bone
Beryllium	0.036	—	<0.00052	—	75% in bone
Arsenic	?18	0.3	2.5	<0.093	Follows phosphorus
Zirconium	420	6.0	13	1.2	67% in fat

[a]Data of Tipton (6) and Bowen (7). From Schroeder and Nason (5).

7.2 PROCEDURES

7.2.1 Determination of Trace Metals in Urine by ICPAES*

The ICP emission spectrometry, with conventional rebulizers, has detection limits that are too poor for determining most trace elements in clinical fluid samples at nanograms per milliliter levels. In recognizing this problem, Barnes and Genna (8) recommend a concentration step using a poly(dithiocarbamate) resin for the analysis of urine. Use of conventional chelating resins, such as Chelex-100, is not possible because the interfering alkali and alkaline earth elements would also be complexed. Thus these authors (8) prepared a poly(dithiocarbamate) resin that extracted copper, cadmium, lead, nickel, mercury, selenium, tin, bismuth, and tellurium from the urine matrix. (They also determined uranium.) The 10 metals are divided into 2 groups for the analysis: (1) Cu, Cd, Pb, Ni, Hg, and U, which require pH 6, and (2) Se, Sn, Bi, and Te, which should be done at pH 1.

The treatment of the urine prior to complexation is relatively simple, involving only collection of the urine sample over HCl and subsequently filtering this mixture. Metals are recovered from the resin by wet ashing with 1:1 NHO_3–H_2SO_4. The detection limits for urine, assuming a 125× concentration factor,

*Reprinted in part with permission from (8), Barnes and Genna, *Anal. Chem.* **51,** 1065. Copyright 1979, American Chemical Society.

are (ng/mL): Cu, 0.05; Cd, 0.04; Pb, 0.3; Ni, 0.06; Hg, 3; Se, 0.3; Sn, 4; Bi, 0.2; and Te, 0.4. recoveries of 90–108% may be expected.

Equipment. Metal determinations were performed using two ICP systems. The experimental facilities and the operating conditions employed are given in Table 7.4. The analysis wavelengths used with both ICP systems are listed in Table 7.5.

Resin columns were made with disposable Pasteur pipettes (Fisher No. 13-678-6B), and sample reservoirs were 1 L, 500 mL, and 250 mL separatory funnels that had been modified by sealing a 1 in. piece of 7 mm od tubing to the funnel outlets.

Reagents. Reagents for the synthesis of the poly(dithiocarbamate) resin included polyethyleneimine, 1800 molecular weight (PEI–18) (Dow Chemical Co., Midland, MI, or Polysciences, Warrington, PA), polymethylenepolyphenyl isocyanate (PAPI) (Upjohn Polymer Chemicals, Kalamazoo, MI), ACS reagent grade ammonium hydroxide, carbon disulfide (Fisher Scientific Co.), and pyridine (Eastman Kodak Co., Rochester, NY). ACS reagent grade metals, salts, acids, and bases were used throughout.

Poly(dithiocarbamate) Resin Synthesis. The poly(dithiocarbamate) chelating resin was synthesized by reacting polyethyleneimine (PEI-18) cross-linked with polymethylenepolyphenyl isocyanate (PAPI) followed by treatment with carbon disulfide and isopropyl alcohol. The PEI-18/PAPI (8.0–1.0 molar equivalent ratio) resin was prepared according to Hackett and Siggia (9) using 72.12 g PEI-18, 27.88 g PAPI, dioxane, 200 mL ammonium hydroxide, 300 mL carbon disulfide, and 800 mL isopropyl alcohol.

Resin Column Preparation. The 8.5 cm long × 0.5 cm o.d. disposable pipette columns were silanized following the procedure described by Hackett (10) with a 15% solution of methylchlorosilane in toluene to which 25 mL of pyridine was added. Without silanization of all the column and container glass surfaces, some metals were lost from the samples. A small plug of glass wool was packed into each column to hold the resin in place, and the empty columns and glass wool plug were soaked in concentrated HNO_3 until ready for use. The columns were drained and rinsed several times with distilled, deionized water prior to filling with resin. After sieving, 100 mg of the 60–80 mesh size resin was slurried with 2 mL distilled, deionized water. The rinsed columns were filled with the resin slurry, and half of the water was drained. The columns were attached to the outlet of the sample reservoir with a Tygon tubing sleeve. The sleeve and the column were then filled with distilled, deionized water.

Table 7.4 Experimental Facilities and Operating Conditions

Inductively Coupled Plasma	System I	System II
Radio-frequency generator	Plasma-Therm model HFS-5000 D generator, 40 MHz, 1000 W	Forrest Electronics ICP generator, 26.2 MHz, 850 W (18)
Power load coil	3 turns, $\frac{1}{8}$ in. diameter	1.5 turns, $\frac{1}{8}$ in. diameter
Aerosol generator	Cross-flow nebulizer, Plasma–Therm model TN-1	Cross-flow nebulizer as described by Scott et al. (19)
Nebulizer chamber	Barrel spray chamber, Plasma-Therm model SC-2.	Scott-type chamber as described by Scott et al. (20)
Plasma torch assembly	Quartz torch, Plasma-Therm model T1.5	Quartz torch. Outer tube 18 mm i.d. Aerosol tube 1.5 mm i.d. (20)
Gas flows:		
Plasma argon flow rate	15 L/min	10.4 L/min
Auxiliary argon	None	None
Aerosol Argon	Flow rate, 1.0 L/min; pressure 12 psi	Flow rate, 0.65 L/min; pressure, 21 psi
Average sample uptake rate	1.6 mL/min	1.5 mL/min
Observation height of plasma	10 mm above the top of the load coil	15 mm above the top of the load coil
Imaging optics	Quartz lens, 2 in. diameter, 200 mm focal length (Oriel. #A-11-661-37), 1:1 image.	Quartz lens, 2 in. diameter, 200 mm focal length (Oriel. #A-11-661-37), 1:1 image
Scanning monochromator	Heath model EU-700-56 (Benton Harbor, MI)	Model EU-700-56 (Heath, Benton Harbor, MI)
Grating	1180 lines/mm	1180 lines/mm
Slit width	30 μm	75 μm
Slit height	5 mm	5 mm
Photometer	Heath model EU-703-31	Heath model EU-703-31
Photomultiplier	Heath model EU-701-30	Heath model EU-701-30
Digital readout	Hewlett-Packard 3420A digital voltmeter	Hewlett-Packard 3430A digital voltmeter

Table 7.5 ICP–AES Elements and Wavelengths

Element	Wavelength (nm)	Element	Wavelength (nm)
Bi(I)	289.80	Pb(I)	405.78
Cd(II)	226.50	Se(I)	196.09
Cu(I)	324.75	Sn(I)	303.41
Hg(I)	253.65	Te(I)	238.58
Ni(I)	341.48		

Procedure. Collect urine samples over concentrated HCl (50 mL/L urine) and then filter through membrane filters (Gelman, Fishere Catalogue No. 09-735 and 09-730-26) and refrigerate. Adjust the pH of 250 mL samples for Cu, Cd, Pb, Ni, and Hg to 6. For Se, Sn, Bi, and Te adjust the pH to 1. Place the sample in silanized glass reservoirs attached to the prepared resin columns. Allow the samples to pass through at about 2.5 mL/min. After passage of the urine through the resin, rinse the reservoirs and the columns with 250 mL of distilled, deionized water. Transfer the resin containing the sequestered metals from the glass column into a 10 mL beaker by forcing air from a pipette bulb placed at the constricted end of the column. Add 1 mL of 1:1 v/v HNO_3–H_2SO_4 to the resin. Cover the beaker with a watchglass. Gently heat the resin on a hot plate until a clear solution is obtained (usually 2–3 min). Transfer the solution to a volumetric flask and dilute to volume with distilled, deionized water. When 2 mL final volumes are used, the reference solutions are prepared to contain 50% v/v 1:1 HNO_3–H_2SO_4 mixture.

The samples and standards are run on the ICP using the parameters listed in Tables 7.4 and 7.5.

7.2.2 IUPAC Reference Method for Determination of Nickel in Serum and Urine by Furnace AAS*

This method is based on the collaboration and recommendation of a group of 24 scientists from 13 countries. The authors emphasize the importance of precautions to avoid nickel contamination. They also stress the following:

1. The laboratory should be well equipped and solely devoted to trace analysis.
2. The analyst should be well versed in trace analysis techniques.
3. An up-to-date electrothermal AAS system must be available and kept in top adjustment and condition.

*Procedure from Brown et al. (11).

The method is based on a digestion of serum or urine using a mixture of HNO_3, H_2SO_4, and $HClO_4$. Subsequently the pH is adjusted to 7 and the nickel extracted with ammonium tetramethylenedithiocarbamate into methyl isobutyl ketone. The nickel is determined by electrothermal AAS.

The procedure as given is not suitable for the analysis of samples high in iron, for example, whole blood. Such samples must be treated for iron removal prior to the determinative step (e.g., addition of HCl to form a chloro complex of iron and then extraction into MIBK). This method, given below, was evaluated by a number of analysts using recovery studies and by standard addition. Coefficients of variation were found to be in the 4–8% range. Recoveries were better than 95%.

Equipment.

Digestion Tubes. The digestion is performed in borosililcate glass tubes (18 mm od; 150 mm length; 25 mL capacity). When new tubes are to be used for the first time, they must be cleaned by adding 1 mL of acid digestion mixture into each tube. The tubes are placed in the digestion aparatus and the contents heated at 300°C for 1 hr. The tubes are then cooled, residual H_2SO_4 is rinsed out with water and the tubes are washed according to the routine procedure. The tubes are discarded as soon as they become etched (e.g., after 50–75 analyses.

Blood Collection Apparatus. Teflon–polyethylene intravenous canulae (20 gauge, 5 cm length), polypropylene syringes (10 mL capacity), and polyethylene test tubes (10 mL capacity) are used for blood collection and serum storage.

Polyethylene Pipette Tips. Polyethylene tips used to dispense samples, standards, and APDC solution by means of piston-displacement pipettes must be acid washed before use, as described below.

Polyethylene Tubes. Polyethylene tubes with attached polyethylene stoppers, narrow form (0.75 mL capacity), must be acid washed as described below.

Washing of Glassware and Plasticware. Before each use, the digestion tubes, and pipette tips are scrubbed in hot detergent solution (the detergent solution should be tested to ensure that it does not contain detectable nickel) and rinsed in tap water. The materials are placed in 2 L polyethylene canisters and washed in batch fashion by filling and decanting with deionized water without any contact of the contents with the analyst's hands. After 6 rinses with deionized water, water is completely drained. Fifty milliliters of concentrated HCl is poured into the canister and the canister is capped tightly. The contents are mixed so that

HCl fumes percolate over the surfaces of the contents. The canister is allowed to stand at room temperature for 1 hr. The canister is then filled with deionized water, shaken, and allowed to stand for 20 min. The contents are allowed to drain. The contents are then rinsed 5 times with deionized water and twice with ultrapure water. The canister is placed with the lid ajar in an oven at 110°C until the contents are dry.

Piston-Displacement Pipettes. Two pipettes are needed: one with a dispensing volume of 2 mL to measure samples of urine, serum, and the calibration solutions and one of 0.5 mL to pipette the APDC solution.

Digestion Apparatus. The digestion apparatus is electrically heated with (1) an aluminum block that contains at least 42 holes (20 mm diameter, 80 mm depth) to accommodate digestion tubes; and (2) a continuously variable temperature regulator (to 300°C). Aluminum foil is packed in the bottom of each hole to facilitate heat transfer to the hemispherical base of the digestion tube and thereby to minimize bumping. The digestion apparatus, placed in a fume hood, is shielded with a glass or lucite canopy. The canopy promotes uniform cooling of the tops of the digestion tubes by the hood air draft and minimizes contamination of the tubes by dust. The temperature during digestion is monitored by a thermometer suspended in a digestion tube that contains 5 ml of concentrated H_2SO_4.

Atomic Absorption Spectrometer with Graphite Furnace Atomizer. Electrothermal atomic absorption spectrometers with background correction capability of various models and manufacturers have been found by a panel of evaluators to be satisfactory for analyses of nickel in serum and urine by this method. Illustrative operating conditions are listed in Table 7.6. Accessories for the electrothermal atomic absorption spectrometer may include (1) automatic sampling system; (2) temperature programming system with ramp modes; (3) optical pyrometer to monitor the atomization temperature; and (4) strip-chart recorder. Ultrapure argon of 99.99% is used to flush the electrothermal atomizer.

Reagents. All reagents were of the highest purity commercial available (e.g., those in the highest purity classification available from Merck and Fisher Scientific). Water was purified by double distillation (all glass still) and ion exchange procedures.

Acid Digestion Mixture. Into a glass-stoppered borosilicate glass bottle (250 mL capacity) are placed successively 120 mL NH_3 (650 g/kg, 910 g/L relative density 1.40); 40 mL H_2SO_4 (960 g/kg, 1766 g/L, relative density 1.84); and 40

Table 7.6 Analytical Conditions for Electrothermal Atomic Absorption Spectrometry of Nickel in MIBK Extracts of Wet-Digested Serum or Urine Samples

	Evaluator						
Parameter	A[a]	B[b]	C[c]	D[d]	E[e]	F[f]	G[g]
Volume of MIBK extract (μL)	20	20	20	50	50	50	20
Background corrector	D_2	D_2	D_2	D_2	D_2	None	None
Temperature program	(a) 25 sec ramp 25–120°C	(a) 10 sec ramp 25.150°C	(a) 20 sec at 100°C	(a) 20 sec at 90°C	(a) 70 sec ramp 15–120°C	(a) 60 sec ramp 25–150°C	(a) 30 sec at 180°C
	(b) 10 sec at 120°C	(b) 20 sec at 150°C	(b) 15 sec ramp 100–1100°C	(b) 50 sec ramp 90–1200°C	(b) 10 sec at 120°C	(b) 5 sec at 150°C	(b) 2 sec ramp 180–800°C
	(c) 45 sec ramp 120–1040°C	(c) 10 sec ramp 150–1000°C	(c) 25 sec at 1100°C	(c) 35 sec at 1200°C	(c) 45 sec ramp 120–1000°C	(c) 55 sec ramp 150–1000°C	(c) 13 sec at 800°C

	(d) 10 sec at 1040°C (e) 7 sec at 2700°C (f) 4 sec at 2750°C	(d) 40 sec at 1000°C (e) 7 sec at 2500°C	(d) 7 sec at 2600°C	(d) 12 sec at 2700°C	(d) 15 sec at 1000°C (e) 7 sec at 2700°C	(d) 15 sec at 1000°C (e) 7 sec at 2600°C	(d) 2 sec at 2200°C
Program for Ar flow (mL/min)	(Steps a–d) 300 (Step e) 10 (Step f) 300	(Steps a–d) 40 (Step e) 0	(Steps a–c) 300 (Step d) 50	(Steps a–c) 300 (Step d) 0	(Steps a–d) 300 (Step e) 0	(Steps a–d) 300 (Step e) 0	Not relevant

[a] Model AA-5000 spectrometer with HGA-500 electrothermal atomizer and pyrolytic graphite cuvette (Perkin-Elmer Co., Norwalk, CT). Step (f) is used to clean the cuvette between samples.
[b] Model AA-503 spectrometer with HGA-500 electrothermal atomizer (Perkin-Elmer Co.).
[c] Model 170-50 spectrometer (Hitachi Instrument Co., Tokyo, Japan) with HGA-2200 electrothermal atomizer and pyrolytic graphite cuvette (Perkin-Elmer Co.).
[d] Model AA-306 spectrometer with HGA-74 electrothermal atomizer (Perkin-Elmer Co.).
[e] Model AA-372 spectrometer with HGA-76B electrothermal atomizer (Perkin-Elmer Co.).
[f] Model AA-703 spectrometer with HGA-500 electrothermal atomizer and pyrolytic graphite cuvette (Perkin-Elmer Co.).
[g] Model 775 spectrometer with CRA-90 electrothermal atomizer (Varian Pty., Canberra, Australia).

mL $HClO_4$ (760 g/kg, 1169 g/L, relative density 1.67). The acids are thoroughly mixed, and the bottle is shielded from dust by a polyethylene outer cap.

Water for Sample Dilution. A screw-capped polyethylene bottle (1 L capacity) is filled with ultrapure water. A screw-capped piston-type dispenser (constructed entirely of glass) is fitted on the bottle in order to dispense 3-mL volumes of water. The bottle and attached dispenser are shielded from dust by a polyethylene outer cap.

Concentrated Ammonium Hydroxide Solution. Into a screw-capped polyethylene wash bottle (250 mL capacity) is placed 150 mL of ultrapure ammonium hydroxide solution (250 g/kg, 228 g/L, relative density 0.91). A screw cap with a fine-tipped delivery tube is fitted onto the bottle to permit dropwise delivery of the contents. The bottle and attached delivery tube are shielded from dust by a polyethylene outer cap.

Dilute Ammonium Hydroxide Solution. Into a screw-capped polyethylene wash bottle (250 mL capacity) are placed 50 mL of ultrapure concentrated ammonium hydroxide solution and 80 mL of ultrapure water. A screw cap with a fine-tipped polyethylene delivery tube is fitted onto the bottle to permit dropwise delivery of the contents. The bottle and attached delivery tube are shielded from dust by a polyethylene outer cap.

Dilute Nitric Acid Solution. Into a polypropylene volumetric flask (250 mL capacity) is placed 1 mL of ultrapure concentrated HNO_3. This is diluted to volume with ultrapure water and transferred to a polyethylene bottle with utmost precaution against nickel contamination. The bottle is shielded from dust by a polyethylene outer cap. This solution is used for preparation of nickel working calibration solutions.

Bromothymol Blue Indicator Solution. Into a screw-capped polyethylene drop-dispenser bottle (60 mL capacity) are palced 20 mg of bromothymol blue (certified reagent) and 1 mL of dilute ammonium hydroxide solution. The contents are diluted to 50 mL with ultrapure water. A screw cap with a fine-tipped polyethylene delivery tube that is fitted onto the bottle and an attached delivery tube are shielded from dust by a polyethylene outer cap.

4–Methylpentane–2–one (Methyl Isobutyl Ketone, MIBK). Ultrapure MIBK is placed in a borosilicate glass bottle (250 mL capacity) with a screw-capped piston-displacement dispenser set to deliver 0.7 mL volume. The bottle and attached dispenser are shielded from dush by a polyethylene outer cap.

Ammonium Tetramethylenedithiocarbamate Solution (Ammonium Pyrrolidinedithiocarbamate Solution, APDC). Into a screw-capped polypropylene graduated tube (25 mL capacity) is palced 0.5 g of APDC (1-pyrrolidinecarbodithioic acid, ammonium salt). This is dissolved in 25 mL of ultrapure water, and the solution is extracted at least three times with 1.4 mL portions of MIBK. The first 2 washings are aspirated and discarded. The last washing is analyzed by electrothermal atomic absorption spectrometry to verify that it contains no detectable nickel. The final aqueous solution of APDC should be colorless and free from precipitate. This solution is prepared immediately before use.

Potassium Phosphate Buffer (1.0 *M*, pH 7). Into a 250 mL volumetric flask are transferred 17.0 g of anhydrous KH_2PO_4 and 21.8 g of anhydrous KH_2PO_4. The contents are dissolved in ultrapure water and diluted to the volume with ultrapure water. The solution is transferred to a 250 mL separatory funnel. An 5 mL APDC solution is added, and the mixture is extracted at least three times with 10 mL portions of ultrapure chloroform. The last $CHCl_3$ washing is analyzed by electrothermal atomic absorption spectrometry to verify that it contains no detectable nickel. The buffer solution is transferred to a polyethylene bottle fitted with a fine-tipped polyethylene delivery tube to permit dropwise delivery of the contents. The bottle and attached delivery tube are shielded from dust by a polyethylene outer cap.

Nickel Stock Calibration Solution (100 mg/L Ni, 1.70 mmol/L). Into a tared borosilicate glass beaker (25 mL capacity) is weighed 50 mg of nickel powder. Ultrapure water of 5 mL and 5 mL ultrapure concentrated NH_3 are added and the nickel powder is dissolved by cautiously warming the beaker. The cooled solution is quantitatively transferred to a polypropylene volumetric flask (500 mL capacity) and diluted to volume with ultrapure water. This solution is stored in a screw-capped polyethylene bottle and is stable for at least 1 yr.

Nickel Intermediate Calibration Solution (400 μg/L Ni, 6.84 μmol/L). Into a polypropylene volumetric flask (500 mL capacity) are pipetted 2 mL of nickel stock standard solution and 2 mL of ultrapure concentrated HNO_3. The contents are diluted to volume with ultrapure water and transferred to a screw-capped polyethylene bottle. This solution is prepared fresh every 3 months.

Nickel Working Calibration Solutions. Into six polypropylene volumetric flasks (100 mL capacity) are accurately pipetted, respectively 0 (blank), 0.5, 1, 2, 3, and 4 mL of nickel intermediate standard solution and 4, 3.5, 3, 2, 1 and 0 mL of the dilute HNO_3 solution. Ten milliliters of Fe–Cu–Zn matrix solution are added to each flask. The contents are diluted to volume with ultrapure water.

These solutions contain 0, 2, 4, 8, 12, and 16 μg/L Ni (0, 34, 68, 136, 204, and 273 nmol/L, respectively), and they are prepared every 2 weeks. The nickel calibration solutions are used for construction of the calibration curve that is prepared with each set of nickel analyses.

Urine Sample for Quality Control. A 24-h urine specimen (containing 4 to 6 μg/L Ni) is acidified by addition of 10 mL of concentrated HNO_3/L. The acidified urine speciment is distributed in 5 mL aliquots in screw-capped polypropylene tubes that are stored at −20°C. One tube is thawed for inclusion in each set of nickel analyses.

Urine Nickel Recovery Sample for Quality Control. The same 24-h specimen of urine that is used for the quality control sample is also used to determine the recovery of nickel. Into a volumetric flask (250 mL capacity) is transferred 5 mL of nickel intermediate standard solution. The contents are diluted to volume with the acidified urine specimen. The spiked urine sample is distributed in 5 mL aliquots in polypropylene tubes, and the tubes are stored at −20°C. One tube is thawed for inclusion in each set of nickel analysis. The nickel concentration obtained by analysis of the unspiked urine sample is subtracted from that obtained by analysis of the spiked urine sample. The difference obtained is divided by the net concentration of added nickel (8 μg Ni/L), and the dividend is multiplied by 100 to yield the percentage recovery of added nickel.

Procedure.

Venipucture and Serum Separation. The antecubital fossa of the arm is cleansed with ethanol and allowed to dry by evaporation. A tourniquet is applied while a polyethylene intravenous catheter is inserted into an antecubital vein. The stylus of the catheter is removed, and the cather is flushed with >2 mL of blood, which is discarded. A polypropylene syringe is then used to collect 10 mL of blood. The blood is placed in a polyethylene test tube and capped. The blood is allowed to clot for 45 min at room temperature. The test tube is centrifuged at 900*g* for 15 min. By use of an acid-washed polyethylene dropper, serum is transferred to a screw-capped polytethylene test tube. The serum specimen is rejected if there is visible hemolysis, lipemia, or turbidity. The serum is stored at 4 or −20°C until the time of analysis.

Digestion Procedure. Samples of 2 mL of serum or urine are transferred to duplicate digestion tubes. Into six more pairs of duplicate digestion tubes are transferred 2 mL of the nickel calibration solutions that contain 0, 2, 4, 8, 12, and 16 μg Ni/L. Nothing is added to another pair of duplicate digestion tubes. Acid digestion mixture (2 mL) is dispensed into all the tubes and the tubes are placed in the digestion apparatus at ambient temperature. The tubes are heated initially at 110°C for 1 hr. This is the most critical stage of the digestion because of the possibility of sample loss by foaming. The digestion then proceeds stepwise as follows: (1) 2 hr at 140°C; (2) 30 min at 190°C; and finally (3) 1 hr at 300°C. At the conclusion of the 4.5 hr digestion period, all HNO_3 and $HClO_4$ should have evaporated. The contents of the tubes should be perfectly clear and colorless, and the sample volume should be approximately 0.2 mL (corresponding to the volume of the residual H_2SO_4).

Extraction of Nickel. After the tubes have cooled to ambient temperature, bromothymol blue indicator solution (3 drops) and ultrapure water (3 mL) are added to each digestion tube with care to rinse down the walls. Concentrated ammonium hydroxide solution is added dropwise with constant swirling until the color begins to change to blue. Phosphate buffer (2 drops) is added. Dilute ammonium hydroxide solution is added dropwise until the color is light blue–green. The APDC solution (0.5 mL) is added to each tube, and the contents are mixed for 10 sec with a vortex-type mixer. The samples are allowed to stand for 5 min; MIBK (0.7 mL) is added to each tube, and the contents of the tube are mixed for 40 sec with a vortex-type mixer. The aqueous and MIBK phases separate without centrifugation. The MIBK extracts are yellow due to the extraction of copper and iron as well as nickel. Approximately 0.5 mL of each supernatant MIBK extract is transferred to a polyethylene tube (0.75 mL capacity) using a Pasteur pipette with care to avoid transfer of aqueous phase. The tube is sealed with the attached polyethylene stopper. If necessary, the analysis can be interrupted at this point, and the MIBK extracts can be stored overnight at 4°C.

Atomic Absorption Spectrometry. Samples of the MIBK extracts are pipetted into the graphite cuvette, and atomic absorption is measured at the nickel absorbance line (232.0 nm). Each analyst should determine for the particular instrument (1) the optimal volume of MIBK extract; (2) the optimal instrumental parameters; (3) the optimal temperature program; and (4) whether or not background correction is advantageous (see Table 7.6).

Computations. The heights or areas of peaks on the recorder tracings are measured. The mean reading obtained for duplicate 0 $\mu g/L$ Ni samples in the set of calibration solution is subtracted from the mean readings obtained with the other calibration solutions, and the resultant values are plotted to prepare the calibration curve. The mean reading obtained with blanks prepared by analysis of duplicate empty digestion tubes is subtracted from the mean readings obtained with the serum or urine samples, and the nickel concentrations in the samples are estimated by reference to the calibration curve.

7.2.3 Determination of Fe in Serum by Furnace AAS*

This method allows direct determination of iron in serum after dilution 40 times with water. Other available methods require a chemical pretreatment of a precipitation step to remove interferences due to protein. Because no chemical digestion is necessary, contamination is kept to a minimum. Standardization is done by direct comparison with aqueous standards. The results from the proposed method were checked against those obtained by a previously established wet digestion method and the values were found to agree satisfactorily.

Equipment. A Model 460 atomic absorption spectrometer (Perkin-Elmer Corp., Norwalk, CT) with a deuterium arc background corrector in conjunction with an HGA-2100 graphite furnace (Perkin-Elmer) was used. The atomizer consisted of a cylindrical graphite tube (6 mm i.d., 28 mm long). A wavelength of 248.6 nm with a spectral bandwidth of 0.2 nm was used. A continuous flow of nitrogen (47.7 mL/min, setting 30 on the HGA controller) was directed through the tube. After the tube was carefully aligned to give a maximum signal, it was cleaned by making a few firings until no peaks were observed on the recorder. Samples or standard solutions were added to the graphite tube directly with an Eppendorf micropipette with disposable tips. Pipette tips were soaked for at least one day in 1 *M* HCl, then thoroughly rinsed with distilled water. A new tip was used for each sample; 50 μL aliquots were used throughout.

Reagents. All chemicals were analytical grade. Standard Fe (III) chloride (1000 ppm Fe) solution (in 0.1 *M* HCl; Wako Chemicals) was used to prepare working standards by dilution with 0.1 *M* HCl. Deionized, distilled water, in which no Fe could be detected, was used throughout.

Procedures. Dilute 25 μL of serum and mix with 1 mL of water and inject into the graphite tube. After drying at 100°C for 50 sec, and charring for 55 sec, including a ramp time of 30 sec, up to 1000°C, atomize at 2400°C for 8 sec. Measure the Fe atomic absorption signals with background correction, using 8 sec integration.

7.2.4 Determination of Zn in Serum by Flame AAS†

The level of zinc in serum is high enough (the sensitivity of zinc by AAS is very good) that flame atomic absorption can be used to determine zinc directly in serum samples diluted by a factor of 10 with water. The proposed method was

*Procedure from Nakamura et al. (12).

†Procedure from Momcilovic et al. (13).

Table 7.7 Instrumental Parameters Used in Determination of Zinc in Serum

Instrument	Varian-Techtron AA5
Burner	AB 51 (single slot)
Wavelength	213.9 nm
Lamp current	6 mA
Slit width	100 μm
Scale expansion	3
Damping	B
Acetylene flow	0.3 L/min
Air flow	7.4 L/min
Recorder	Varian G-2000
Recorder range	10 mV

tested against a wet digestion procedure using nitric acid, and satisfactory agreement was obtained. Standards were prepared in dilute nitric acid solution.

Equipment. A Varian-Techtron AA-5 atomic absorption spectrophotometer was used with a Varian G-2000 strip-chart recorder (both from Varian-Techtron Pty., Ltd., Melbourne, Australia) to measure zinc concentrations in serum with an air–acetylene flame. Table 7.7 lists the instrumental settings and variables.

Standard glass volumetric flasks and pipettes were used. The glassware was washed in an automatic washer and soaked in dilute (100 mL/L) HNO_3 and rinsed thoroughly with distilled demineralized water. An ion exchange column (demineralizer column, Research Model No. 1; IWT, Rockford, IL 61105) was used to produce the demineralized water used in all these experiments where water is mentioned.

Reagents. All chemicals used were reagent grade ($ZnCl_2$, granular, reagent grade, from Anachemia Chemicals Ltd., Montreal, Toronto, Canada; HCl and HNO_3, Baker analyzed reagent from J.T. Baker Chemical Co., Phillipsburg, NJ 08865). The stock standard, containing 1000 μg/mL Zn and dilute (10 mL/L) HCl, was prepared from $ZnCl_2$ and kept in a plastic bottle (Nalgene, Labware Division, Rochester, NY 14602). The stock standard was diluted with water or dilute (10 mL/L) HNO_3 to obtain two sets of working standards containing 500, 1000, 2000, and 4000 μg/L Zn.

The standards in HNO_3 were prepared by adding 1 mL of dilute (100 mL/L) HNO_3 working standard and diluting to 10 mL with water. Four final concentrations of 50, 100, 200, and 400 μg/L Zn were used. Reagent blanks were prepared for each set of standards.

Procedure. Expose each vial of serum to room temperature overnight (about 16 h). Then mix thoroughly before taking samples for zinc determination. Dilute to 10 mL in glass volumetric flasks identical to those used for the standards. Run by flame AAS using instrument parameters given in Table 7.7.

7.2.5 Determination of Zn in Serum and Urine by Furnace AAS*

A furnace AAS procedure for determining zinc in serum requiring only a 10 μL sample is used. A strong emphasis is placed on the need to use ultrapure reagents and water. Working standards are prepared fresh each day in zinc-free dialysis fluid to obviate matrix problems due to differences in sample and standard composition. The proposed method was tested against an established flame atomic method and good agreement was obtained. Note that Taylor and Bryant (15) report a large scatter of zinc results by furnace atomic absorption, but they did not test the following procedure.

Equipment. All determinations were made with a Model 5000 atomic absorption spectrometer equipped with an HGA-2100 graphite furnace with ramp accessory and a Mode 056 strip-chart recorder (all from Perkin-Elmer Corp., Norwalk, CT 06856). A Perkin-Elmer hollow-cathode zinc lamp was used as the source at a current of 15 mA.

The spectrometer was operated at 213.9 nm, in the peak height mode, and with a 0.7 nm low slit width. Graphite furnace conditions, established according to Fernandez and Innarone (16), were: dry for 60 sec with 10 sec ramp to 95°C; char for 30 sec with 15 sec ramp to 450°C; and atomize for 6 sec at 2400°C.

Argon was used as the purge gas. The flow rate was adjusted to 60 flow meter divisions (corresponding to 110 mL/min) during atomization, to reduce zinc sensitivity threefold. This avoided excessive dilution of the sample, which can result in significant extraneous zinc contamination. The final sensitivity for a typical zinc assay at these settings is 37 pg/0.1 absorbance unit (A).

A new Monoject needle was used to free the clot from the Falcon labware tubes (Nos. 2059 and 2063; Falcon, Oxnard, CA 93030), thus avoiding zinc contamination from wooden applicators. Disposable Eppendorf, tray-mounted, polypropylene micropipette tips, supplied in covered boxes of 100 tips, were used to dilute samples, to prepare standard curves, and to introduce specimens into the graphite furnace. To eliminate a small but variable degree of zinc contamination from these tips, immediately before use each tip was rinsed twice with an 0.8 *M* ultrapure NHO_3 solution and then three times with water in three different plastic tubes (to minimize contamination from repetitive rinsing in a single tube), which was a procedure found to be quick and adequate.

*Procedure from Kayne et al. (14).

Reagents. All water used in these studies was processed through deionizers manufactured and maintained by Hydroservice and Supplies, Inc., Durham, NC 27705; it then contained <15 ng Zn/L. All chemicals used were analytical reagent grade (J.T. Baker Chemical Co., Phillipsburg, NJ 08865) unless otherwise noted. Ultrapure HNO_3 was used in standard curve and sample preparation (J.T. Baker, Ultrex grade). Certified Atomic Absorption Reference Solution containing 1 mg/L Zn (Fisher Scientific Co., Fairlawn, NJ 07140) was used to prepare the standard curve.

Deionized water stored in the tubes for several days did not acquire any detectable zinc.

For preparation of large volumes of zinc-free solutions, glassware was immersed in 1.6 *M*/HNO_3 for 24 h, rinsed 4 times with water, and allowed to dry in a basket that was lined and covered with absorbent paper shown not to contaminate the glassware with zinc. Such glassware was handled only with plastic gloves that were first rinsed in nitric acid and then in water to avoid contamination from the relatively high amounts of zinc present on the surface of the skin.

Procedure. Wherever possible, disposable plasticware (polypropylene or polyethylene) is used to collect specimens and to prepare them for analysis.

Allow blood samples to clot in Falcon labware tubes. Store the serum in the same tubes. These tubes should also be used for all sample dilutions and solutions used in preparing standard curves.

Dilute serum 100-fold by adding 10 μL of serum to 990 μL of water (10 μL of this solution gives readings of about 0.2 A). To obtain complete and reproducible delivery of dilutions of serum, fill a rinsed pipette tip and dry the outside carefully with a paper wipe to prevent the aliquot from sticking to and climbing up the tip upon expulsion because of the surface-tension-reducing properties of protein solutions.

Introduce 10 μL samples into the furnace, using a new, rinsed pipette tip for each sample. Use three identical dilutions of each serum sample to provide three readings. Assay the dilution resulting in the median reading twice more; the median of the three readings on this dilution is regarded as the most precise estimate of the true serum value.

7.2.6 Determination of Cu in Serum by Flame AAS*

The following flame atomic absorption method was chosen on the basis of a collaborative study of methods for determining copper in serum. Twelve collaborating laboratories were involved.

*Procedure from Osheim (17).

The serum sample is diluted 1 : 1 with water and the viscosity of the standards adjusted to match that of the samples using 100% glycerin. An external commercially available copper serum control sample is run with each sample set. The coefficient of variations for intralaboratory and interlaboratory results was found to be in the range of 2.2 to 4.4% and 2.6 to 6.1%, respectively.

Equipment. The atomic absorption unit was equipped with a nebulizer and an air–C_2H_2 burner head. Performance was monitored by assuring that a 4.0 mg/L standard produces a response of ≥ 0.2 absorbance unit.

Reagents. External control: Precilip, Catalogue No. 125067 (Bio-Dynamics/bmc, 9115 Hague Rd., Indianapolis, IN 46250) or equivalent with established value for copper. Dilute according to label.

Glycerol USP: 10% v/v aqueous solution.

Copper standard solutions: stock standard solution, 1000 mg/L. Dissolve 1.000 g Cu metal in a minimum volume of (1 + 1) HNO_3–H_2O. Dilute to 1000 mL with 1% HNO_3.

Intermediate standard solution, 100 mg/L: Dilute 10 mL stock solution to 100 mL with H_2O.

Working standard solutions: Dilute 0.0, 0.25, 0.5, 1.0, 2.0, and 4.0 mL intermediate solutions to 100 mL with 10% glycerol to give standard solutions 0.0, 0.25, 0.5, 1.0, 2.0, and 4.0 mg Cu/L.

Procedure. Rinse all glassware used with 2 *N* HCl. Mix samples thoroughly before pipetting. Using a Mohr pipette, transfer 1.0 mL serum and 1.0 mL Precilip (external control) to separate test tubes. Add 1.0 mL H_2O to each and mix 5 sec on vortex mixer or capped tubes and shake 10 sec. Use 1 external control for each 10 samples or fraction thereof.

Analyze by AAS using the following conditions: wavelength, 324.7 nm; slit, 0.7 nm; flame, air–C_2H_2 (lean blue). Aspirate a series of working standard solutions, external control solution, and sample dilutions. Repeat analysis if copper value in external control solution is not within accepted range. Prepare standard curve of concentration in milligrams of copper per liter versus A and determine concentration of the sample. Multiply the result by 200 to account for sample dilution and to convert the result to μg Cu/100 mL.

7.2.7 Determination of Cu in Serum by Furnace AAS*

Taylor and Bryant (15) studied procedures for the determination of copper and zinc in serum by AAS. They found that low results are obtained for copper using electrothermal AAS when inorganic copper standard working solutions are em-

*Procedure from Evenson and Warren (18).

ployed. Evenson and Warren (18) (whose procedure follows) however, report the use of such solutions in serum analysis as giving accurate results (tested on NBS standards and by flame AAS) and in fact list this to be an advantage of their procedure.

Sample preparation is simple, involving 10-fold dilution of the sample with 10 *M* HNO_3. It is important that the final pH be about 3 for best accuracy.

Equipment. All analyses were performed on a Model 303 atomic instrument (Perkin-Elmer Corp., Norwalk, CT 06856) equipped with a Model No. HGA-2000 graphite cuvette (Perkin-Elmer Corp.) for volatilization and atomization in place of the flame. The signal produced was recorded on a 10 mV strip-chart recorder (Model PWA; Texas Instruments, Houston, TX 77006).

Vacutainer tubes (No. 3200; Becton-Dickinson, Rutherford, NJ 07070) were used to draw blood samples.

A 50 μL Pressure-Lok pipette (Precision Sampling Corp., Baton Rouge, LA 70815) with a Teflon-tipped plunger was used for delivery of samples, standards, and controls into the graphite cuvette. (The Teflon tip on the plunger helps minimize trace metal contamination between plunger and sample.)

All tubes, volumetric flasks, and pipettes were soaked in 2*M* NHO_3 for 24 h, then rinsed with doubly deionized water about 10 times. The Eppendorf pipette tips were also acid washed before use. Plastic urine cups, Auto-Analyzer cups, and individually wrapped 5.0 mL sterile culture tubes were found to contain less copper than could be detected.

Reagents. Working standards were prepared to contain 0.05, 0.1, 0.2, 0.3, 0.4 and 0.5 mg/L Cu in 10 m*M* HNO_3. The standards were run in sequence; the serum samples were separated by a cup of 10 m*M* HNO_3. An NBS SRM 157a was analyzed as control. All standards, controls, and sera were measured in triplicate, with the mean of the last two measurements recorded. All reagents were analytical reagent grade (unless otherwise noted).

Procedure. Pipette 2.0 mL of the 10-fold dilution of serum and 20 μL of 1.0 m*M* HNO_3 into acid-washed 10 mL glass tubes.

Using the 50 μL Pressure-Lok pipette, introduce 5, 10, 15, 20, 25, and 30 μL aliquots of the standard into the cuvette for the standard curve. Analyze each standard in triplicate using the instrument settings in Table 7.9. When the concentration of the 10-fold diluted serum samples are read from the standard curve, the standards (multiplied by 10, the dilution factor) are 0.33, 0.67, 1.0, 1.33, 1.67, and 2.0 mg Cu/L.

Pipette serum samples in triplicate, using 15 μL aliquots, and rinse the pipette with distilled water between each serum dilution. Run the samples using the instrument settings given in Table 7.8. Average the absorbance values of the last

Table 7.8 Instrument Settings

HGA-2000 Controller	
Dry,	30 sec at 125°C
Char,	35 sec at 700°C
Atom,	12 sec at 2000°C
Perkin-Elmer Model 303	
Hollow cathode,	15 mA
Slit,	6
Wavelength,	324.7 nm
Recorder	
Scale expansion	X1
Noise suppression	X1

two injections and determine the copper concentration from the standard working curve. For samples that have an abnormally high copper content, use 5.0 μL aliquots and multiply the answer by three. Thus the standard curve can be used for samples containing as much as 6.0 mg Cu/L.

7.2.8 Determination of Cr in Serum and Urine by Furnace AAS*

Perhaps the greatest difficulty in determining chromium in urine and serum by electrothermal AAS is compensation for background effects. Many AAS instruments still do not have the capability of doing background correction at the 357.9 nm resonance line of chromium. There are several methods of handling background problems in this spectral region, such as the following:

1. Use an instrument encorporating Zeeman background correction.
2. Do background correction with a tungsten filament lamp.
3. Choose a sample preparation, furnace program (dry, ash, and atomize), and matrix modification regimen so that the need for background correction is minimal.

The following procedure by Kayne et al. (19) employs method (2), but reference is also made to the procedural parameters used by Routh (20) to minimize problems due to background interferences. A 5-fold dilution with water is used for sample preparation. The detection limit is 0.1 μg/L.

Equipment. The deuterium lamp assembly was removed from a Perkin-Elmer Model 603 atomic absorption spectrophotometer and a plane front surface mir-

*Procedure from Kayne et al. (19).

ror was substituted to direct light through the chopper in the usual background-correction optical path. An air-cooled 100 W (12 V) tungsten–halogen lamp was mounted on the outside of the instrument, with the light directed down through a shielded port cut in the mirror housing and outside case. A Model 6324 lamp housing with a blower (Oriel Corp., Stamford, CT 06902) was used to shield and cool the lamp as well as to facilitate alignment. An observation port on the side of the mirror box allowed the lamp image to be seen on the monochromator entrance slit. The resulting configuration yields very even illumination through the furnace tube of sufficient intensity to balance fully the photon signal produced by the hollow-cathode lamp up to its maximum operating current. An Oriel G–774 3300 filter in the optical path was used to remove stray light from this source. For stability and versatility of control, the tungsten lamp was powered by a Hewlett-Packard 6267 B power supply operating in the constant current mode. The operation of the instrument in the background correction mode is exactly the same as the unmodified version. To allow this, a shutter was placed in the reference light path and a source of 60–120 V dc was applied to the deuterium lamp terminals to energize the background correction circuit through the optically coupled switch that is part of the instrument circuit. For convenience, the deuterium supply can simply be left connected and used to switch on the circuit.

For sample atomization a Perkin-Elmer HGA 2200 graphite furnace unit equipped with the temperature ramping accessory and the optical temperature sensor was used. Instrument parameters are given in Table 7.9.

Reagents. Purified HNO_3 was obtained from NBS (prepared by subboiling distillation) and J.T. Baker Chemical Co. (Ultrex grade) and 30% H_2O_2 from Fisher Scientific Co. For $K_2Cr_2O_7$, Fisher reagents and both the solid analytical reagent grade and the 1000 mg/L certified atomic absorption standard were used. The water used for all procedures and washing was first deionized, then distilled

Table 7.9 Chromium Analysis Conditions

Dry, 100 sec; ramp, 80 sec; 110°C
Char, 60 sec; no ramp; 1100°C
Atomizer, 8 sec; no ramp; 2500°C
Argon gas flow, 40 mL/min
Flow time, 3 sec, stop
Temperature sensor, on
Maximum power, on
Wavelength, 357.9 nm
Slit, 0.7 nm
Lamp current, 20 mA

Table 7.10 Instrument Parameters for Determination of Cr in Urine

Wavelength	357.9 nm
Slit	0.5 nm
Dry	100°C/40 sec
Ash	1000°C/50 sec
Atomize	2300°C/15 sec

in glass, and finally passed through a Nanopure 3 ion exchange and ultrafiltration system (Barnstead Co., Boston, MA 02132). Any Chromium in the water supply was below the level of detection. Solutions were kept in polyethylene or polystyrene containers that were washed in HNO_3 and checked for contamination by chromium.

Procedure. Place 0.5 mL serum or urine into 16 × 100 mm fused silica test tubes. After overnight incubation at 80°C, remove the tubes and allow them to cool to room temperature. Dilute the contents of the tubes to a final volume of 2.5 mL with deionized, distilled water. Take 20 μL portions of the solution with Eppendorf automatic pipettes and inject into a graphite furnace. Run replicate samples and use the peak-height measurements. Standard Cr solutions were prepared in an acid–H_2O_2 matrix and then incubated at 80°C prior to analysis.

For urine Routh (20) diluted aliquots of acidified urine (to 1 *N* Ultrex HNO_3) 1:1 with blank or standard Cr solution (the method of standard additions was used). Then 20 μL aliquots were injected into the furnace. The instrument parameters are given in Table 7.10. Background correction was used. Routh (20) found that losses of Cr were occasional at ash temperatures above 1000°C. Background correction was used, and the background signals found to be easily correctable using standard instrumentation (Varian Techtron AA-775).

7.2.9 Determination of Al in Serum and Urine by Furnace AAS*

Contamination can be a very serious problem in determining aluminum at the nanograms per gram level in most laboratories. Toda et al. (22) and Smeyers-Verbeke et al. (21) both had problems in this regard. The latter authors suggest rinsing of apparatus with aluminum-free nitric acid. The level of aluminum in reagents used must be ascertained and if necessary purification should be undertaken. Syringes used for sample collection were often found to be contaminated with aluminum. To minimize contamination, samples for analysis were not treated with a preservative but were simply stored in their containers in a freezer until used.

*Procedure from Smeyers-Verbeke et al. (21).

It is essential to use the method of standard additions for calibration. Background correction is also employed.

Equipment. All determinations were performed using a Perkin-Elmer atomic absorption spectrometer Model 460 together with a Perkin-Elmer graphite furnace HGA 76B. Argon was used as the purging gas. Solutions were injected manually into the furnace using Eppendorf pipettes.

Reagents. Except for preparing the aluminum standards, the use of glass materials was avoided. All materials were rinsed before use with diluted aluminum-free HNO_3 (Merck), followed by several rinses with double-distilled water. After this washing procedure all containers were tested for contamination.

To prepare the standard solutions and the sample dilutions double-distilled water, prepared just before use in a quartz device, was used. No detectable level of aluminum was found.

The standards were prepared just before use from an aluminum chloride standard solution (1000 mg/L, provided by Merck) by simple dilution with double-distilled water.

Urine samples were collected in thoroughly rinsed polyethylene or polystrol containers. For the smaller serum samples, PTFE material was preferred. The PTFE containers were always used for the dilution of urine and serum and for the preparation of the samples for standard addition. For the collection of blood samples a 10 mL sterile syringe (Sarstedt Monovette Co.) was used. The system, which was found to be free of detectable aluminum, allows direct centrifugation of the blood in the syringe.

Procedure. Urine samples are either used undiluted or are diluted 1:2 with double-distilled water, depending on the aluminum concentration. Serum samples are diluted 1:3 with double-distilled water. The dilution is adapted to the aluminum in the sample. The standard addition method is used for both samples and measurements performed using the furnace conditions given in Table 7.11.

7.2.10 Determination of Cd in Blood and Urine by Graphite Furnace AAS*

Severe problems due to high levels of nonspecific background signals are encountered in the direct determination of cadmium in blood and urine by furnace atomic absorption. Pleban and Pearson (24) published a method requiring no wet ashing step, but Zeeman background correction was necessary to compensate for the background signal. This method is not given here because most workers do not have a Zeeman instrument.

*Procedure from Perry et al. (23).

Table 7.11 Furnace Conditions for the Analysis of Al in Serum and Urine

Urine	
Volume	25 or 50 μL
Drying	100°C/60 sec
Ashing	300°C/60 sec
Ramp ashing	300–1400°C
	30°C/sec = 1 × 30[a]
Atomization with gas interrupt	2650°C/8 sec
Serum	
Volume	25 μL
Drying	100°C/60 sec
Ramp ashing	100–600°C
	6°C/s = 2 × 30[a]
Ashing	1400°C/15 sec
Atomization with gas interrupt	2650°C/sec

[a]Indicates HGA 76B furnace setting.

Perry et al (23) wet ashed blood and urine samples with a mixture of nitric acid and hydrogen peroxide prior to analysis by graphite furnace AAS. Both background absorption and matrix interferences were reduced to tolerable levels by this treatment. The detection limit for both samples is 2 pg. A coefficient of variation at the 2.8 μg/L level was 7%. National Bureau of Standards Bovine Liver was used to access accuracy. The results obtained were 0.27 μg/g $\pm$0.01 (accepted value is 0.27 μg/g $\pm$0.4).

Equipment. A Model 403 atomic absorption spectrophotometer with attached deuterium arc for background correction, equipped with a Model HGA 2000 graphite furnace, was used. A Model 165 recorder recorded the absorption peaks. All this equipment was obtained from Perkin-Elmer Corp., Norwalk, CT 06856. Disposable polyethylene containers Catalogue No. B7938 for urine were purchased from Scientific Products, McGraw Park IL 60085.

Reagents. All water used for preparation of standards, dilution of samples, and washing of glassware was deionized and had a minimum resistance of 5 MΩ. Redistilled HNO_3 and $HClO_4$ (G. Frederick Smith Chemical Co., Columbus, OH) were used throughout. Hydrogen peroxide (30%) was purchased from Fisher Scientific Co., St. Louis, MO 63032. Pyrex digestion tubes were decontaminated by treating each tube with 6 blank digestions using 0.5 mL of 70% $HClO_4$ and rinsing 6 times with deionized water between each digestion. (Decontaminated tubes are available from Environmental Science Associates, Burlington, MA 01803.)

Standard solutions of cadmium were freshly prepared each week and carefully checked daily for constancy of absorption in the concentration range of 1 to 5 μg/L from a 1000 mg/L reference standard (Fisher Scientific Co.) in dilute (10 mL/L) HNO_3.

Procedure—Blood. Immediately after collection, transfer 0.5 mL of blood to digestion tubes and add 1 mL concentrated HNO_3. Place the tubes in a heating block and digest the blood slowly for 3 h at a temperature just below boiling. When the volume has been reduced to about a third, add 0.4 mL 30% H_2O_2, evaporate the sample at the same temperature, remove from the heating block, and dissolve the residue in 5.0 mL HNO_3 (10 mL/L) to provide a sample solution for analysis. Plasma is handled in the same manner as blood after the cells are removed by centrifugation.

Procedure—Urine. Twenty-four-hour urine samples are collected from the patients directly into disposable polyethylene containers to which had been added 20 mL of 3 *M* HCl acid. Urine samples are collected in plastic containers containing 3 *M* HCl. Place a 1.0 mL aliquot of urine in the digestion tube and add 0.2 mL concentrated HNO_3. Digest the urine as described for blood. After digestion, dissolve the dry sample in 2.0 mL HNO_3 (10 mL/L) for assay. All reagents used in sample collection and preparation must be carried through in the blank. The reagent blank consistently corresponds to 50 ng/L.

Procedure for Analysis. Transfer 20 μL of the sample solution to a standard graphite tube with an Oxford pipette. Optimum temperature and time for drying, charring, and atomizing are 150°C for 30 sec, 300°C for 60 sec, and 1950°C for 8 sec, respectively. Purge the chamber with argon. Operate the furnace in the interrupt mode so that the flow of purge gas to the furnace can be stopped during the atomization step.

7.2.11 Determination of Pb in Blood and Urine

The determination of lead in blood and urine is one of the most important trace analyses that is done. Despite this there is no general agreement on methodology. The reader's attention is drawn to a study of the determination of lead in blood by Boone et al. (3) mentioned previously. This 1979 study showed that present methods gave positive biases at the low concentration end and negative biases at the high concentration levels. Good agreement was obtained at only about 40 μg/100 mL as compared with isotope dilution mass spectrometry.

There have been many publications (mostly from hospital laboratories) dealing with the determination of lead in body fluids by the Delves cup approach (4). As indicated, the results from these procedures show a great dependence on the repeatability of the pretreatment step and on the exact positioning of the cup in the flame. Because I feel that the electrothermal AAS methods are generally superior to the Delves cup approaches, the latter are not given here.

Electrothermal atomization methods abound for the determination of lead in blood and urine. Much research has been done to investigate the complex interferences that occur in determining lead in such atomizers. Slavin and coworkers (25) in a series of papers have clearly demonstrated the importance of employing the L'Vov platform when performing determinations of the more volatile elements such as lead in complex matricies. Using conventional electrothermal atomization from the furnace wall, some lead compounds, particularly the chloride, can be lost before attaining the desired atomization temperature of lead. The L'vov platform delays the volatilization of volatile lead compounds (the temperature of the platform follows more closely the gas temperature until the atomization temperature has been reached). Thus the atomization takes place under isothermal condition and little problem of lead loss occurs.

A variety of matrix modifications have also been suggested to stabilize lead during thermal pretreatment steps. For example, the addition of ammonium nitrate to a sample high in chloride prevents the formation of volatile lead chloride by removing the chloride as ammonium chloride at temperatures lower than those at which lead chloride is lost.

*7.2.11.1 Determination of Pb in Blood and Urine by Flame AAS**

The following flame AAS method for the determination of lead in blood and urine was published by Berman in 1964 (26). A study done by Kopito et al. (27) in 1974 showed that the method gives good lead recoveries (about 96%). This reliable method is included for those workers who do not have electrothermal atomic absorption equipment.

The proteins in the blood samples are precipitated using trichloroacetic acid and then the lead is extracted using ammonium pyrrolidinedithiocarbamate–methyl isobutyl ketone (APDC–MIBK). For urine, acidification only is used prior to the APDC–MIBK extraction.

Reagents and Equipment. Trichloroacetic acid (TCA, 5%) and ammonium pyrrolidinedithiocarbamate (APDC, 1%) were used. Work was done on a Perkin-Elmer Model 214 unit using an air–acetylene flame. The 217.0 nm line was employed.

Procedure.

Blood. Add 10 mL of 5% TCA to 5 mL of whole blood. Allow the samples to stand for 1 h and stir occasionally with a glass rod. Centrifuge the supernatant

*Procedure from Berman (26).

liquid and decant. Add 10 mL of distilled water to the residue, stir the sample, and centrifuge again. Decant the supernatant liquid and mix with the previous liquid. Adjust the pH 2.2–2.8 by the addition of about 0.6 mL of 0.5 *N* NaOH Add 1 mL of a 1% solution of APDC and 5 mL of MIBK to this mixture. Shake the samples for 2 min manually or 10 min mechanically, and determine the Pb in the organic phase. Use standards containing 0, 0.2, 0.5, 1, and 2 μg/mL Pb in 5% TCA similarly extracted into MIBK. If an emulsion forms when the ketone is added, centrifuge the sample to obtain complete separation of the layers.

Urine. Acidify 30 mL urine to pH 2.2–2.8 with 5% TCA. Add 1 mL 1% solution of APDC and 5 mL MIBK. Shake the samples for 2 min manually or 10 min mechanically and determine the Pb in the organic phase against similarly extracted standards of 0, 0.2, 0.5, 1, and 2 g/mL Pb. If an emulsion forms at the interface of the layers, centrifuge the sample for 10 min.

*7.2.11.2 Determination of Pb in Blood by Furnace AAS**

This method is the modification of a previous method that involved dilution of the sample in Triton-X 100 followed by atomization from the wall of a furnace atomizer. Improvements include use of the L'vov platform, peak area measurements, and use of 0.2% $NH_4H_2PO_4$ as a matrix modifier. Standard lead solutions containing the matrix modifier are used for calibration. A charring step at 600°C is possible with the $NH_4H_2PO_4$, which reduces the residue to minimal amounts and results in a background signal easily correctable with the deteurium arc.

Equipment. The authors have used both analog or microprocessor-equipped instrumentation. Pyrolytic-coated grooved tubes and solid pyrolytic graphite platforms were employed. Argon was used as the purge gas.

Reagents.

Blood Dilutant Solution. A mixture of 0.2% $NH_4H_2PO_4$ and 0.5% Triton-X 100 is prepared. Standards are prepared in 0.2% $NH_4H_2PO_4$ and 0.5% HNO_3.

Procedure. Pipette 400 μL of a diluent solution containing 0.2% $NH_4H_2PO_4$ and 0.5% Triton-X 100 into a 1500 μL Eppendorf microcentrifuge tube. Pipette 100 μL of blood into the tube and flush pipette tip 3–4 times with the dilutent solution to minimize transfer error. Cap tube and shake sample well. Prepare Pb standards of 0.05, 0.1, 0.15, and 0.2 mg/L in 0.5% HNO_3 (Ultrex grade) and 0.2% $NH_4H_2PO_4$. With the 1:4 dilution ratio these standards correspond to

*Procedure from Fernandez and Hilligoss (2).

Table 7.12 Instrument Parameters

Atomic Absorption Parameters	
Wavelength	Pb 283.3 nm
Slit	0.7 nm
Calibration mode	Peak area
Integration time	5
Operating mode	AA–BG
HGA Parameters	
Step 1. Dry 130°C; ramp 10 sec, hold 5 sec	
Step 2. Dry 200°C; ramp 15 sec, hold 20 sec	
Step 3. Char 600°C; ramp 15 sec, hold 45 sec	
Step 4. Atom 1700°C; ramp 0 sec, hold 6 sec (argon 40 mL/min) (maximum power heating model)	
Step 5. Clean-out 2500°C; ramp 1 sec, hold 4 sec	

blood-lead concentrations of 250, 500, 750, and 1000 μg/L. Determine the Pb concentration in the blood samples by direct comparison to the Pb standards prepared in 0.5% HNO_3 and 0.2% $NH_4H_2PO_4$. Use 10 μL aliquots. Use the experimental conditions in Table 7.12. Prepare and analyze blank solutions of 0.5% HNO_3, 0.2% $NH_4H_2PO_4$, and 0.5% Triton-X 100 to verify that no Pb is present in these solutions.

*7.2.11.3 Determination of Pb in Urine by Furnace ASS**

A carbon rod atomizer is employed in this method. Matrix modification is accomplished using H_3PO_4. Iodine is also added as an oxidant to prevent the loss of organic lead compounds. To improve the life of the graphite tubes they are coated with molybdenum. The procedure is suitable for the determination of from 5 to 200 μg/L lead.

Equipment. A Varian 175B atomic absorption spectrometer was used together with a CRA-90 atomizer fitted with an ASD-53 automatic injection system. The diluted samples were mixed using a Denley Spiramix 5. The warming plate was constructed from an aluminum block with 48 indents to hold the cups. Electric heating elements were placed in the base of the block and the temperature was controlled using a Pye Ether Mini. The block was maintained at a surface temperature of 40°C. Gilson adjustable-volume and Eppendorf fixed-volume disposable tip pipettes were used for transfer of samples and reagents. Borosilicate glass cups fitted with polyethylene caps were used for sample preparation.

*Procedure from Hodges and Skelding (28).

Reagents. All reagents were at least Analar (British Drug House) grade. The orthophosphoric acid was Aristar grade (BDH Chemicals).

AMMONIUM MOLYBDATE SOLUTION, 5% BY WEIGHT. Dissolve 5 g of ammonium molybdate in approximately 60 mL of deionized water, add 5 mL of orthophosphoric acid, mix, transfer into a 100 mL calibrated flask, and make up to volume with deionized water.

AMMONIUM MOLYBDATE SOLUTION, 1% BY WEIGHT. Dissolve 1 g of ammonium molybdate in about 60 mL of deionized water, add 1 mL of orthophosphoric acid, mix, and make up to 100 mL in a calibrated flask.

ASCORBIC ACID SOLUTION, 25% BY WEIGHT. Dissolve 6.25 g of ascorbic acid in deionized water and make up to 25 mL in a calibrated flask.

URINE DILUENT. To approximately 60 mL of deionized water contained in a 100 mL calibrated flask add 5 mL of 1% m/v ammonium molybdate solution and 2 mL of orthophosphoric acid. Swirl to mix and then add 1 mL of ascorbic acid solution. Dilute to volume with deionized water and shake to mix. The solution will turn a royal blue color on standing.

IODINE SOLUTION, 1.0 *N*. Dissolve 16.6 g of potassium iodide in approximately 80 mL of deionized water contained in a 100 mL calibrated flask; add 12.7 g of iodine and shake to dissolve. Make up to 100 mL with deionized water and shake well to mix.

Standard Lead Solution. Dissolve 0.1599 g lead(II) nitrate in approximately 80 mL of deionized water; add 1 mL of distilled concentrated nitric acid. Transfer the solution quantitatively into a 100 mL calibrated flask, make up to volume with deionized water, and shake to mix. This stock solution contains 1 mg/mL Pb.

DILUTE STANDARD LEAD SOLUTION. Transfer 125 μL of the stock solution into a 25 mL calibrated flask containing approximately 20 mL of deionized water. Make up to volume with deionized water and shake thoroughly to mix. This solution contains 5 μg/L Pb.

Procedure. ***Tube Conditioning (Mo Coating).*** Fit a standard (pyrolytically coated) carbon tube to the CRA90 work head. Set up the instrument as shown in Table 7.13, except use a heating rate of 50°C/s to atomize when conditioning and use 5 injections.

Transfer into a clean glass cup approximately 1.5 mL of 5% m/v ammonium

Table 7.13 Instrument Parameters

Varian 175B spectrometer	
Wavelength	283.3 nm
Lead hollow-cathode	lamp 5.0 mA
Hydrogen lamp	Intensity to balance
Expansion	Low 5
Mode	Peak concentration
CRA-90 atomizer	
Dry	100°C, 60 sec
Ash	900°C, 40 sec
Atomizer	2100°C, hold time 0.5 sec ramp rate 300°C/sec
ASD-53 autosample dispenser	
Cam	5 μL
No. of injections	3 [see (1) below]

molybdate solution, place in position 24 of the carousel, and initiate the program start. Set the instrument to read background absorbance. In the course of replicate injections the background signal will stabilize at a value close to zero. Reset the instrument to read background-corrected absorbance.

Standards. Into five clean glass cups pipette 0, 5, 10, 15, and 20 μL of the dilute standard lead solution, representing additions of 0, 50, 100, 150, and 200 μg Pb/L of urine, respectively. Into each cup, pipette 20 μL iodine solution and 500 μL of a normal low-lead-content urine sample. Place the cups in the indents on the warming block and allow to stand for 5–10 min. Remove the cups from the block and add 1000 μL of urine diluent to each cup. Stopper the cups with polyethylene overcaps and place on the Denley mixer for 10 min. Remove the caps, place sequentially in the carousel of the autosample dispenser, and initiate the program start. Use instrument parameters listed in Table 7.14. Carry out reagent blanks by substituting deionized water for the urine.

Plot the readings obtained against μg/L Pb added and extrapolate the line to obtain the Pb content of the urine sample. Using this value, construct a calibration graph so that the line obtained passes through the origin.

Samples. Pipette 20 μL of iodine solution into clean glass cups and add 500 μL of each of the samples. Place the clearly marked cups in the indents on the warming block and allow to stand for 5–10 min. Remove the cups from the block and add 1000 μL of diluent to each cup. Stopper the cups with the polyethylene overcaps and place on the Denley mixer for 10 min. Remove the caps, place in the carousel of the autosample dispenser, and initiate the program. Use instrument parameters given in Table 7.14. Read off the lead content of the sample from the calibration graph.

Samples more than 24 h old and samples containing precipitate must be acidified by adding 10% v/v of distilled concentrated HNO_3. Transfer 500 μL of the acidified urine into a clean glass cup and add 65 μL of ammonia solution (specific gravity 0.88) to neutralize the HNO_3, then add 20 μL of iodine and proceed as above.

This dilution must be taken into consideration when calculating the lead content by multiplying the number of micrograms of lead per liter (read off their calibration graph) by 1.16.

Note the following:

1. Better replication can be achieved by making a large number of injections (nine) of the urine blank prior to the remainder of the calibration.
2. When large numbers of samples are to be analyzed, the calibration should be repeated at frequent intervals; this is conveniently carried out after each 25 samples.
3. To obtain the full benefit of the orthophosphoric acid treatment, it is essential that an ashing temperature of at least 800°C be used.

7.2.12 Determination of Pb and Cd in Blood and Urine by Furnace AAS*

Lead and cadmium are determined in blood and urine without use of acid decomposition and oxidation. Samples are dry ashed in the oven on microboats at 500°C prior to atomization in a graphite furnace. Loss of cadmium is prevented by the addition of $(NH_4)_2HPO_4$ for matrix modification. Samples are also diluted 1:2 in Triton-X 100 to allow hemolysis of the erythrocytes. The results obtained by the proposed procedure were checked by a reliable flame method in the case of lead in blood and good agreement was obtained. The coefficient of variation was better than 7%.

Equipment. The analyses were done with an IL 151 (Instrumentation Laboratory) atomic absorption spectrophotometer equipped with an IL 555 graphite furnace and Philips PM 8202 recorder. The spectrophotometer is a single-beam instrument with which either peak height or peak area can be measured and digitally recorded. The instrument was equipped with a deuterium lamp for background correction. An IL 555 furnace cell may be used with two different graphite cuvettes: (1) a conventional cylindrical tube for direct sample injection and (2) a rectangular-shaped cuvette used in conjunction with samples placed

*Procedure from Lagesson and Andrasko (29).

on graphite microboats. In this study the samples were analyzed in microboats and were preashed outside the graphite furnace. A special probe plate for ashing in microboats was constructed of titanium sized to fit into an ordinary laboratory oven (Sola Basic/Lindberg).

Reagents. All water used to prepare standards and matrix solutions was distilled and deionized. Each of the matrix solutions—NH_4NO_3, NH_4F, and $(NH_4)_2HPO_4$—in a concentration of 10 g/L was treated with ammonium pyrrolidine dithiocarbamate as a complexing agent and extracted with methyl isobutyl ketone. No lead or cadmium could be detected in these matrix solutions after the extraction procedure. Triton-X 100 surfactant (pa grade) was purchased from Kebo-Grave and used without further purification.

Procedure. Because Pb and Cd are relatively volatile elements, it is difficult to remove sample matrix without losses of these elements. Charring time and temperature must be optimized to facilitate maximum pyrolysis of matrix while retaining Pb or Cd.

Lead in Blood. Add 20 μL of Triton-X 100 (10 mL/L, in water) to 10 mL blood samples before the analyses to complete the hemolysis of erythrocytes. Blood samples supplemented with a known amount of lead are used as standards. For the atomic absorption measurements use the rectangular graphite cuvette and preash the samples in microboats. Pipette 25 μL of the solution of Triton-X 100 into the microboats and add a 2 μL blood sample. Place the titanium plate, holding 16 microboats, in the laboratory oven. Dry samples at 100°C and ash at 500°C (these two processes required about 5 min). Then analyze in the graphite furnace according to the temperature program listed in Table 7.14. Some carbon residue appears in the microboats after several analyses, but no detectable effect of this on the results was noted with succeeding analyses.

Lead in Urine. The organic content of urine is not as high as that of blood and thus no time-consuming preashing is necessary. However, the high salt content will interfere at the atomization step. To prevent salt interference add 10 μL NH_3NO_4 matrix solution to the graphite microboats together with 10 μL urine sample. Place the microboats in the laboratory oven, dry the samples at 100°C, ash at 350°C, and then analyze in the graphite furnace according to the temperature program shown in Table 7.14.

Cadmium in Blood. The volatility of cadmium gives rise to difficulties in such a complex medium as blood. Cadmium metal has a vapor pressure of 133 Pa (1.0 mm Hg) at 394°C, which is approximately the temperature required for burning away the organic matrix. Pipette 10 μL $(NH_3)_2HPO_4$ matrix solution

Table 7.14 Instrumental Conditions for Determination of Pb and Cd in Blood and Urine

	Spectrophotometer			Flameless Sampler[a]			
	Slit width (μm)	Wavelength (nm)	Recording	Pressure	Purge Gas	Gas Flow Rate (SCFH)[c]	Temperature program[c]
Pb in blood	320	217.0	Peak height or peak area (digital)[d]	Atmospheric	Argon	15	A
Pb in urine	320	217.0	Peak height	Atmospheric	Argon	20	B
Cd in blood	320	228.8	Peak height	173 Pa (25 psi)	Argon	20	C
Cd in urine	320	228.8	Peak height	Atmospheric	Argon	20	D

[a]Rectangular graphite tube used for all four analyses.

[b] Indicated temp. (°C) / Time setting (×5 sec):

A						
Indicated temp. (°C)	0–280	280–400	400	400–1900	1900	
Time setting (×5 sec)	1	1	1	2	1	

B						
Indicated temp. (°C)	0–125	125–300	300	300	300–2000	2000
Time setting (×5 sec)	1	1	4	4	2	2

C						
Indicated temp. (°C)	0–125	125–280	280–525	525	525–1900	1900
Time setting (×5 sec)	1	3	5	5	2	2

D						
Indicated temp. (°C)	0–125	125–280	280–320	320	320–850	850–1700
Time setting (×5 sec)	1	1	3	3	0	4

[c]SCFH, standard cubic foot per hour (1 SCFH = 0.47 L/h).

[d]Integrating time, 16 sec.

into the graphite microboats with 10 μL of blood sample and place the microboats in the laboratory oven. Dry the samples at 100°C, ash at 300°C, and then analyze with use of the instrumental settings listed in Table 7.14.

Cadmium in Urine. Mix 10 μL urine with 10 μL NH_4F matrix solution (10 g/L) in the microboats and place in the laboratory oven. Dry the samples at 100°C, ash at 300°C, and analyze in the graphite furnace. Table 7.15 shows the instrumental settings and the temperature program for the analysis.

7.2.13 Determination of Cd and Pb in Human Urine by Furnace AAS*

This method is rapid and relatively easy to use. Reagent additions to the urine are diammonium hydrogen phosphate and nitric acid. No prior decomposition is employed. In the case of lead, matrix matched calibration is necessary (blank urine). Calibration using single aqueous standards is satisfactory for cadmium. Pyrolytically coated graphite tubes are used. Detection limits for lead and cadmium are 4.1 and 0.9 ng/mL, respectively. Accuracy was assessed for lead by spiking. Cadmium accuracy was also assessed in this manner, but in addition checks of the cadmium concentration were verified using two other procedures. The method is capable of handling 25 samples per hour.

Reagents. Certified atomic absorption standards containing 1000 mg each of Cd(II) and Pb(II) per liter were obtained from Fisher Scientific. Fresh working standards of lower concentrations were prepared daily by serial dilution of the stock solutions in high-purity water.

Ten percent aqueous solutions of $(NH_4)HPO_4$ and NH_4NO_3 (Baker analyzed reagents, J. T. Baker Chemical, Phillipsburg, NJ) were prepared. Each solution was purified by extraction with ammonium pyrrolidinedithiocarbamate and water-saturated methyl isobutyl ketone. The aqueous phase in each instance was stored separately in 1 L Nalgene linear polyethylene bottles.

All other reagents and solutions used were of the highest purity available. In addition, prior to use all the glass and plastic labwares were acid washed.

Procedure. ***Cadmium.*** To 2.5 mL of urine contained in a 5 mL Pyrex volumetric flask add 100 μL of 10% $(NH_4)_2HPO_4$ and 100 μL of 10% HNO_3. Adjust the volume to 5 mL with high-purity water. Stopper the flask and agitate the solution vigorously for 20 s by using the Fisher Scientific Vortex-Genie. Inject a 10 μL aliquot of this solution into a pyrocoated graphite tube. Use the program of the HGA-500 shown in Table 7.15. Obtain the average absorbance values of 4–5 injections and correct for the reagent blank. The amount of cad-

*Reprinted in part with permission from (30), Subramanian, Meranger, and Mackeen, *Anal. Chem.* **55,** 1060. Copyright 1983, American Chemical Society.

Table 7.15 Optimized Instrumental Parameters for the Determination of Cd and Pb in Human Urine

Setting[a]	For Cd	For Pb
Hollow-cathode lamp current (mA)	6.0	7.0
Wavelength (nm)	228.8	283.3
Slit (nm)	0.7	0.7
Nitrogen flow[a] (mL/min)	300	300
Integration time (sec)	5.0	5.0
Drying temp (°C)	120	120
Drying time (ramp/hold) (sec)	30–20	30–20
Ashing temp (°C)	500	500
Ashing time (ramp/hold) (sec)	10–30	10–30
Atomization temp (°C)	1500	2300
Atomization time (ramp/hold) (sec)	1–3	1–3
Cleaning temp (°C)	2700	2700
Cleaning time (ramp/hold) (sec)	1–3	1–3

[a]Temperatures given represent the digital display on the control panel of the HGA-500.

mium in the sample is calculated by reference to linear working curves prepared from fresh aqueous standards in 0.2% $(NH_4)_2HPO_4$ and 0.2% HNO_3.

Lead. The procedure for Pb is the same as above except for the following: to 2.5 mL of the sample add 250 μL of a 10% solution of NH_4NO_3 and 500 μL of 10% HNO_3. The Pb content of the urine sample is obtained by reference to a calibration plot constructed from a human urine control (Product No. 2934-80, Level I, Fisher Scientific) fortified with varying amounts of the element.

The internal purge gas flow was operational only during the drying and ashing cycles; there was no gas flow during the atomization cycle.

7.2.14 Determination of Hg in Urine and Blood by Cold Vapor AAS*

Inorganic and organic mercury in blood can be differentiated by this method. The method is automated, but manual operation can be easily done if desired. Blood samples must be pretreated with trichloroacetic acid. Total mercury is obtained by treatment with a mixture of 10% $SnCl_2$–$CdCl_2$, whereas inorganic mercury is determined using 10% $SnCl_2$ only. Organic mercury is calculated by subtraction. Standardization is accomplished using inorganic and methylmercury chloride standard solution.

*Reprinted in part with permission from (31), Coyle and Hartley, *Anal. Chem.* **53,** 354. Copyright 1981, American Chemical Society.

Equipment. The atomic absorption unit used was a Model 403 Perkin-Elmer double-beam spectrophotometer fitted with a Perkin-Elmer Model 56 recorder. A mercury hollow-cathode lamp was used as the light source. The gas flow cell was a Corning Eel design 7 mm i.d. and 100 mm long with quartz glass end windows.

A Technicon Auto-Analyzer II proportioning pump was used to pump sample and reagents through the manifold illustrated in Fig. 7.1. Standard Technicon flow-rated pump and transmission tubings were used throughout. A Hook and Tucker A40 Autosampler II presented samples and wash solution to the manifold for 30 and 75 sec, respectively. The wash solution was 0.1% L-cysteine in 2 *M* HCl. The vapor–liquid separator (see Fig. 7.2) was assembled from Quickfit glassware and plastic tubing. The sample stream was injected through a right-angled glass tube (4 mm o.d., 2 mm i.d.), and the nitrogen stream was injected though a glass nozzle with an internal tip diameter of 0.8 mm. Both these pieces were mounted in a length of rigid polyethylene 9 mm i.d. and 50 mm long.

Reagents. All chemicals were analytical grade. The manifold reagents for total mercury measurements were made up at the concentrations detailed in Figure 7.1. The sodium hydroxide and L-cysteine were made up in glass-distilled water,

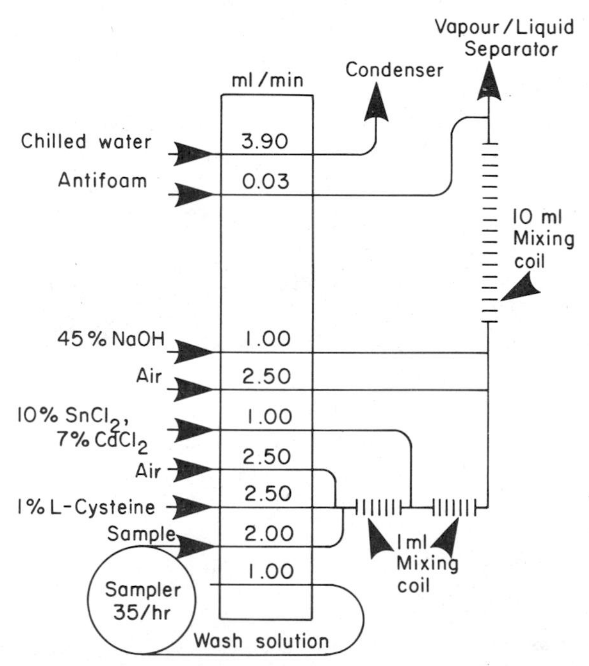

Figure 7.1 Autosampler manifold diagram for Hg (31).

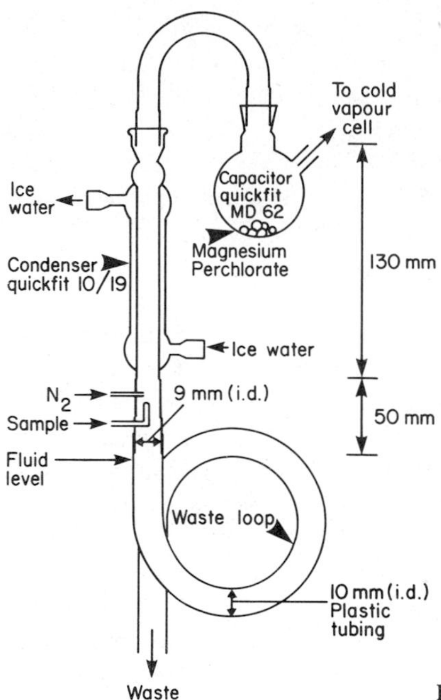

Figure 7.2 Vapor–liquid separator (31).

the 10% $SnCl_2$–7% $CdCl_2$ was prepared in 2 *M* HCl, and the wash solution was a 0.1% solution of L-cysteine in 2 *M* HCl. The antifoam was prepared as a 10% solution of 1-octanol in ethanol. When the manifold was used for the selective measurement of inorganic mercury, the 10% $SnCl_2$–$CdCl_2$ reagent was replaced by a 10% solution of $SnCl_2$ in 2 *M* HCl acid.

A stock 1000 μmol/L inorganic mercury standard was prepared by dissolving 0.2715 g HgCl in 5% v/v H_2SO_4. A 1 g sample of L-cysteine was then added and dissolved before making up to a final volume of 1 L with the H_2SO_4. Portions of this stock standard were then diluted with glass-distilled water to give a series of working standards containing 0.25, 0.50, 0.75, and 1.00 μmol/L Hg.

A stock 1000 μmol/L methylmercury chloride standard was prepared by dissolving 0.2511 g of the anhydrous solid in 100 mL acetone. The final volume of this solution was adjusted to 1 L with glass-distilled water. Working standards were prepared at the same concentrations as the inorganic mercury standards.

Procedure. ***Determination of Total Mercury in Urine.*** Urine samples and standard solutions are analyzed directly. Place 1 mL aliquots in sample cups on the Autosampler turntable. Use the $SnCl_2$–$CdCl_2$ reagent in the manifold.

Determination of Total Mercury in Blood. The blood samples are pretreated as follows. Add 1 mL of a 6% L-cysteine solution to 1 mL of the whole blood or aqueous standard and mix. Add 1 mL of 40% trichloroacetic acid to each tube and shake the tubes vigorously. Centrifuge the specimen for 5 min at 3000 rpm. Place the clear protein-free supernatant in sample cups on the Autosampler. Use the $SnCl_2$–$CdCl_2$ reagent in the manifold.

Determination of Inorganic Mercury in Blood. Treat specimens in the same manner as in the total mercury in blood method; however, replace the $SnCl_2$–$CdCl_2$ manifold reagent with a 10% SnCl solution.

Treatment of Above Solutions. The final reaction in which the mercuric ions are reduced to mercury occurs in the large mixing coil. From there the stream is pumped into a specially designed vapor–liquid separator where it is nebulized with a fine jet of nitrogen flowing at 1 L/min. This allows for a complete diffusion of mercury from the liquid phase into the gas phase. The surplus fluid is allowed to drain to waste through a waste trap, and the vapor is carried by the nitrogen flow through a condenser. This is fed with ice-chilled water, and here the majority of water in the vapor condenses the runs down to waste. The dried mercury vapor then passes into a round-bottom capacitor flask that contains about 5 g of magnesium perchlorate as a desiccant. In addition to removing any surplus water vapor, this flask is important because its capacitor action damps out pulsations in the vapor flow before the vapor passes into the analysis cell. This ensures that the subsequent atomic absorption signal monitored on the chart recorder increases and decreases smoothly.

The signal noise from the instrument is dampened (recorder response 3, which is equivalent to a time constant of 4 s) at the recorder by a 32 μF capacitor wired across its input terminals.

The signal is amplified by using the scale expansion control both on the atomic absorption instrument and on the recorder. (Consequently an aqueous inorganic mercury standard of 1 μmol/L, which gave an absorbance of approximately 0.160, gave an almost full-scale deflection of the 10 in. chart recorder.) During blood analyses a further threefold expansion is necessary to compensate for the dilution factor inherent in the preparation state.

Draw a concentration versus peak-height curve from the recorder tracing, and then read concentrations of unknowns directly from this curve.

7.2.15 Determination of Se in Blood Serum and Seminal Fluid by Furnace AAS*

Phosphate was found to cause a spectral interference in this application whereby overcompensation results if a deuterium arc background-correction system is employed. To overcome this problem trichloroacetic acid is used to precipitate

*Procedure from Saeed and Thomassen (32).

the proteins. The precipitated proteins are then dissolved in 10% NH_3. Silver is added to the sample in the furnace to stabilize selenium thermally. A recovery of 97.4% selenium was obtained using this approach.

Equipment. A Perkin-Elmer Model 5000 atomic absorption spectrometer equipped with selenium electrodeless and deuterium arc lamps, an HGA-500 graphite furnace, an AS-40 automatic sampler, and a Perkin-Elmer Model 56 recorder were used. The furnace was purged with argon.

Reagents. The reagents employed were analytical reagent grade. A commercially available 1000 μg/mL standard solution of selenium was used. A stock 1000 μg/mL solution of silver was obtained by dissolving a suitable amount of silver nitrate in distilled water. Stock solutions of 50% trichloroacetic acid and 10% ammonia were prepared in distilled water. Working solutions were prepared daily.

Procedure. After sampling, the semen is centrifuged for 10 min at 1000 rpm. The supernatant fluid is then kept frozen until required.

To a test tube (graduated at exactly 2 mL) containing 1 mL of 10% trichloroacetic acid solution add dropwise 0.2 mL of the serum or seminal fluid. Separate the supernatant liquid carefully from the denatured proteins by centrifugation at 3700 rpm for 10 min. Add a 0.5 mL portion of 10% ammonia solution and shake the mixture vigorously until the precipitate dissolves completely. Dilute sample to volume (2 mL) with distilled water. Inject 20 μL of solution into the furnace, followed by 30 μL of 0.1% silver nitrate solution. Run the furnace program as shown in Table 7.16. Repeat for solutions to which standard additions of selenium have been made.

Table 7.16 Furnace Program[a]

	Temp. (°C)	Ramp/hold (sec)
Dry	100	5/60
	140	20/10
	170	10/5
Ash	800	5/15
		Baseline-5
Atomize	2200	0/5
	Recorder-6	Baseline-1
Clean-out	2700	1/1

[a]Pyrolytically coated tubes were used with an internal argon flow of 20 mL/min during atomization.

Table 7.17 Instrumental Conditions for Atomic Absorption Measurements and Hydride Generation

Atomic Absorption Measurements	
Instrument	Perkin-Elmer 560
Radiation source	Hollow-cathode lamp, 16 mA
Wavelength	196.0 nm
Spectral band pass	2.0 nm, reduced slit height
Measurement mode	Integrated peak area 20 sec, at ×1 scale expansion
Recorder	Tekman 220/2, continuous recording at 2 mV fsd, i.e., ×5 scale expansion
Hydride Generation	
Instrument	Perkin-Elmer MHS 20
Tube temperature	900°C
Purge I (argon)	42 sec
Purge II (argon)	30 sec
Reaction time	10 sec
Reaction volume	20 mL

7.2.16 Determination of Se in Clinical Samples by Hydride Generation AAS*

Samples analyzed by the method are blood, plasma, and erythrocytes. Samples are decomposed and oxidized using a mixture of HNO_3 and H_2SO_4 (HNO_3–H_2O_2 mixtures were not adequate for this purpose). Hydride generation then follows using sodium borohydride reduction. Accuracy was evaluated using spiked plasma and blood samples and through an interlaboratory comparison study. Spiked samples gave 90–100.6% recoveries.

Equipment. A Techne Dri Block heater, Model DB 3H, was used for ashing the samples. The heater was fitted with three alloy blocks, each drilled with 12 holes to a depth of 48 mm and a diameter of 16.75 mm. Borosilicate test tubes (16 × 110 mm) were used for sample oxidation.

Atomic absorption spectroscopic measurements were carried out on a Perkin-Elmer Model 560 atomic absorption spectrometer and a Perkin-Elmer MHS 20 hydride generation system. Instrumental conditions are given in Table 7.17.

Soak all glassware in 2% v/v Decon 75 for a minimum of 2 h, rinse well with deionized water, then immerse in 1.6 *M* HNO_3. After 4 h, wash the glassware thoroughly in deionized water and dry in a hot air oven.

*Procedure from Lloyd et al. (33).

Reagents. The antifoam emulsion was obtained from Dow Corning. All other reagents were AnalaR grade materials from BDH Chemicals. The reagents required are Decon 75, 2% v/v in deionized water; HNO_3, 16 *M* HCl; 6 *M* H_2SO_4; 4.5 *M* NaOH solution; 0.25 *M* sodium tetrahydroborate (III), 6% m/v solution in 0.25 *M* NaOH hydroxide solution; antifoam emulsion DB 110 A, 1% v/v in deionized water; 0.15 *M* NaCl solution; and 0.15 *M* Selenous acid standard solution, 1 mg/mL Se.

Working Stock Standard Solution (10 μg/mL). Transfer 1 mL of the selenous acid standard solution into a 100 mL calibrated flask. Add 1 mL 16 *M* HNO_3 and dilute to volume with deionized water.

WORKING STANDARD SOLUTIONS. Transfer 0, 0.5, 1.0, 1.5, 2.0, 2.5, and 3.0 mL of the working stock standard solution of selenous acid into 100 mL calibrated flasks. Add 1 mL of 16 *M* HNO_3 to each flask and dilute to volume with deionized water. These solutions contain 0, 50, 100, 150, 200, 250, and 300 ng/mL of selenium, respectively.

Procedure. Blood samples were collected by venipuncture from an antecubital vein and transferred into polycarbonate tubes containing either K_2H_2EDTA or lithium heparin as an anticoagulant. All samples should be stored at 4°C prior to separation of plasma from red cells. Plasma and erythrocytes were separated by centrifugation from whole blood and transferred to trace-metal-free polycarbonate tubes. The red cells were washed three times with 0.15 *M* NaCl solution and reconstituted to the original volume of blood using this solution. All samples were stored at −20°C until required for analysis.

Transfer duplicate 100 μL portions of whole blood, plasma, or washed red cells into borosilicate glass tubes. Add 1 mL of 16 *M* HNO_3 and 1 mL of 4.5 *M* H_2SO_4 and place the tubes into the reheated aluminum blocks at 155°C for 60 min. Remove the tubes from the block, allow them to cool to room temperature, then add 2 mL of 6 *M* HCl. Mix the solutions by tapping the test tubes before returning them to the aluminum block, preheated at 95°C, for 30 min. Remove the tubes from the block, allow them to cool to room temperature, then transfer the contents quantitatively into the reaction vessels for hydride formation and dilute to a final volume of 20 mL with deionized water. Add 100 μL of antifoam reagent and mix. Use the instrumental conditions given in Table 7.18 to determine the selenium concentrations in the solutions of the ashed samples. Establish a calibration graph by adding duplicate 100 μL volumes of the selenium working standard solutions to reaction vessels containing 1 mL of 16 *M* HNO_3, 1 mL of 4.5 *M* H_2SO_4, and 2 mL of 6 *M* HCl. Dilute to a final volume of 20 mL with deionized water and add 100 μL of antifoam reagent. Analyze the sample solutions and blanks in duplicate and run a working standard solution after every five samples. Calculate the selenium concentrations from the integrated absorbance values and the calibration graphs. The continuous chart recordings are only used as a visual check that the analysis is proceeding satisfactorily.

7.2.17 Determination of As and Se in Liver and Kidney by Hydride Generation AAS*

Semiautomated hydride evolution is used together with quartz tube atomization for the determination of arsenic and selenium in human liver and kidney. Liver and kidney decomposition is done using nitric and perchloric acids and the sample is diluted to contain 5% hydrochloric acid. The determination range is 50–500 ng/g.

The system is automated using a Technicon autosampler. Reagent streams include 2× HCl, 1% sodium borohydride, and sample together with two air streams.

Equipment. A Varian Techtron Model AA-5 atomic absorption spectrophotometer equipped with a Cathodeon arsenic (or selenium) hollow-cathode lamp was used. An electrically heated open-ended silica tube, 15 cm long and 1.2 cm i.d. with a 4 mm diameter inlet tube fused in the middle, was used for atomizing arsenic and selenium in the gaseous stream. The argon flow rate was regulated by a calibrated flow meter. The temperature of the silica tube was regulated by means of a 0–110 V Variac. The furnace was mounted on the burner head and aligned in the usual manner to let maximum light from the hollow-cathode lamp reach the detector. A Technicon Autosampler with 40-sample capacity, a proportionating pump, and manifold were used in conjunction with a 10-mV strip-chart recorder. The manifold is represented schematically in Fig. 7.3.

Reagents. ARSENIC(III) STOCK SOLUTION, 1000 mg/L. Dissolve exactly 0.1320 g of high-purity arsenic(III) oxide (NBS SRM 83c, dried at 100°C for 2 h) in 2 mL of 1 *M* NaOH solution. Add 25 mL of water followed by 4 mL of 1 *M* HCl, dilute to 100 mL with high-purity water, and store the solution in a precleaned polyethylene bottle.

SELENIUM(IV) STOCK SOLUTION, 1000 mg/L. Dissolve exactly 1.406 g of analytical reagent grade selenium(IV) oxide (BDH Chemicals) in the minimum amount of 10% ultrapure HCl. Dilute to 1 L with 10% HCl and store in a 1-L screw-capped linear polyethylene bottle. For both of these reagents, solutions of lower concentrations were prepared fresh daily by serial dilution of the stock solutions.

*Procedure from Subramian and Meranger (34).

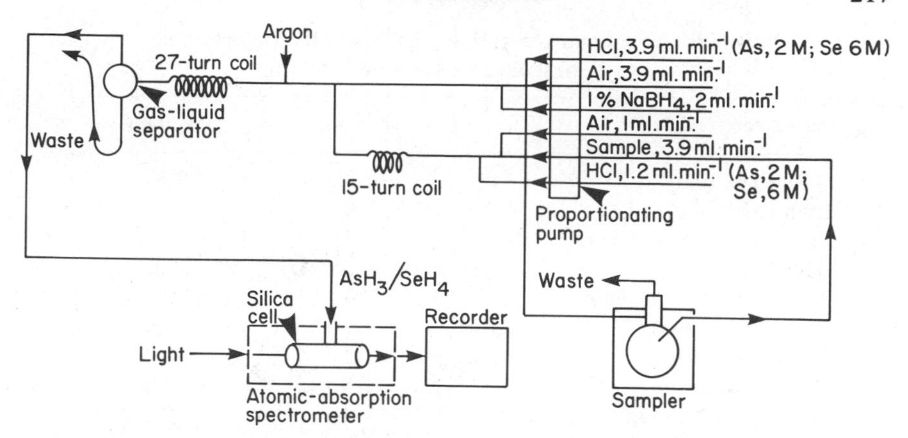

Figure 7.3 Autosampler manifold hydride system for As–Se determination (34).

SODIUM TETRAHYDROBORATE(III) SOLUTION, 1% m/v. Dissolve 10 g of sodium tetrahydroborate(III) (Alfa Inorganics) and 2 g NaOH in 1 L of high-purity water. Keep refrigerated in a tightly capped linear polyethylene container until ready for use. The solution is stable for at least 1 week under these conditions. All other reagents and solutions used were of the highest purity available (e.g., HNO_3 Baker Ultrex and $HClO_4$ BDH Aristar).

Procedure. Weigh out (0.90–1.3 g, wet weight) of sample accurately into 50 ml Pyrex beakers that have been cleaned with ultrapure HNO_3. (Samples were stored at −20°C prior to analysis.) Digest the samples with 8 mL concentrated HNO_3 and 2 mL concentrated $HClO_4$ on a sand bath until a clear, colorless solution is obtained. Evaporate the digestate to fumes of $HClO_4$ and dilute to 10 mL with 0.5 *M* HCl. Transfer about 3–4 mL of the solution into the 5 mL disposable polypropylene cups placed in the Autosampler. Take blanks and the reference standards (NBS SRM 1577 Bovine Liver) through the same procedure.

Allow the atomic absorption spectrophotometer to warm up and the silica furnace to attain thermal equilibrium (usually 30 min) under the optimized operating conditions given in Table 7.18. Close the vent to the lowest flow to prevent air turbulence in front of the cell opening, thereby minimizing baseline drift. Turn on argon at the predetermined flow rate. Insert the manifold tubes into the reagent solutions, HCl, and sodium tetrahydroborate(III) solution. Start the proportionating pump. When the system has attained equilibrium (in about 20 min) as indicated by minimum baseline noise on the strip-chart recorder, the spectrophotometer is adjusted to zero absorbance. Switch on the Autosampler containing the test solution and the arsenic(III) hydride (or selenium(IV) hydride) vapor is carried by the argon stream through a U-shaped spray trap into

Table 7.18 Optimum Operating Conditions for Determining Arsenic and Selenium

Parameter	Arsenic	Selenium
Wavelength (nm)	193.7	196.0
Band width (nm)	0.3	0.3
Hollow-cathode lamp current (mA)	7.0	4.5
Damping	Maximum (D)	Maximum (D)
Argon flow rate (mL/min)	450	400
Sample time (min)	1.0	1.0
Wash time (min)	2.0	2.0
Recorder span, full scale (mv)	5.0	5.0
Chart speed (cm/min)	0.5	0.5

the resistance-heated silica tube where it is atomized. Record the atomic absorption peaks of the blanks, calibration standards, reference standard, and samples.

Equipment. A Perkin-Elmer Model 603 atomic absorption spectrophotometer equipped with a HGA-500 graphite furnace and a deuterium arc background corrector and Varian Techtron hollow-cathode lamps were used for determining cadmium and lead.

7.2.18 Determination of Cd, Cr, Hg, Pb, and Zn in Hair by AAS*

Cadmium, chromium, lead, and mercury are important toxic trace elements that can be readily monitored in hair. Zinc is an essential trace element that shows relatively uniform distribution in hair with time. Because of this, zinc is a useful reference element.

In the following an electrothermal atomic absorption procedure is presented for the determination of zinc, chromium, cadmium, and lead. Mercury is determined by the cold vapor approach of Hatch and Ott (36). The proposed procedures were checked by comparing the results with those obtained using (1) instrumental neutron activation analysis (Hg, Cr, and Zn); (2) induced proton activation analysis (Cd); and (3) anodic stripping voltametry (Pb). Acceptable agreement was obtained. With chromium the level was so near the detection limit of both techniques that the comparison was not entirely conclusive.

Equipment. A Perkin-Elmer (PE) Model 305 atomic absorption spectrometer equipped with Intensitron hollow-cathode lamps and deuterium arc lamp, on line with a PE mercury analysis system and an HGA-76 graphite furnace

*Procedure from Bagliano et al. (35).

equipped with an AS-1 automatic sampler and a burner with a 4 in. single-slot head, were assembled in turn with the spectrometer. A microball mill (Model II, Braum Melsungen) equipped with a cylindrical Teflon vessel tightly closed with a Teflon lid and containing a Teflon-coated steel ball was used to pulverize and homogenize the hair samples. A digestion bomb (PE autoclave-1) heated on a thermostatically controlled ceramic hot plate was used for the preliminary tests to establish the optimal conditions for the dissolution of the hair samples. Home-made screw-capped Teflon beakers (50 mm o.d., 60 mm height) heated in a conventional oven were later used routinely for the decomposition of the samples.

Two resin exchangers (a Seradest Seral in line with an Elgastat) were used to prepare very pure double-deionized water. All glassware was stored in a dithizone solution in carbontetrachloride; before use it was rinsed throughly with the deionized water and drained until dry at room temperature. The digested hair samples were diluted with the deionized water and stored in polyethylene volumetric flasks. Eppendorf pipettes were used to prepare the standard solutions, which were stored in polyethylene bottles.

Reagents. The above double-deionized water was used for all solutions and dilutions. Stock solutions (100 μg/mL) for Cd, Cr, Pb, and Zn were prepared by diluting the contents of Titrisol ampuls (Merck) to 1:1 with water. A Fixanal ampul (Riedel de Haen) was used to prepare a 100 μg/mL Hg stock solution. Two sets of standard solutions were prepared by diluting the stock solutions with two different acid mixtures: (1) a H_2SO_4 (39.4 g/L)–HNO_3 (12 g/L) and (2) $HClO_4$ (8.4 g/L)–HNO_3 (35 g/L). The concentrated acids used to prepare the mixtures were of Suprapure quality (Merck).

All the reagents used to determine mercury by the Hatch and Ott procedure (36) were of guaranteed low mercury content (PE mercury reagent kit). During the preliminary tests, lead was extracted with a 2% w/v solution of ammonium pyrrolidinecarbodithioate (Merck–Schuchart) in methyl isobutyl ketone (Merck). Analytical grade acetone (Merck) was used for washing the hair. Welding grade acetylene (Messer–Griesheim) was used as the fuel gas for zinc determinations, and 99.9% argon (Messer–Griesheim) was used as the purge gas with the graphite furnace.

Preparation of Hair Samples. Place triplicate subsamples of about 100 mg of hair in Teflon beakers and stir magnetically for 10 min periods successively with acetone, three portions of water, and again with acetone; in each wash 125 mL of solvent is used and decanted off. Wrap the hair sample in a sheet of blue-band filter paper and air dry at room temperature in a clean dust-free room for one day. Transfer the wrapped sample to a closed balance room and leave overnight to equilibrate with atomospheric moisture. Weigh two subsamples to a tenth of a milligram and dry the third subsample at 110°C for 2 h. Measure the weight loss to a tenth of a milligram and use to refer all concentrations to a dried sample basis. Powdered hair can be prepared as follows. With a tantalum knife cut the bundles into lengths of 2–3 mm, which are then washed and dried as described above. Place 1 g of hair into a tightly capped Teflon vessel containing a Teflon-coated ball. Cool in liquid nitrogen for a few minutes and vibrate for 1–2 min with the microball mill. Obtain a fine powder by repeating the grinding two or three times. Repeat this whole procedure for each fraction as often as needed to pulverize all the available hair material. Sieve through a 125 μm nylon sieve to provide homogeneous powders.

Procedure. Weigh about 100 mg of hair sample into the dissolution beaker. Add 3 mL of a mixture (5 + 1, v/v) of HNO_3 (d = 1.40) and $HClO_4$ (d = 1.67). Screw the cap on tightly and heat at 100°C for 45 min in a drying oven. Cool to room temperature in a clean, dust-free room and transfer the digestion solution to a 100 mL polyethylene volumetric flask. Dilute to the mark with the double-deionized water. Prepare a blank in parallel. Pipette an aliquot of the solution containing no more than 0.25 μg Hg, and transfer it to the aeration flask of the Hg analysis system previously assembled and aligned with the spectrometer. Select the 253.7 nm wavelength and a 0.7 nm band pass. Dilute the aliquot with 100 mL of the deionized water and follow the method recommended by the manufacturer. Follow the same procedure with aliquots of Hg standard solutions. Empty and rinse the aeration flask carefully between measurements. Perform duplicate measurements on each solution and run one standard aliquot every 10 measurements. Measure the height of the peaks recorded. Remove the Hg absorption cell.

Table 7.19 Instrumental Conditions[a]

			Drying[b]		Charring[c]		Atomization[d]		Cleanings	
Element	Wavelength (nm)	Background Correction	Temp. (°C)	Time (sec)	Temp. (°C)	Time (sec)	Temp. (°C)	Time (sec)	Temp. (°C)	Time (sec)
Hg	253.7	No	NA[e]	NA	NA	NA	NA	NA	NA	NA
Cd	228.8	Yes	300	20	500	20	1900	5	2500	4
Cr	357.9	No	300	20	900	20	2600	6	2700	4
Pb	283.3	Yes	300	10	700	20	1950	5	2700	4
Zn	213.9	Yes	NA	NA	NA	NA	NA	NA	NA	NA

[a]Argon was used as purge gas with a flow rate of 145 mL/min; in all cases, the band pass was 0.7 nm.
[b]Ramp heating rate 3.
[c]Ramp heating rate 2.
[d]Gas stop for 5 s.
[e]Not applicable.

Mount the HGA-76 graphite furnace within the burner compartment of the spectrometer. Inject 20 μL aliquots of sample or standard solutions into the furnace atomizer and determine Cd, Cr, and Pb following the operating conditions given in Table 7.19; use a pyrolytic tube atomizer for Pb. Inject at least 3 aliquots of each solution and calculate the mean peak height in each instance.

Replace the graphite furnace with the slot burner, and spray the sample or standard solutions into an air–acetylene (lean) flame; measure the peak height for Zn at the 213.9 nm wavelength using a 0.7 nm band pass. Carry out 2 determinations for each solution.

Determine the amount of each element by comparing the peak height of the samples with a linear regression calibration curve obtained from the standard solutions. Determine the content of the five elements in a blank prepared for each daily set of digested samples.

7.2.19 Determination of Cu, Cd, and Zn in Liver by AAS*

As small a sample as 15 mg of human liver can be analyzed by the following procedure for copper, cadmium, and zinc. Copper and cadmium are determined by furnace AAS, whereas zinc is in high enough concentration to be handled by flame AAS. The sample is dissolved in nitric acid. Accuracy is assessed by determination of the analyte metals in NBS Bovine liver.

Equipment. A model 303 atomic absorption spectrophotometer and a graphite tube (HGA-2000) atomizer (both from Perkin-Elmer, Norwalk, CT 06852) were used for all Cd and Cu analyses. Flame atomization with a tantalum three-slot burner (air–acetylene flame) was used for Zn analysis (see Table 7.20) for specific instrumental settings).

Samples were injected into the graphite furnace atomizer with 50 and 10 μL syringes (Precision Sampling Corp., Baton Rouge, LA 70815) with a Teflon-tipped plunger, which eliminates measurable Cu and Cd contamination of the sample from the syringe.

All glassware was allowed to soak overnight in 2.0 *M* HNO_3, then rinsed with doubly deionized water. The water used in this study contained no Cu, Zn, and Cd detectable by this analytical method and neither did the 10 mmol/L solution of HNO_3.

Reagents. All reagents were analytical grade. Potassium chloride, KNO_3, $NaNO_3$, and NaCl were obtained from Mallinckrodt Chemical Works, St. Louis, MO 63160. Phosphoric acid, HCl, and HNO_3 were obtained from du Pont Wilmington, DL 19898. Analytical reagent grade HNO_3 was used in preparing so-

*Procedure from Evenson and Anderson (37).

Table 7.20 Instrumental Settings for Atomic Absorption Spectrophotometers

	Cu	Cd	Zn
Perkin-Elmer 303			
Wavelength (nm)	324.7	228.8	213.9
HGA-2000 Atomizer			
Purge gas	N_2	N_2	
Purge gas flow rate(L/min)	5	5	
Drying time (sec)	10	10	
Drying temp (°C)	125	125	
Charring time (sec)	15	15	
Charring temp (°C)	600	600	
Atomization time (sec)	10	10	
Atomization temp (°C)	2500	1950	
Flame atomizer			
Gas			Air–acetylene
Air flow L/min			34
Acetylene flow (L/min)			6
Aspiration rate (mL/min)			5

lutions and in the acid-digestion step. At 1.0 *M*, such HNO_3 is sufficiently free of Cd, Cu, and Zn to contribute no measurable contamination to the analysis.

For Cu and Zn, analytical standards ("Dilute-It"; J. T. Baker Chemical Co., Phillipsburg, NJ 0.8865) were diluted with 10 mmol/L HNO_3 to a final concentration of 10,000 mg Cu and Zn per litre. Working standards were made by diluting an aliquot of primary standard with 10 mmol/L HNO_3.

The Cd standard was Harleco American Public Health Association standard (Scientific Products, McGraw Park, IL 60085), the pH of which is 1.5. According to information provided by Harleco this standard is made by dissolving Cd metal in HCl. The working standard was prepared by diluting this stock standard with 10 mmol/L HNO_3.

Procedure. Place the sample (human liver or NBS Bovine Liver) in a dried, preweighed, acid-cleaned 3 mL test tube and obtain the wet weight of the tissue. Then place the tubes in a beaker on a hot plate, and cover the beaker with an inverted larger beaker. The air temperature inside the larger beaker should be 80°C. Dry the tissue for 24 h to constant weight, cool in a desiccator, and weigh. About 70% of the wet weight is lost, leaving about 5.0 mg of dry liver tissue. Add 1 mL of 1.0 *M* HNO_3 and digest the dried sample on a hot plate at 80°C for another 24 h. The 1.0 *M*/HNO_3 is then slowly evaporated on a hot plate and

2.0 mL of 10 mmol/L HNO_3 is added with vigorous mixing. The liver tissue is almost completely dissolved after this treatment. Incomplete solubilization did not alter the analytical results. Delay analysis for 4–6 h after reconstitution. The concentration of the three metals in the tissue is such that only 30 μL of the reconstituted solution of liver tissue is needed for Cd and Cu analysis. Zinc is determined in the remaining solution (1–2 mL) by flame atomization.

7.2.20 Determination of Trace Elements in Bone by ICPAES*

The following trace elements can be determined after dissolution of bone samples by direct aspiration of sample solutions: Fe, Zn, Ba, Al, Mo, and Pb. Arsenic, Se, and Hg are done directly on the sample solution using hydride generation and elemental mercury generation, respectively. Nickel, Co, Ti, V, Mo, Cd, and Pb at low concentrations are determined after preconcentration on a poly(dithiocarbomate) resin.

Samples are first dried at 100°C in an oven. They are then weighed into Teflon beakers. This is followed by a HNO_3–H_2O_2 digestion. The procedure was evaluated using International Atomic Energy (IAEA H-5) standard bone sample. Good agreement with accepted values was obtained.

Equipment. The instrumental facilities and operating conditions used are summarized in Tables 7.21 and 7.22. Simplex optimization was used to establish optimum operating conditions for each determination.

Reagents. Stock solutions were prepared from high-purity and reagent-grade metals or salts with distilled, deionized water and high-purity acids. The poly(dithiocarbamate) resin was synthesized as described earlier (8). Calcium solution and other reagents were purified by treatment with poly(dithiocarbamate) resin to minimize reagent blanks. Standard solutions were prepared by serial dilution of 1000 μg/mL stock solutions with distilled water, and the concentration of calcium in each reference solution was held constant.

Procedure. Dry powdered bone samples in an oven at 100°C for 2 h. Weigh 1–3 g portions of dried powdered bone sample into 100 mL polytetrafluoroethylene(IPTFE). Add 15 mL concentrated HNO_3. Heat the mixture slowly on a hot plate. Add 30% H_2O_2 dropwise until all organic material is destroyed. Approximately 15–30 min will be required for complete dissolution. After cooling, transfer the solution to a 50 mL volumetric flask and dilute to volume with distilled water.

For As, Se, and Hg determinations in bone, hydride and elemental generation techniques are used. For the As and Hg determinations, dissolve 2 g of bone

*Procedure from Mahanti and Barnes (38).

Table 7.21 Instrumentation and Operating Conditions

A. Instrumentation

Generator	Plasma-Therm Model HFS-5000D, 40.68 MHz with 3-turn ($\frac{1}{8}$ in. copper) load coil
Nebulizer	Modified Babington with double-barrel glass spray chamber and sample uptake of 1.2 mL/min
Plasma torch	Conventional 18 mm i.d. quartz with 1.5 mm i.d. injector orifice
Direction	Minuteman monochromator Model 310-SMP, Czerny-Turner with 1200 groove/mm grating. Slit widths 40–60 m, slit height 5 mm; 1 : 1 image formed by quartz lens (Oriel A-11-661-37). RCA IP28 photomultiplier (−700 V), Keithley 411 picoammeter, Heath EU-201V log/linear recorder

B. ICPES Operating Conditions

Power (kW)	0.70[a]
Outer gas flow (L/min)	16
Intermediate gas flow (L/min)	0
Aerosol flow rates (L/min)	0.8[b]
Nebulizer back pressures (psig)	5 mm × 50 m from 13.5 to 18.5 mm above the induction coil[c]

[a]Except for Cu and Fe (0.8 kW), Na (0.6 kW), Ni (0.5 kW), P. (0.9 kW), and Se (0.75 kW).
[b]Except for Ca, P, and Zn at 0.9 L/min and As, Se, and Hg carrier gas flow of 0.5 L/min and auxiliary gas flow of 1 L/min.
[c]Except for Ba, Co, Ni, and Sr (11.5–16.5 mm above induction coil).

sample slowly at 70°C by heating with concentrated HNO_3 and H_2O_2. The final solution volume is 5 mL. To a 1 mL aliquot of sample solution for each determination, add 0.5 mL concentrated HCl and 0.5 mL of freshly prepared 2% soldium tetrahydroborate. For the arsenic determination, add 0.1 mL of 10% KI solution. For the Se determination, add 2 mL of concentrated HCl, and heat the solution slowly on a hot plate at 70°C. The final volume is 5 mL. A 1 mL aliquot of sample solution is used for the Se determination, and add 0.5 mL of concentrated HCl and 0.5 mL of 2% sodium tetrahydroborate.

Poly(dithiocarbomate) Resin Treatement. Dissolve a 3 g bone sample as described. Add 120 mL of 0.2 MEDTA, purified earlier with the resin, and adjust the pH to 5 with ammonia solution. Pass the solution through a column containing 80 mg of 70–80 mesh resin at 1 mL/min. After washing, digest the resin with 30% H_2O_2 and 2 mL of concentrated HNO_3 by heating slowly on a hot plate. Add dilute (3 + 1) HNO_3 to obtain a final volume of 5 mL. The concentrations of trace elements are quantified by using acid-matched calibration stan-

Table 7.22 Wavelengths and Figures of Merit[a]

Element and Wavelength (nm)	Detection Limit (ng/mL)	IEC (mg/L)	BEC (mg/L)	LQD Bone (μg/g)
Al, 236.705	50	0.02	12.5	
Ba, 455.40	0.6	0.014	0.02	0.15
Ca, 422.67	7	—	0.20	1.75
Cd, 228.80	2	0.12	0.10	0.5 (0.0165)[b]
Cu, 324.75	6	0.06	0.26	1.5 (0.05)
Co, 345.35	10	0.06	0.35	2.5 (0.083)
Fe, 259.94	6	0.12	0.26	1.5
Mg, 179.55	0.5	0.10	0.12	0.125
Mn, 257.61	2.8	0.04	0.13	0.7
Mo, 202,03	10	0.08	0.40	2.5 (0.083)
Na, 588.995	11	—	0.46	2.75
Ni, 341.45	12	0.06	0.44	3 (0.10)
P, 214.91	150	3.0	5.0	37.5
Pb, 220.35	100	3.0	4.0	2.5 (0.833)
Sr, 407.77	0.8	0.014	0.03	0.2
Ti, 334.94	4	—	0.22	1.0 (0.033)
V, 292.4	10	0.03	0.4	2.5 (0.083)
Zn, 213.86	9	0.12	0.42	2.25
As, 193.69[c]	1	—	0.04	0.0125
Se, 196.02[c]	1	—	0.04	0.0125
Hg, 253.65[d]	0.05	—	0.02	0.006

[a]Detection limit in a calcium solution equivalent to a 2% bone solution is defined as the concentration giving a signal equal to three times the standard deviation of the background. The interference equivalent concentation (IEC) is the equivalent concentration at each element wavelength corresponding to the spectral interference from a 5400 mg/L Ca solution. The background equivalent concentration (BEC) is the analyte concentration equivalent of the background level at the analyte wavelength. The LQD is the lowest quantitatively determinable concentration calculated from five times the limit of detection with a 1 g sample dissolved in 50 mL after resin preconcentration. For As and Se by hydride generation and Hg by elemental Hg generation, LQD is calculated for a 2 g sample in 5 mL of solution.

[b]With poly(dithiocarbamate) resin preconcentration.

[c]With hydride generation.

[d]With elemental Hg generation.

dards. A concentration factor of 30 is obtained beginning with a 150 mL sample solution.

7.2.21 Determination of Pb in Bone Ash by Furnace AAS*

Lead concentrates to a large extent in the bones. Thus exposure to lead can be assessed through analysis of bone. The following procedure was developed for the analysis of milligram-sized samples of museum specimens.

Enough bone must be used to give bone ash residue of at least 20 to 25 mg. The material is then dissolved in 0.5 mL of 9.4 *M* HNO_3. Lanthanum solution is added to overcome interferences in the furnace atomic absorption determinative step. The procedure was assessed by evaluating the percentage recovery of lead added to standard bone samples in solution. The method of standard additions is employed.

Equipment. A Perkin-Elmer 460 atomic absorption spectrophotometer equipped with deuterium arc background corrector was used. A HGA 2100 graphite furnace and AS-1 Autosampler were used. Graphite tubes were first pyrolytically coated with graphite and then coated with tungsten carbide to prolong the furnace life and increase analytic consistency.

The transfer of small volumes of standards, samples, and diluting fluids was accomplished with Eppendorf pipettes with an absolute error of less than 1%. Statistical and linear regressions analyses were performed with a HP-97 programmable calculator.

Reagents. All water used in preparation of samples and standards was deionized and redistilled from glass. Electronic-grade concentrated (15.6 *M*) HNO_3 was employed.

Stock solutions of lanthanum were prepared by dissolving 2.000 g of La_2O_3 in 160 mL of concentrated HNO_3 and diluting to 1 L with distilled water to yield a final concentration of 1705 μg/mL. Working solutions of 853 μg/mL were prepared by diluting the stock solution with distilled water.

Lead Standard. Commercially prepared standard solutions of 1000 (±1%) μg/mL available from Fisher Scientific were used. This standard was prepared from $Pb(NO_3)_2$ and standarized with secondary standard EDTA (primary standard $CaCO_3$). This stock solution was used to make up all standard curve samples and was the source of lead for the standard addition analysis.

*Procedure from Wittmers et al. (39).

Procedure.

Bone Standard Preparation. Scrape bone samples to remove adhering tissue, periosteum, and marrow. Freeze bone samples in dry ice and break into managable pieces. Weigh samples, place in Vycor crucibles, and dry at 110°C for 20 h to constant weight. Ash the samples in a muffle furnace at 450°C for 48 h (or until completely white) and cool in a desiccator. (The ashing temperature was selected to avoid loss of lead by volatilization of the sulfide or chloride.) Record the ash weight and then grind the samples in an agate mortar to a fine powder. Return the sample to a desiccator until analysis.

At the time of analysis completely dissolve 20–25 mg of bone ash in 0.5 mL of 9.4 *M* HNO_3 and dilute with 1.5 mL of water. Take a 0.2 mL aliquot of this dissolved bone solution and transfer to a sampling cup. Dilute with 1 mL of water containing 853 μg/mL La ion. (All glassware, crucibles, and sample cups that contact the samples and standards must be soaked in 7.8 *M* HNO_3 for 2–4 h during the cleaning process.) Prepare 0, 2, 4, and 8 μg/mL Pb stock standard by pipetting 0, 0.2, 0.4, and 0.8 mL, respectively of the 1000 μg/mL Pb commercial standard into 100 mL volumetric flasks, adding 1 mL of 15.6 *M* HNO_3, and diluting to volume with water. Then prepare running standard solutions by transfering 20 μL of each into sampling cups and adding 1 mL of La working solution. The resulting standards contain 0.0000, 0.0392, 0.0784, and 0.157 μg/mL, respectively.

For each bone sample to be analyzed by the method of standard additions, place 0.2 mL of the dissolved bone ash in each of four sample cups. Add to three sample cups 20, 40, and 60 μL of 2 μg/mL Pb standard. Dilute the solution in all cups to equal volume (1 mL) with La working solution. Run the solutions using the manufacturer's recommended operating parameters.

REFERENCES

1. P. Bailey and T. A. Kilroe-Smith, *Anal. Chim. Acta* **77,** 29 (1975).
2. F. J. Fernandez and D. Hilligoss, *At. Spectrosc.* **3,** 130 (1982).
3. J. Boone, T. Hearn, and S. Lewis, *Clin. Chem.* **25,** 389 (1979).
4. H. T. Delves, *Anayst* **95,** 431 (1970).
5. H. A. Schroeder and A. P. Nason, *Clin. Chem.* **17,** 461 (1971).
6. I. H. Tipton, *Gross and Elemental Composition of Reference Man.* ICRP Handbook, Report of Subcommittee II. Pergamon Press, New York 1972, Chapter 2.
7. H. M. Bowen, *Trace Elements in Biochemistry.* Academic Press, London, 1966.
8. R. M. Barnes and J. S. Genna, *Anal. Chem.* **51,** 1065 (1979).
9. D. S. Hacket and S. Siggia (G. W. Ewing, Ed.), *Environmental Analysis.* Academic Press New York, 1977, p. 253.

10. D. S. Hackett, *Diss. Abstr. Int. B.* **37,** 4430 (1977).
11. S. S. Brown, S. Nomoto, M. Stoeppler, and F. W. Sanderman, Jr., *Pure and Applied Chem.* **33,** 773 (1981).
12. K. Nakamura, H. Wantanabe, and H. Orii, *Anal. Chim. Acta* **120,** 158 (1980).
13. B. Momcilovic, B. Belonje, and B. Shah, *Clin. Chem.* **21,** 588 (1975).
14. N. E. Vieira and J. W. Hansen, *Clin. Chem.* **27,** 73 (1981).
15. A. J. Taylor and T. N. Bryant, *Clin. Chim. Acta* **110,** 83 (1981).
16. F. J. Fernandez and J. Innarone, *At. Absorpt. Newslet.* **17,** 117 (1978).
17. D. L. Osheim J., *Assoc. Off. Anal. Chem.* **66,** 1140 (1983).
18. M. A. Evenson and B. L. Warren, *Clin. Chem.* **21,** 619 (1975).
19. F. J. Kayne, G. Komar, H. Laboda, and R. E. Vanderlinde, *Clin. Chem.* **24,** 2151 (1978).
20. M. K. Routh, *Anal. Chem.* **52,** 182 (1980).
21. J. Smeyers-Verbeke, D. Verbelen, and D. L. Massart, *Clin. Chim. Acta,* **108,** 67 (1980).
22. W. Toda, J. Lux, and J. C. Van Loon, *Anal. Lett.* **13,** 1105 (1980).
23. E. F. Perry, S. R. Koirtyohann, and H. M. Perry, *Clin. Chem.* **21,** 626 (1979).
24. D. E. Pleban and K. H. Pearson, *Clin. Chim. Acta* **99,** 276 (1979).
25. W. Slavin, G. R. Carnick, D. C. Manning, and E. Prusekowska, *At. Spectros.* **3,** 69 (1983).
26. E. Berman, *At. Absorpt. Newslett.* **3,** 111 (1964).
27. L. E. Kopito, M. A. Davis, and H. Schwachnan, *Clin. Chem.* **20,** 205 (1974).
28. D. J. Hodges and D. Skelding, *Analyst,* **106,** 299 (1981).
29. V. Lagesson and L. Andrasko, *Clin. Chem.* **25,** 1948 (1979).
30. K. S. Subramanian, J. C. Meranger, and J. E. Mackeen, *Anal. Chem.* **55,** 1064 (1983).
31. P. Coyle and T. Hartley, *Anal. Chem.* **53,** 354 (1981).
32. K. Saeed and Y. Thomassen, *Anal. Chim. Acta* **143,** 223 (1982).
33. B. Lloyd, P. Holt, and H. T. Delves, *Analyst* **107,** 927 (1982).
34. K. S. Subramanian and J. C. Meranger, *Analyst* **107,** 157 (1982).
35. G. Bagliano, F. Benischek, and H. Huber, *Anal. Chim. Acta* **123,** 45 (1981).
36. W. R. Hatch and W. L. Ott, *Anal. Chem.* **40,** 2085 (1968).
37. M. A. Evenson and C. T. Anderson, *Clin. Chem.* **21,** 537 (1975).
38. H. S. Mahanti and R. M. Barnes, *Anal. Chim. Acta* **151,** 409 (1983).
39. L. E. Wittmers, A. Aliche, and A. C. Aufderheide, *Amer. J. Clin. Path.* **75,** 81 (1981).

CHAPTER

8

WATER SAMPLES

8.1 INTRODUCTION

It would seem at first glance that the analysis of water would be the simplest of all analytical tasks. Although this may be correct for distilled water solutions containing an easily detectable amount of a single inorganic salt, it is far from the truth for natural water. Natural waters are dynamic systems containing living as well as nonliving, organic together with inorganic, and dissolved as well as insoluble substances. Events can occur during or after sampling that may change the sample composition drastically from its original form. These are the result of a variety of causes, principal among which are contamination and loss of a substance due to precipitation, complexation, adsorption, or ion exchange on the container wall.

Natural waters vary greatly in matrix constituents. A laboratory may be required to analyze samples ranging from rainwater and soft water rivers and lakes, seawater and estuarian waters, to highly polluted effluents. A single analytical method for a metal will not necessarily be applicable to such a wide range of compositions.

Table 8.1 lists the concentrations of some selected trace elements in normal seawater (1). Obviously, for most elements the levels are below the detection limit for direct analysis by any of the available methods. Thus some method of preconcentration and separation will often be necessary.

Compounding these difficulties is the question of sampling. Little attention is generally given to the problem and hence it is common for sampling to be the largest source of error in the final result.

8.1.1 Sampling

No agreement can be found in the literature on sampling methods. Generally three types of sampling are done on bodies of water: grab, composite, or continuous. Samples of precipitation may be short interval, event, or composite. Samples are taken with a fixed volume sampling device or by pumping.

Natural waters can be heterogeneous vertically, horizontally, and with time. This is due not only to man-made pollution but also to natural phenomena such as erosion, currents, thermoclines, and precipitation washout of dust.

Table 8.1 Concentration of Selected Trace Elements in Seawater[a]

Element	Chemical Form	Concentration (μg/L)
Ag	$AgCl_2^-$	0.0003
Al		0.01
As	AsO_4H^{2-}	0.003
Au	$AuCl_4^-$	0.000011
B	$B(OH)_3$	4.6
Be		0.0000006
Bi		0.000017
Cd	Cd^{2+}	0.00011
Co	Co^{2+}	0.00027
Cr		0.00005
Cu	Cu^{2+}	0.003
Fe	$Fe(OH)_3$	0.01
Hg	$HgCl_4^{2-}$	0.00003
Li	Li^+	0.18
Mn	Mn^{2+}	0.002
Mo	MoO_4^{2-}	0.01
Ni	Ni^{2+}	0.0054
Sb		0.00033
Se		0.00009
Si	$Si(OH)_4$	3
Sn		0.003
Ti		0.001
U	$UO_2(CO_3)_3^{4-}$	0.003
V	$VO_5H_3^{2-}$	0.002
Zn	Zn^{2+}	0.01

[a]From Bowen (1).

Grab, short interval, and event sampling can thus lead to erroneous estimates of water composition. A series of such samples taken with time and, in the case of rivers and lakes, vertically and horizontally, may give a useful result.

Composite samples are obtained by combining a large number of samples taken at different intervals. The intervals may be either time or spacial or both. In this instance an average composition can be obtained. A distinct disadvantage of this approach is that any important compositional variations within the water system become undetectable.

In designing a sampling strategy, the objectives of the work must be born in mind together with the problems mentioned above. A detailed discussion of sampling, however, is beyond the scope of this book.

8.1.2 Contamination

For many of the parameters that must be measured, contamination of the water is an ever-present danger. Thiers (2) classifies the problem as follows:

1. Positive contamination results from the addition of contaminants to the sample.
2. Negative contamination occurs when losses of the substance in question (analyte) occur.
3. Pseudocontamination is an error introduced by the presence of a substance other than the analyte.

Positive and negative contamination pose the most serious problems. Causes of positive contamination are more obvious and occur more often. Negative contamination arises frequently from precipitation and adsorption or ion exchange on the surface of containers. Adsorption of analyte on particulate matter, if present, may also pose a problem.

8.1.3 Sample Containers

Presently, plastic bottles have almost entirely superseded glass for the storage of waters for inorganic analysis. Compared to borosilicate glass, most plastic bottles are inexpensive, less fragile, and have minimal problems due to the ion exchange. On the negative side, most plastics have a porosity that allows samples to evaporate slightly over long storage periods.

The most practical among the plastics is high density linear polyethylene. Compared with other softer plastic containers, these are rugged and suffer less from evaporation and adsorption–ion exchange problems.

8.1.4 Bottle Cleaning

Plastic bottles, including Teflon, may contain impurities that will result in sample contamination. Metallic impurities present as a result of the fabrication process can cause serious positive contamination in trace metal analysis. Other additives used as stabilizers may leach out causing negative or pseudocontamination problems.

When trace metals are to be analyzed, an acid soaking following the detergent wash is essential. When an acid treatment is needed, there is no general agreement in the literature on the type or strength of acid and the method of treatment. I use warm 1:1 nitric or hydrochloric acid. This mixture is added and followed by a period of vigorous shaking. Subsequently the container is rinsed thoroughly with distilled or deionized water.

There is a considerable disagreement on the nature and treatment of storage containers for waters to be analyzed for mercury. Some researchers recommend glass bottles and others plastic. An acid wash as described above is recommended. However, several workers suggest rinsing with oxidizing reagents such as permanganate or persulfate as an additional precaution.

8.1.5 Preservation

For the trace metals except mercury, only acid is commonly added as a preservative. No general agreement exists on the nature and final concentration of the acid. I recommend that the water be brought to pH 1 or lower with high-purity nitric acid. This procedure will result in metals being leached from particulate material. A prior filtration of the sample through prewashed members of 0.45 μm will minimize this effect.

8.1.6 Filtration

Depending upon the nature of the sample—the parameter to be measured and the data desired—it may be necessary to perform a filtration. This is usually essential for trace metals.

This treatment may aid in maintaining sample integrity because of the resulting reduced bacterial activity. As indicated in the preceding section, prior to the acidification of waters for heavy metal analysis, a filtration through 0.45 μm porosity membrane filters is commonly done. In this way a distinction can be made between dissolved and total metal in a water sample. Although it is a generally accepted procedure to use 0.45 μm porosity size to make this distinction, I have evidence that in many cases particulate passes through such a filter. Most commonly, membrane type filters are employed. These must be acid washed and distilled water rinsed to eliminate metal contaminants.

8.1.7 Separation and Preconcentration

Ion exchange chromatography and solvent extraction are the most frequently used methods for preconcentration and/or separation of metals from matrix constituents in waters. In my laboratory solvent extraction has been found the most satisfactory approach. In addition to yielding a satisfactory separation, solvent extraction has the advantage that the metals of interest are in an organic solvent when flame AAS is used. If the proper solvent is chosen, an enhanced signal compared with water solutions is obtained.

8.1.8 Plasma Emission Spectrometry

The detection limits obtained with present commercial ICP and DCP equipment are in most cases too poor for the direct determination of trace metals in natural waters. For example, Nygaard (3) in 1979 concluded that the DCP was not suitable for the direct analysis of seawaters for trace metals. In seawater the precision was poor at four times the detection limit.

Enhancement effects have been reported for trace metal signals caused by alkali metals in high salinity waters. In this regard Eastwood et al (4) studied the effect of 0%, 30%, 50%, and 100% salinity relative to fresh waters on the DCP emission of a number of elements. To illustrate, manganese emission intensity at 30% salinity had increased by 100% over the signal in fresh water (2576.10-Å line). This contrasts with zinc where the increase was less than 30% even at 100% salinity (2137.56-Å line).

A number of modifications to the sample introduction module of plasma emission equipment have been made that have resulted in detection limits that are adequate for determining trace metals in natural waters. The two main approaches adopted are a change in nebulizer type of use or an electrothermal atomizer to volatilize discrete samples into the plasma. For example, Garbarino and Taylor (5) describe a Babington-type nebulizer that gave equivalent or better detection limits (compared with conventional pneumatic nebulizers) for several metals. In addition, this nebulizer was found to be relatively insensitive to suspend material in waters.

Ultrasonic nebulizers have been recommended by several workers to improve detection limits. These devices are suitable for waters with relatively low total dissolved solids contents. When using ultrasonic nebulization, desolvation must be employed.

Nixon et al. (6) and Gunn et al. (7) used an electrothermal atomizer (graphite rod or metal filament) of the type employed for atomic absorption to volatilize microliter-sized aliquots of solutions into an ICP. For example, a carbon rod was mounted on electrodes under a cylindrical glass manifold (7). The sample placed on the rod was volatilized into an argon stream that formed part of the sample carrier gas stream. Linear calibration graphs over four orders of magnitude were obtained. Table 8.2 (7) gives detection limits for 16 elements.

Unfortunately no comparison is given to detection limits obtained by conventional nebulization. It is my experience that the detection limits in Table 8.2 are about one order of magnitude better. For refractory elements (e.g., boron and zirconium) addition of freon to the argon carrier gas greatly enhances the signal. This is due to the increased volatility of the metal halides thus produced.

The covalent hydride forming elements will be detectable in many waters because the hydride evolution method greatly enhances the detection limits of these elements compared with values obtained using conventional nebulization. Hydride methodology is now available for use with the ICP.

Table 8.2[a]

Element	Line (nm)	Detection limit µg/L (10 µL)	(pg)
Ag	I 328.1	0.0001	1
As	I 228.8	0.2	2000
Au	I 267.6	0.001	10
Be	I 234.9	0.0001	1
Cd	I 228.8	0.003	30
Ga	I 417.2	0.001	10
Hg	I 253.7	0.006	60
In	I 325.6	0.002	20
Li	I 670.8	0.0004	4
Mn	II 257.7	0.0001	1
P	I 213.6	0.02	200
Pb	I 405.8	0.01	100
Re	I 346.0	0.01	100
Sb	I 259.8	0.03	300
Tl	I 535.0	0.006	60
Zn	I 213.8	0.002	20

[a]From Gunn et al. (7).

8.1.9 Interlaboratory Comparison of the Analysis of Fresh Waters for Cu, Zn, Cd, and Pb.

Ihnat et al. (8) report the results of an interlaboratory comparison of the analysis of fresh water for copper, zinc, cadmium, and lead. The methods employed are listed in Table 8.3 with the detection limits of each technique. Both unfiltered and filtered samples were run by these techniques. Remember that the various techniques may determine different metal forms in waters, and this may account for some of the analytical variation reported. Techniques were used under routine operating conditions.

Copper. Agreement among the techniques was generally good. Positive bias was obtained with evaporation AAS and furnace AAS, whereas differential pulse anodic stripping yielded a negative bias. The carbon rod AAS technique gave a negative bais with distilled water and a positive bias with natural waters (species dependency?). Solvent extraction AAS gave a positive bias at lower concentrations and was close to unit slope for higher levels.

Zinc. Reasonably good agreement was obtained between furnace AAS and evaporation AAS for distilled and filtered natural water samples, with a positive bias obtained for furnace AAS compared with extraction AAS.

Table 8.3[a]

Method	Detection Limit (μg/L)			
	Cu	Zn	Cd	Pb
Evaporate flame AAS	0.8	0.3	0.1–0.4	1.6
Furnace AAS	0.5	—	0.01	0.2
Differential pulse anodic stripping	0.1	—	0.05	0.05
Carbon rod AAS	0.1	0.05	0.005	0.05
Solvent extraction flame AAS	0.5–0.6	0.3–11	0.5–1.1	2.1–5.1

[a]From Ihnat et al. (8).

Cadmium. Agreement was found to be good among the techniques. Differential pluse anodic stripping and carbon rod AAS and furnace AAS had good enough detection limits for Cadmium in natural waters.

Lead. Excellent agreement was obtained for results from differential pulse anodic stripping and carbon rod AAS. Furnace AAS gave a positive bias compared with carbon rod AAS.

8.2 PROCEDURES

8.2.1 Determination of Trace Metals in Fresh Waters by Evaporation and ICPAES*

An evaporation is used to concentrate trace metals in fresh waters to levels that would be detectable by inductively coupled plasma emission spectrometry. The samples were preserved by adding HNO_3 to give a 0.2% v/v solution. Then an evaporation in a silicon dioxide tube (to avoid aluminum contamination) in the presence of HNO_3 is employed. This treatment involving HNO_3 would also destroy any metal compounds.

A heated spray chamber is also used to enhance the emission signal. Detection limits for the metals ranged from 3 μg/L for aluminum to 0.03 mg/L for manganese. The trace elements determined were aluminum, cadmium, cobalt, chromium, copper, iron, manganese, molybdenum, lead, vanadium, and zinc.

*Reprinted in part with permission from (9), Goulden and Anthony, *Anal. Chem.* **54**,1681. Copyright 1982, American Chemical Society.

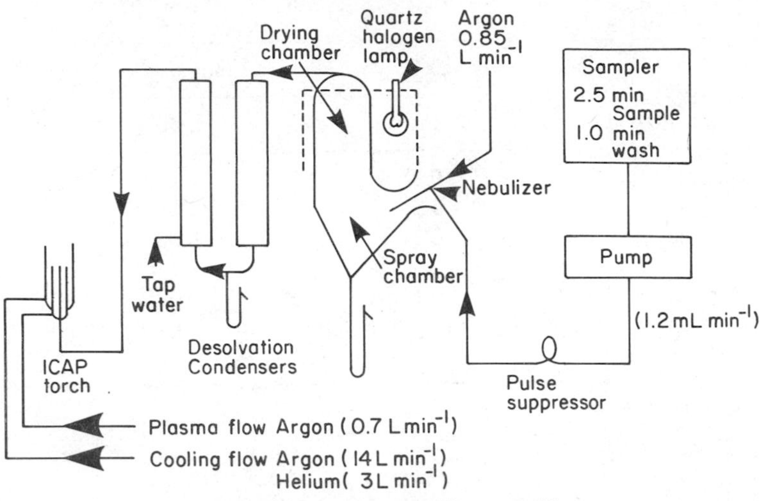

Figure 8.1 Desolvator—ICP system (9).

Equipment. The flow through the plasma system is shown schematically in Fig. 8.1. The sample is pumped to a nebulizer in the spray chamber; the spray escaping from the conical section is heated in the cylindrical portion of the chamber. The resulting aerosol is cooled in the two condensers where most of the water separates out and the desolvated aerosol is passed to the torch in the ICAP system.

When used in the automated mode, a Sampler IV (Technicon Corp.) is used. The pump is a pump I (Technicon Corp.) Between the pump and the nebulizer is 1 m of AWG 30 Teflon spaghetti tubing to act as a pulse suppressor. The nebulizer is a concentric glass nebulizer. Nebulizers from Meinhard Associates have been used, as well as nebulizers of the same type made in the laboratory. The modified spray chamber is made from a 125 mL conical flask to which is joined a piece of glass tubing, 45 mm o.d. The dimensions and the arrangement of the desolvation condensers are shown in Fig. 8.2. The heater for the cylindrical chamber is made from aluminum sheet (as shown in Fig. 8.2) and in this shroud is placed a 600 W, 115 V, quartz–halogen lamp (Cole Parmer, Catalogue No. C-3151-30).

The torch used is similar to that previously described (10); it used 17 mm bore silica tubing for the outer tube. The spraying and desolvation equipment are mounted on the bench beside the torch chamber. Then aerosol is carried to the torch through a 3 mm i.d. glass tube connected to the base of the torch with a short length of silicon rubber tubing to allow the torch to be positioned. The ICAP system is an Applied Research Laboratory ICP torch compartment with

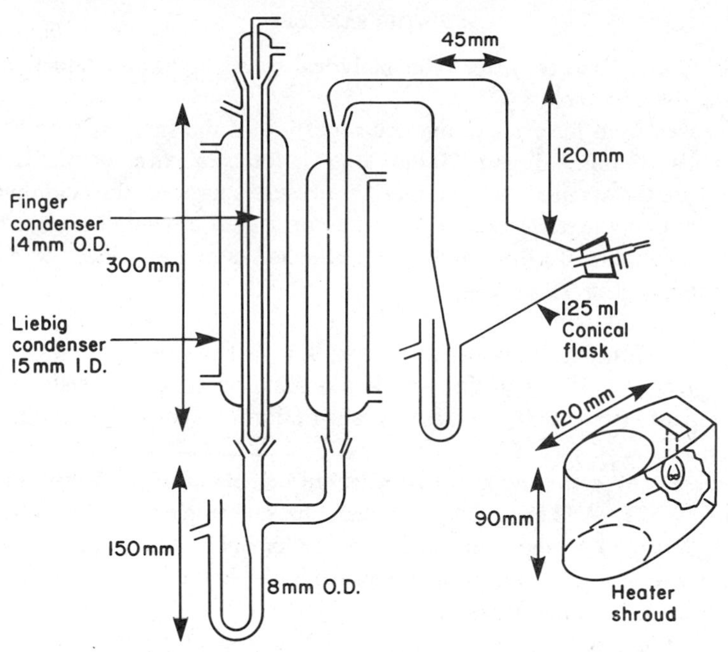

Figure 8.2 Spray chamber and condensers (9).

Table 8.4 Analytical Wavelengths and Detection Limits

Element	Wavelength (nm)	Order	Detection Limits (μg/L)
Al	308.22	2	2.9
Co	238.89	2	0.31
Cr	267.72	2	0.27
Cu	324.72	2	0.57
Fe	259.94	2	0.19
Mn	257.61	2	0.026
Mo	202.03	2	0.28
Ni	231.60	2	0.30
Pb	220.35	3	0.59
V	311.07	2	0.22
Zn	213.86	3	0.067
Ba	455.40	1	0.24
Ca	317.93	2	0.36
K	766.49	1	4.8
Mg	279.08	2	0.97
Na	589.00	1	4.6
Sr	407.77	1	0.03

a Model QA-137 spectrometer. The analytical wavelengths and detection limits used are given in Table 8.4.

The tubes used for evaporation and digestion of the samples are ultraviolet grade silica, 23 mm o.d. and 170 mm long. A forced air oven at 180°C is used to evaporate the samples; an aluminum hot block is used for the acid digestion. The antibumping system used in the oven consists of a manifold connection to a nitrogen cylinder. To the manifold are attached numerous pieces of AWG 30 Teflon tubing, each 35 cm long.

Reagents. Ultrapure nitric acid (Ultrex, J.T. Baker Chemical) was used for sample preservation and in the sample digestion procedure. Metals standards were prepared from 1000 mg/L atomic absorption standards (Fisher Scientific).

Procedure. Preserve the samples by acidifying them to 0.2% v/v with concentrated HNO_3. Add about 35 g of a well-shaken sample to a silica tube. Place the tube in a rack in the oven, insert the Teflon tube, and adjust the nitrogen pressure so that about 1 mL/min is blown into the liquid. Evaporate the sample to about 2 mL and add 0.5 mL concentrated HNO_3. Transfer the tube to the aluminum hot block and evaporate the solution to near dryness. Add 1 mL of water and evaporate the solution to dryness. Repeat this twice. Dissolve the solid in a mixture of 0.35 mL concentrated HNO_3 and 1 mL water (heating of this solution may be required in some cases) and make up the weight to 3.5 g. Analyze the concentrated sample in the ICAP system. Dilution factors are calculated from the weights and the measured gravities of matrix-matched standards.

The ICAP system is set in operation with 0.2% HNO_3 pumped through the nebulizer. Turn on the heating lamp and when conditions have stabilized, which takes about 30 min, pump the samples and standards to the spray chamber. The operating conditions are as follows: observation height is 2 cm above top of the radio frequency coil; plasma forward power, 1600 W; plasma reflected power, less than 10 W; argon cooling flow, 14 L/min; helium cooling flow, 3 L/min; argon plasma flow, 0.7 L/min; nebulizer gas flow, 0.85 L/min; nebulizer liquid flow, 1.2 mL/min; and lamp power, 180 W (55 V). The calibration of the instrument is made with a matrix-matched solution containing 1 mg/L of each of the metals to selected samples. The zero for each standard and sample is determined by making measurements off-peak by moving the spectrometer primary slit. The wavelength shift used is 0.023 nm below the peak of the elements measured in the third order and 0.034 nm for those measured in the second order. The off-peak and on-peak measurements are made with a to sec integration time. Aspirate the sample for 2.5 min. Between each sample use a 1 min wash of 0.2% HNO_3.

Table 8.5

Element	Determination limits (μg/L)
Pb	2.5
Zn	0.13
Cd	0.25
Ni	0.5
Mn	0.06
Fe	0.25
V	0.38
Cu	0.5

8.2.2 Determination of Trace Elements in Seawater by Solvent Extraction and ICPAES*

Diethyldithiocarbamate is used to extract lead, zinc, cadmium, nickel, mangenese, iron, vanadium, and copper into chloroform from seawater for determination by inductively coupled plasma emission spectrometry. To obtain extraction of manganese a pH of 6.2 is essential. At this pH, citrate must be present to prevent precipitation of iron as hydroxide. Under these conditions lead, zinc, cadmium, nickel, manganese, iron, and copper are quantitatively extracted, but a vanadium extraction of 83% is obtained. For this reason standards must be treated in a closely similar manner to the samples.

Because the chloroform layer cannot be nebulized satisfactorily, it is evaporated and the residue taken up in hydrochloric and nitric acids. The determination limits for the elements are shown in Table 8.5.

Equipment. An Applied Research Laboratories (ARL) ICPQ system (based on a QA 137 polychromator) was used. The system consists of a high-frequency generator, a plasma torch, a sample nebulizer, and the direct reading polychromator. The experimental faculties and the pertinent plasma operating conditions are summarized in Table 8.6.

Reagents. For the stock standards of each of the elements (100 μg/mL), dissolve a weighed portion of the high-purity metal or salt in 10 mL HNO_3 and dilute to volume in a volumetric flask. For the acetate buffer, pH 6.2, dissolve 60 g of sodium acetate in 1 L of distilled water and adjust the pH to 6.2 by adding 1 *M* acetic acid.

*Procedure from Sugimae (11).

Table 8.6 Plasma Emission Spectrometer and Plasma Operating Conditions

Spectrometer	ARL QA 137 1 m Pachen–Runge mounting, grating ruled 1920 lines/mm, 0.52 nm/mm reciprocal linear dispersion (1st order), primary slit width 20 m, secondary slits 50 μm, Hamamatsu R-300 photomultipliers.
Programmed wavelengths	Pb(II), 220.35 nm; Zn(II), 202.55 nm; Cd(II), 226.50 nm; NI(II), 231.60 nm; Mn(II), 257.61 nm; FE(II), 259.94 nm; V(II), 311.07 nm; Cu(I), 324.75 nm
Readout	Digital readout of integrated signal
Radio-frequency generator	Henry Radio 3000 PGC/27; frequency 27.12 MHz, crystal controlled
Plasma torch assembly	Fused quartz with capillary injector
Nebulizer and spray chamber	Glass pneumatic nebulizer into dual-tube spray chamber
	Plasma operating conditions
Frequency	27.12 MHz
Forward power	1600 W
Reflected power	< 10 W
Argon coolant gas flow rate	11.0 L/min
Argon plasma gas flow rate	1.31 L/min
Argon carrier gas flow rate	1.0 L/min
Observation height	16 mm above load coal

Procedure. Collect 2 L of seawater and filter it through a 0.45 μm membrane filter. Immediately acidify the filtered seawater with HCl to pH 1 to lessen the danger of adsorption or precipitation of colloidal materials on the walls of the container.

Transfer a 1 L aliquot of the acidified water sample to a 1 L beaker and evaporate to about 400 mL on a hot plate. Transfer to a 1 L separatory funnel, rinse with distilled water, dilute to 500 mL with distilled water, and mix. (*Note:* These steps are precautionary measures to prevent emulsification during solvent extraction and are necessary only for the analysis of polluted coastal seawater and estuarine waters that may contain emulsifying agents. Clear phase separation is aided by increasing the concentrations of electrolytes such as sodium chloride, that is, by salting out. A 500 mL aliquot of the acidified water sample is usually measured into the separatory funnel.) Add a few drops of bromocresol green indicator, 25 mL of aqueous 10% w/v ammonium citrate solution, and sufficeint NaOH solution to give a light blue color, indicating pH 5.5–6.0. Add 30 mL of acetate buffer, pH 6.2, which should adjust the pH to about 6.2 Add 20 mL of aqueous 2% w/v sodium diethyldithiocarbamate solution and mix. Add 100 mL of chloroform and shake vigorously for 10 min on a mechanical shaker. Allow the chloroform layer to separate and filter this layer through an 11 cm Toyo No. 5C filter paper into a 200 mL conical flask. Again add 20 mL of chloroform to the aqueous layer, shake for 1 min, and allow the chloroform layer to separate. Combine the two chloroform layers.

(Direct nebulization of the chloroform extract into the plasma affects the resistive impedance of the plasma to a large extent and makes it difficult to couple the radio-frequency energy into the plasma. This leads to complicated changes in signal intensity and considerably reduced sensitivity.)

Evaporate the combined extract almost to dryness by gentle heating. Add 12 mL of (1 + 9) HNO_3 and quantitatively transfer to a 25 mL volumetric flask with distilled water for nebulization.

8.2.3 Determination of Trace Metals in Water by Solvent Extraction AAS*

In the analysis of water samples, a number of ionic species may be encountered at concentration levels far greater than those of the metals being determined. Fluoride, calcium, potassium, magnesium, sodium, phosphate, silicate, and biodegradable detergent are potential interferences present in many waters. These ions were individually added to solutions of the analyte. After extraction the absorbances obtained were compared with those of solutions without interferences. Most of these ions are known masking agents; however, only the biodegradable detergent appeared to affect the absorbances of the metals to any degree.

The following procedure is suitable for the analysis of sea, brackish, and fresh waters, and for most effluents. Elements that can be readily determined are silver, cadmium, cobalt, copper, iron, nickel, lead, and zinc with detection limits of 0.6, 0.8, 1.5, 0.8, 1.3, 2.5, and 0.6 μg/L, respectively.

*Reprinted in part with permission from (12), Kinrade and Van Loon, *Anal. Chem.* **46,** 1896. Copyright 1974, American Chemical Society.

Equipment. An IL 153 (Instrumentation Laboratories) was used for all atomic absorption measurements. The pH measurements were made using a Beckman Expandomatic pH meter.

Reagents. All chemicals used were reagent grade of the highest quality available. Standard metal solutions (1000 μg/mL) were prepared for Ag(I), Co(II), Fe(III), Ni(II), Cu(II), Pb(II), Cd(II), and Zn(II) from high-purity metal. Buffer was prepared by mixing the appropriate amounts of citric, boric, and phosphoric acids in water to give a solution that is 0.5 *M* in each acid.

Prepare a 1% w/v mixed chelating solution of ammonium pyrrolidinedithiocarbamate and diethylammonium diethyldithiocarbamate in water. Extract this twice with methy isobutyl ketone (MIBK).

Procedure.

Extractable Metals. Acidify the water to pH 1 with HNO_3. Place the desired sample (usually 200 mL) in a 250 mL separatory funnel fitted with a Teflon stopcock. Add 4 mL of the buffer. Shake to mix well. The pH should be 4.0 ± 0.1. If the pH must be adjusted, add sufficient 20% NaOH solution to obtain this value. Add 5 mL of 1% mixed chelating agent. Shake briefly. Add 10–20 mL (depending on concentration factor required) of MIBK. Shake vigorously for 60 sec. Allow the layers to separate. Remove the aqueous lower layer. Retain the MIBK layer in tightly capped glass bottles until samples have been made ready for analysis. Prepare standards from multielement stock solution so that the 200 mL of water extracted contains four concentrations within the ranges of Fe 10–20 μg/L; Cu, 5–100 μg/L; Ni, 5–100 μg/L; Cd, 1–20 μg/L; Zn, 10–2000 μg/L; and Pb, 10–2000 μg/L. In this way a direct concentration relationship exists with samples. Run a reagent blank. Use experimental conditions recommended by the atomic absorption manufacturer.

Total Metal. Add 1 mL HNO_3 to the desired sample. Evaporate the sample to dryness on medium heat of a hot plate. Add 2 mL HCl and 1 mL HNO_3 acid. Evaporate to dryness. Add a drop of HCl and dilute to 200 mL. Continue as for extractable metal above. Run a blank containing all reagents.

8.2.4 Determination of Trace Metals in Water by Furnace AAS*

Prior to electrothermal atomic absorption analysis it is essential to remove the major cationic matrix constituents of seawater. Separation techniques for trace metals commonly require large amounts of chemicals that can lead to high

*Reprinted in part with permission from (13), Kingston, Barnes, Brady, Rains, and Champ, *Anal. Chem.* **50**,2064. Copyright 1978, American Chemical Society.

blanks. Of the methods available those employing Chelex-100 give suitably low blank values without excessive purification of reagents.

The trace metals (cadmium, cobalt, copper, iron, manganese, nickel, lead, and zinc) are separated from the major cations of seawater, sodium, potassium, calcium, and manganese using Chelex-100 resin: The pH of the seawater is adjusted to 5.0 to 5.5. It is passed through a column containing Chelex-100 resin. Ammonium acetate is used to elute the alkali and alkaline earth elements, followed by elution of the trace metals using nitric acid. The trace metals are determined by electrothermal atomic absorption spectroscopy.

A radioactive tracer procedure was used to study the separation. The recovery of cadmium, copper, manganese, nickel, and zinc was greater than 99.9% and recovery of cobalt, lead and iron was 99.5, 98.4, and 93.1%, respectively.

Equipment. The instrumental system used in this study consisted of a Perkin-Elmer Model 603 atomic absorption spectrometer with HGA-2100 graphite furnace (GFAAS). An Isolab QS-Q polypropylene column with porous polyethylene resin support was used for 100 mL and 1 L sample volumes. Although the same column was used for both sample volumes, the amount of resin and reservoir system were entirely different. For the 100 mL sample, the QS-S 25 mL conventional polyethylene extension funnel was attached to the column to act as a reservoir for the sample.

For a 1 L sample the reservoir was a 1 L Teflon (FEP) bottle inverted and modified with a machined Teflon (TFE) closure insert containing a microbore venting tube and outlet tube. The outlet was connected to a valve (TFE) by a 1.59 mm ($\frac{1}{16}$ in.) i.d. Teflon (FEB) tubing connector and linked to the reservoir with a specially machined mount (TFE) that sealed the column into the closed system. The mount contained a vent (sealed with a nylon screw that allowed the removal of air from the system) as well as an inlet and this was tightly clamped to the column using the lip on column at point B (see Fig. 8.3). The clamp (a modified glass joint clamp) and mount provided a seal that allowed the reservoir to be raised above the column to obtain enough pressure to control the flow rate using the pressure of the raised reservoir and the valve (Fig. 8.3).

Reagents. High-purity water and nitric and glacial acetic acids were prepared using subboiling distillation at the National Bureau of Standards (NBS). All reagents used in the separation process were prepared and stored in clean FEP Teflon bottles unless otherwise stated.

Ammonium hydroxide was prepared by bubbling filtered ammonia gas through high-purity water until room temperature saturation was achieved. A 1.0 *M* ammonium acetate solution was prepared by mixing 60 g purified glacial acetic acid and 67 g of saturated NH_4OH and diluting to 1 L in a polypropylene volumetric flask. The acidity was adjusted to pH 5.0 by dropwise addition of

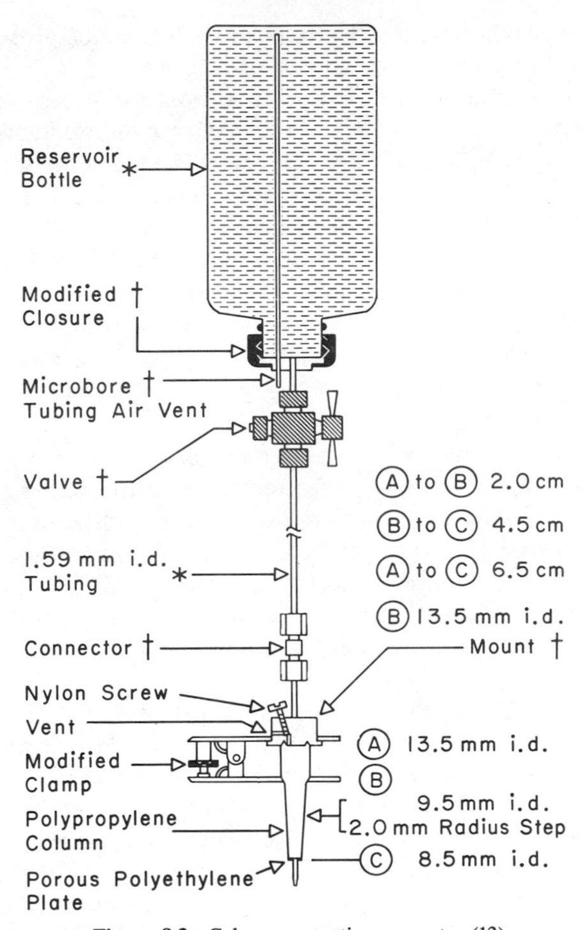

Figure 8.3 Column separation apparatus (13).

HNO_3 and/or NH_4OH. All reagent and sample preparations were done in a Class 100 clean laboratory.

Chelex-100 chelating resin, 200–400 mesh size, was purchased from Bio-Rad Laboratories. All standard stock solutions were prepared from high-purity metals or salts in subboiling distilled NBS acids as described by Dean and Rains (14). Working solutions were prepared as needed.

Seawater. The sample was collected with a submersible pump and plastic tubing permanently submerged approximately 100 m offshore from the Institute. The seawater was pumped directly into a conventional polyethylene drum that had been cleaned first with hydrochloric acid and then with nitric acid and purified water before use. After filtration through a 0.45 μm millipore filter using an all-polypropylene filter apparatus, the seawater was collected in a polyethylene carboy and acidified (to 0.6 M HNO_3) with high-purity HNO_3 to prevent bacterial growth, to stabilize the trace element concentrations, and to strip any trace elements bound by colloidal particles.

Procedure.

Column Preparation and Purification Procedure. The column preparation procedure consisted of precleaning the column in 1:4 HCl and then in 1:4 HNO_3 for one week in each bath and then rinsing the columns with water after each acid wash. Load the column with a slurry of 200–400 mesh size Chelex-100 resin, (sodium form). For 100 mL seawater samples, use 3.2–3.4 mL of resin which covers the lower barrel of the column from point B to point C in Fig. 8.3. For 1 L samples, use 5.8–5.9 mL of resin, which fills the column from point A to point C in Fig. 8.3. Wash the resin with 15 or 20 mL of 2.5 M HNO_3 (for the small and large resin volumes, respectively) in 5 mL portions to elute any trace metal contamination present in the resin. Then rinse with two 5 mL volumes of water. To transform the resin to the NH_4^+ form, add 10–15 mL of 2.0 M NH_4OH in 5 mL volumes. After checking the pH of the effluent to ensure basicity, rinse the column with 10 to 15 mL of water to remove the excess NH_4OH.

Column Preconcentration and Separation Procedure. For the 100 mL seawater sample, weigh 101.8 + 0.2 g directly into a clean 250 mL Teflon (FEP) beaker and adjust the pH to 5.5 with the dropwise addition of NH_4OH. Then add 0.5 mL of 8 M ammonium acetate to aid in buffering the system. Any necessary agitation of the solution is done with a Teflon stirring rod. Add a small amount of the seawater to the reservoir and the column to allow the resin to undergo its natural shrinkage as it changes ionic form and pH. This shrinkage results in a resin volume of approximately one half the original volume. After the completion of this transformation is observed (2–3 min), add the remaining seawater to the reservoir as needed to keep it filled; the flow rate is approximately 0.8 mL/min. To selectively elute Na^+, K^+, Ca^{2+}, and Mg^{+2} replace them with NH_4^+. Add 40 mL of 1.0 M ammonium acetate to the column in 10 mL aliquots. At the completion of the ammonium acetate addition, add 10 mL of water to remove residual ammonium acetate. Elute the transition metals using 7 mL of 2.5 M HNO_3 and collect in clean, preweighed, 10 mL conventional polyethylene bottles. Cap the bottles with clean polyethylene-lined caps and reweigh to determine the weight of the effluent accurately.

The procedure for the 1 L samples is the same as the 100 mL samples with minor alterations because of the apparatus (see above). Weigh the sample (1018.0

Table 8.7 Instrumental Parameters

	P-E 603			HGA-2100						
				Drying		Charring		Atomization		
Element	Wavelength (nm)	SWB (nm)	Scale Expansion	Temp (°C)	Time (sec)	Temp (°C)	Time (sec)	Temp (°C)	Time (sec)	Gas
Cd	228.8	0.7	1	100	30	200	20	2100	7	Ar[a]
Co	240.7	0.2	2	100	30	500	30	2700	7	Ar[a]
Cu	324.7	0.7	1	100	40	700	30	2500	6	Ar[a]
Fe	248.3	0.2	2	100	30	600	30	2700	7	Ar[a]
Mn	279.5	0.7	2	100	30	300	30	2700	7	Ar[a]
Ni	230.0	0.7	5	100	30	1000	30	2700	6	Ar[a]
Pb	283.3	0.7	3	100	40	400	30	2200	7	Ar[b]
Zn	213.9	0.7	0.5	100	30	500	20	2000	7	Ar[b]

[a]Interrupt mode.
[b]Normal mode.

+ 0.5 g) into a 1 L Teflon (FEB) bottle and adjust the pH in the same manner as previously described. The bottle becomes the reservoir and is fitted with a modified closure (see Fig. 8.3). Invert the bottle and purge the air from the system by means of the vent on the column mount. Adjust the flow rate using the valve and the height of the reservoir. Keep the flow rate to less than 0.2 mL/min until the shrinkage of the resin is complete. Then increase the flow rate to 1.0 mL/min and leave overnight to flow through the column. After passing the sample through the column, remove the valve and tubing and the connector above the column mount and replace with a smaller reservoir containing 70 mL of 1.0 *M* ammonium acetate. Adjust the flow rate to 0.5 mL/min until the reagent is exhausted. Wash the resin with 10 mL of water. Elute the transition metals with two 5 mL portions of 2.5 *M* HNO_3 into preweighed polyethylene bottles as previously described.

Introduce a 25 μL sample into the graphite furnace with the autosample. Use the instrument parameters given in Table 8.7.

8.2.5 Determination of Cd, Cu, Ni, and Zn in Seawater by Solvent Extraction and Furnace AAS*

A 50-fold concentration factor is obtained by this procedure. Dithizone is used to extract cadmium, copper, nickel, and zinc from seawater samples into chlo-

*Procedure from Smith and Windon (15).

roform. Back extraction from chloroform into dilute nitric acid follows. Detection limits obtainable with the technique in nanograms per liter are 6.0 Cu, 0.4 Cd, 32 Ni, and 16 Zn. At these levels contamination can be serious and hence actual detection limits obtained may be limited by this problem.

Equipment. Atomic absorption analyses were accomplished with a Perkin-Elmer Model 403 atomic absorption system equipped with a deuterium arc background corrector and a HGA-2200 heated graphite furnace. An AS-1 automatic sampling system was used in conjunction with the heated graphite furnace. Ultrapure water was made from distilled water by redistilling in a quartz subboiling still.

Reagents. A 0.040% w/v dithizone solution was freshly prepared from Suprapur dithizone (EM Laboratories) and high-purity chloroform (Burdick and Jackson). The dithizone reagent may be further purified as described by Patterson and Settle (16). This purification step is only necessary when working at zinc concentrations less than 100 ng/L. An ammonium citrate buffer was made by bubbling ammonia into a 20% w/v ammonium citrate solution until a pH of 7.7 was reached. This solution was then purified by multiple extractions with dithizone in chloroform as described by Patterson and Settle (16). A 1% w/v dimethylglyoxime solution was prepared from Suprapur dimethylglyoxime (EM Laboratories) and ethanol. High-purity nitric acid was obtained from the National Bureau of Standards. Diluted Suprapur ammonium hydroxide was used for sample pH adjustment. All reagents were prepared and stored in Teflon.

Procedure. Copper is extracted from seawater at a pH of 2.0 $\pm$ 0.2 from a 100 mL Teflon bottle to which 2.5 mL of 0.040% dithizone–chloroform and 7.5 mL of chloroform has been added. Equilibrate the aqueous and organic phases on a mechanical shaker for 20 min, and then pour into a 250 mL Teflon separatory funnel. Rinse the empty Teflon bottle with water and discard rinse water. It is important that every drop of rinse water be removed from the bottle and from the bottle cap. After 5 min, withdraw the organic phase from the separatory funnel directly into the rinsed Teflon bottle. Care must be exercised so that no water is drawn into the bottle with the organic phase. Back extract the organic phase with the addition of 200 μL of the NBS HNO_3, using an Eppendorf pipette, and 1.8 mL of high-purity water from an Oxford pipette. The organic phase must change from bluish-green to orange in color after the addition of HNO_3 and prior to the addition of the water. Shake the organic and HNO_3 phases on a mechanical shaker for 5 min to ensure complete back extraction.

Discard the aqueous phase in the separatory funnel and rinse the Teflon separatory funnel with water to prepare it for the next separation. The rinse water is completely removed from the separatory funnel and discarded.

After the 5 min equilibration, pour the back-extracted sample into the Teflon separatory funnel and separate the phases. Collect the acid phase in a 2.0 mL precleaned polyethylene vial and store in a refrigerator until analyzed.

Cadmium and zinc are extracted from seawater at a pH 7.7 ± 0.2 by using the procedure described for copper after the addition of 1.0 mL of the citrate buffer. Nickel is also extracted after the addition of 5.0 mL of 1% dimethylglyoxime in ethanol.

Standards are prepared by spiking preextracted water samples and reextracting as described above. Any residual dithizone remaining in the preextracted sample is removed prior to the spike addition by washing the aqueous phase with two 10 mL portions of chloroform and discarding the wash; 5 min is allowed after each wash for the phases to separate.

Each of the metals is determined by flameless atomic absorption spectrometry. Use the manufacturer's recommended operating conditions for lamp current, wavelength, and slit width. Use the deuterium background corrector for all analyses. Instrumental parameters for the heated graphite furnace are given in Table 8.8.

The sample injection into the heated graphite furnace should be done with an automatic sampling system. Contamination does not appear to be a problem when precleaned polyethylene vials are used as sample containers.

8.2.6 Determination of Pb, Cd, and Cr in Estuarine Waters by Furnace AAS*

This procedure is for the direct determination of lead, cadmium, and chromium in saltwater samples. Nitric acid and ammonium nitrate are added for cadmium and lead determinations to volatilize chloride prior to the atomization cycle. Only nitric acid is added for the chromium determination. Chromium standards contain only chromium and nitric acid. For cadmium and lead artificial seawater matrix is used in addition to acid. Detection limits are as follows in micrograms per liter: 0.1 Cd, 4 Pb, and 0.2 Cr. Repeatabilities expressed as the relative standard deviation are 20, 18, and 25% for cadmium, lead, and chromium, respectively. If salt content is above 18%, the sample must be diluted. These detection limits will be good enough for seawaters containing average levels of the metals. When low levels must be determined, it is necessary to use the more lengthy ion exchange or solvent extraction methods.

Equipment. A Perkin-Elmer heated-graphite atomizer (HGA-2100) equipped with normal (or pyrolytically coated) tubes was used in conjunction with a system composed of a monochromator (Jarrell-Ash, Model 82/410), photomulti-

*Procedure from Stein et al. (17).

Table 8.8 Instrumental Parameters for HGA-2200

Element	Sample Volume (mL)	Gas Flow Rate (cm^3/min)	Dry Cycle	Char Cycle	Atomization Cycle	Recorder Setting[a]
Cd	10	40[b]	15 sec, 100°C	20 sec, 350°C	5 sec, 2000°C	1.0
Cu	20	40	25 sec, 100°C	20 sec, 900°C	5 sec, 2300°C[c]	0.5
Ni	20	40	25 sec, 100°C	20 sec, 900°C	8 sec, 300°C[c]	0.5
Zn	2.0	40	10 sec, 100°C	15 sec, 500°C	7 sec, 2200°C	0.5

[a]Full-scale absorbance setting.

[b]Gas interrupt mode.

[c]Temperature sensor used.

plier (Pacific Photo-optics Instruments, Model 3150), amplifier (Masters Instruments, Model PT 75), and recorder (Environ-Tech Corp., Model Sc-1200R). Sequential background correction was made with a deuterium lamp for lead and cadmium or a quartz-iodine lamp for chromium. Later, a Perkin-Elmer Model 372 atomic absorption spectrophotometer was used with simultaneous deuterium arc background correction. Hollow-cathode lamps were used for all elements. Operating parameters are described in Table 8.9.

Reagents. Reagent grade chemicals were used throughout. All solutions were prepared with demineralized water containing 10 mL concentrated HNO_3/L (demineralized acid water) unless otherwise specified. Artificial seawater containing sodium sulfate and chlorides of sodium, magnesium, calcium, and potassium was prepared according to the Environmental Protection Agency (18), except that sodium hydrogen carbonate, potassium bromide, boric acid, stron-

Table 8.9 Instrument Conditions

Metal	Wavelength (nm)	Gas flow rate[a] (mL/min)	Drying		Charring		Atomization	
			(°C)	(sec)	(°C)	(sec)	(°C)	(sec)
Lead	283.3	20	110	20	700	20	220	5
Chromium	357.9	15	110	20	1200	30	2700	5
Cadmium	228.8	15	110	20	300	20	1900	5

[a]Interrupted gas flow used with Model 372 for measurements at low levels.

tium chloride, and sodium fluoride were omitted. These omissions did not measurably affect the nonspecific attenuation. The solution had a salinity of 36%.

For each metal a stock solution (1 g/L as nitrate) was prepared with demineralized acid water and stored at 22°C for up to 6 months. Intermediate solutions (1 and 10 mg/L) were stored at 22°C for up to 1 month.

Standard solutions were prepared weekly with demineralized acid water or the appropriate artificial seawater dilution and were stored at 22°C for up to 1 month.

Procedure. Determine the salinity of each sample. For Pb analysis, dilute the sample to < 18% salinity with demineralized acid water. In each 100 mL sample to be analyzed for Pb (diluted or undiluted) or Cd, dissolve ammonium nitrate as follows: salinity, 0–9%, 7 g; 10–18%, 15 g; 19–27%, 24 g; 28–36%, 30 g.

Prepare a set of five standard solutions within each analytical range using demineralized acid, water, and the artificial seawater dilution whose salinity is within approximately 5% of that of each diluted or undiluted sample for Cd and Pb. Run appropriate size aliquots (10, 20, or 50 μL) of samples and standards using parameters in Table 8.9.

8.2.7 Determination of Trace Metals in Seawater by Solvent Extraction and Furnace AAS*

The elements covered by this method are cadmium, zinc, lead, copper, iron, manganese, cobalt, chromium, and nickel. The procedure is used when small samples of seawater are to be done. A double chelation system is used consisting of pyrrolidine-*N*-carbodithioate (APDC) and oxinate. Oxinate is used to allow complete extraction of manganese. A further preconcentration is obtained by back extraction of the metals from the methyl isobutyl ketone solvent into an-acidic aqueous phase. Regarding back extraction, the authors argue for the usefulness of such an approach because (1) metal complexes are unstable in MIBK, but metals are stable in acidic aqueous solution; (2) a better concentration ratio can be obtained; and (3) improved behavior of microliter aliquots in the graphite furnace occurs with aqueous as compared to organic solvent solutions.

It is important to reduce the reagent blank as much as possible. In this regard purified MIBK and APDC as normally obtained from the supplier must be further decontaminated.

Equipment. A Varian Techtron atomic absorption spectrophotometer Model AA-5 fitted with a Perkin-Elmer heated graphite atomizer Model 2200 (HGA-2200) and temperature-ramp accessory was used for all trace metal measure-

*Reprinted with permission from (19), Sturgeon, Berman, Desaulniers, and Russell, *Talanta* **27,** 85. Copyright 1980, Pergamon Press, Ltd.

ments. The base of the HGA-2200 furnace was modified so that it could be fitted to the optical rail of the spectrophotometer. A simultaneous background-correction system was used. The continuum source was a Hamamatsu deuterium hollow-cathode lamp. Sample solutions were delivered to the furnace in either 10 or 20 μL volumes with a Perkin-Elmer AS-1 autosampler, and absorbance peaks were recorded on a fast-response (< 300 msec full scale) Servo II strip-chart recorder (Esterline Corp., Indianapolis, IN). The instrumental settings and furnace thermal programs, optimized for each element, were those suggested by Perkins-Elmer.

A Beckman Century SS-1 pH meter with a glass electrode was used for pH measurement. Polypropylene beakers and borosilicate Squibb-type separatory funnels fitted with Teflon stopcocks and polypropylene tops were used for sample preparations and extractions.

Reagents. Stock standard solutions (1000 mg/L) of the elements of interest were prepared by dissolution of the pure metals or their salts. Following dilution with high-purity distilled demineralized water, their concentrations were verified by standardization against NBS SRM 1643, Trace Elements in Water.

All reagents were purified prior to use. Concentrated nitric, hydrochloric, and acetic acids were prepared by subboiling distillation in a quartz still from reagent grade feed stocks.

A saturated solution of ammonia (25%) was prepared by bubbling gaseous ammonia (generated by evaporation of liquid ammonia) through a trap containing a basic solution of EDTA prior to passage into chilled high-purity water.

Methyl isobutyl ketone was purified by subboiling distillation.

Fresh 5% aqueous solutions of ammonium pyrrolidine-*N*-carbodithioate (APDC, Baker Analyzed) were filtered through a 0.3 μm membrane filter to remove insoluble material and stripped free of metal impurities by repeated extraction with distilled MIBK. In order to stabilize the reagent, the solutions were adjusted to a pH of about 9 with ammonia.

The Baker Analyzed 8-hydroxyquinoline was purified by vacuum sublimation at 120° onto a cold finger (metal oxinate impurities are not volatilized below 160°) and 2% solutions of 8-hydroxyquinoline (oxine) in dilute HCl were prepared.

All reagents were analyzed for trace metal content prior to use. Subboiling distilled MIBK was analyzed by evaporating a 100 mL aliquot to 5 mL in a polypropylene beaker, adding 1 mL concentrated HNO_3, and completing the evaporation. The residue was dissolved in 2 mL 0.3 *M* HNO_3 and analyzed by GFAAS. For analysis of the purified APDC a 20 mL aliquot of a 5% solution was evaporated to dryness in a Vycor crucible, 1 mL concentrated HNO_3 was added, and the evaporation was completed. The residue was then ashed at low temperature in a muffle furnace and the ash taken up in 2 mL 0.3 *M* HNO_3 and analyzed by GFAAS. Oxine and acetate buffer solutions were analyzed directly by furnace atomic absorption. Distilled demineralized water and concentrated HNO_3 were analyzed by spark-source mass spectrometry.

All evaporations were done on a porcelain hot plate in a "clean" fume hood. Coastal seawater was obtained from the Atlantic Regional Laboratory of the National Research Council of Canada in Halifax, Nova Scotia. The samples had been filtered through a nominally 0.45 μm membrane filter, acidified to pH 1.6, and stored in polyethylene bottles.

Procedure. All sample preparations and analyses by the authors were done in a "clean" laboratory equipped with laminar-flow "clean" benches and fume cupboards, thus providing a Class 100 working environment. All labware was thoroughly cleaned in (1 + 1) HNO_3, rinsed with water, further cleaned with hot aqueous 0.05% APDC solution, and again rinsed with water. Once in use, beakers and separatory funnels were washed only with water between runs because frequent thorough cleaning was found to produce highly variable analytical blanks.

Seawater samples are analyzed by the method of standard additions, with successively larger metal spikes added to 100 mL samples of seawater contained in 250 mL polypropylene beakers, with 1.0 mL of 1 *N* ammonium acetate buffer, 0.5 mL of 5% APDC solution, and 0.5 mL of 2% oxine solution added to each sample. Two reagent blanks are also prepared, consisting of the added chelating agents and buffer as well as 15 mL of water to act as a physical carrier.

Adjust the samples and blanks to pH 4.0 and heat to 80° on a water bath for 10 min in order to complex Cr. Transfer the solutions to separatory funnels and cool to room temperature. Add 15 mL of MIBK to each. Shake each mixture vigorously for 3 min and allow the phases to separate. After the aqueous phase has been drained from the separatory funnel and its pH adjusted to 9.2 with ammonia, it is recombined with the MIBK phase and the mixture shaken for a further 3 min in order to extract Mn. Following phase separation, discard the aqueous phase and rinse the separatory funnel with a small volume of water to remove the film of seawater adhering to the glass surface. The metal complexes are then back extracted by shaking the MIBK phase with 300 μL of concentrated HNO_3 for 1 min, then adding 2.7 mL of water and shaking for a further 2 min. Following phase separation, drain the 3.0 mL acid layer (1.5 *M* HNO_3) into a 30 mL screw-capped polypropylene bottle. This procedure provides a 33-fold preconcentration of the sample. The MIBK phase is analyzed for Co, for this element is not efficiently back extracted.

Coastal seawater was analyzed for Fe and Mn by direct injection of 10 μL volumes into the atomizer. A 5 μL volume of 25% ammonium nitrate solution as matrix modifier was then added and the sample subjected to the following atomization program: ramp dry to 110° during 40 sec, hold for 20 sec; ramp ash to 1000° during 15 sec, hold for 5 sec; atomize by heating at maximum rate to 2700° in 5 sec (using temperature-controlled heating). Nitrogen gas, at a flow rate of 40 mL/min, was used to purge the interior of the furnace during atomization.

Background correction was used during analysis of the nitric acid back extracts from the solvent-extraction samples and also during direct analysis of seawater. All "standard additions" curves were analyzed by regression procedures to obtain the intercepts.

8.2.8 Determination of Metals in Sewage Effluent by AAS*

This method is for the determination of zinc, copper, iron, lead, nickel, and cadmium. A mixed chelating solution of APDC and DDDC is used to extract the metals into MIBK prior to flame atomic absorption determination. To ensure that the metals are in an extractable form, a portion of the effluent is evaporated in the presence of nitric acid.

Equipment. All-glass apparatus should be used and should be cleaned by boiling with 1% v/v HNO_3 (where possible). Adequate rinsing with deionized water is required before use. It is advisable to set aside apparatus specifically for metals analysis in order to minimize contamination. A Pye-Unican SP 90 Series 2 atomic absorption spectrometer was used.

Reagents. Reagents should be the highest grade possible. The following are required: nitric acid (SG 1.42 Aristar grade); 4-methylpentan-2-one (MIBK); ammonium tetramethylenedithiocarbamate (APDC); diethyldiammonium dithiocarbamate (DDDC); sodium citrate (Analar); citric acid (Analar); ammonia solution (SG 0.88).

The chelating solution consists of 1% w/v each of APDC and DDDC in water prepared daily. The buffer solution is 1.2 *M* sodium citrate and 0.7 *M* citric acid. Prior to use, it is necessary to purify the citrate buffer by extracting, using 50 mL of MIBK and approximately 0.5 g of APDC/DDDC, until the extracts are no longer green (this is usually achieved by two extractions).

Procedure. Take 500 mL of effluent, add 5 mL HNO_3, and boil down to approximately 50 mL in a Kjeldahl flask. When cool make up to the original volume. Take 250 mL of the treated sample and pour into a separation funnel. Neutralize to pH 7 by cautious dropwise addition of ammonia solution (approximately 2.5 mL). Add 5 mL of citrate buffer and check that the pH is around 5.0. Add 10 mL of chelating solution. Add 35 mL of MIBK. Shake the funnel

*Procedure from Webster (20).

for 1 min and then allow to stand for 5–10 min to facilitate separation of the two phases. Discard the aqueous phase and pass the organic phase through phase-separating paper (Whatman 1PS) into a 50 mL volumetric flask. Make up to 50 mL with MIBK. The solution is now ready for analysis. The samples are run using the experimental conditions recommended by the manufacturer.

8.2.9 Determination of Pb in Drinking Water by Solvent Extraction and Furnace AAS*

The direct determination of lead in waters by furnace AAS is generally not possible because of loss of lead, usually as the chloride, during ashing or during the heating to the atomization temperature and because of its very low levels in drinking water. The former can be minimized by using a L'vov platform, matrix modification, and by a rapid heating rate (>2000°C/sec.).

In the present method, a solvent extraction using ammonium tetramethylene-dithiocarbamate (APDC) in 4-methylpentan-2-one (MIBK) is used to separate lead from the interfering matrix. An aliquot of the solvent is pipetted directly from the tube in which the extraction occurs into the furnace.

Equipment. The atomic absorption spectrophotometer used was a Perkin-Elmer 360 fitted with a deuterium arc background corrector. This instrument was fitted with a special band pass mode for carbon furnace work that consists of a slit of reduced height. The graphite furnace was an HGA 74 and nitrogen (with a low oxygen content) was used as the purge gas. This particular furnace can operate in three modes of gas flow during atomization; for this work, however, the miniflow approach was used. In this mode the purge gas flow is reduced from the normal rate of 4 mL/sec to 0.5 mL/sec for 8 sec of atomization. This results in an enchancement of sensitivity. All furnace injections were made with Oxford micropipettes utilizing disposable plastic tips. The extraction vessels were glass specimen tubes, 25 mL in capacity, with plastic snap-on caps. A Perkin-Elmer 56 flat-bed recorder was used to trace the absorbance readings at a chart speed of 0.10 cm/sec.

Operating conditions for the atomic absorption spectrophotometer were those recommended by the manufacturer. The resonance line used was 283.3 nm and the band pass was 0.7 nm (carbon furnace mode).

The operating conditions of the furnace had to be established because the stability of the lead chelate was unknown. The initial drying stage was set at 100°C for 30 sec. This temperature was selected because it is slightly lower than the boiling point of 4-methylpentan-2-one and would thus avoid splattering within the tube during drying. The time was sufficient to evaporate off the solvent completely. The atomization temperature was set at 2000°C. The time for

*Procedure from Mitcham (21).

Table 8.10 Miniflow Program

	Drying	Ashing	Atomization
Temperature (°C)	100	700	2000
Time (sec)	30	15	10

atomization was set at 10 sec, which is 2 sec longer than the program for miniflow operation. The ideal ashing temperature for the organic extract is 700°C for 15 sec.

Reagents. Atomic spectroscopy grade ammonium tetramethylenedithiocarbamate was used. Laboratory reagent grade concentrated HNO_3 was used. This is employed as a 1% m/v solution in distilled water and is filtered prior to use. Analytical reagent grade 4-methylpentan-2-one was used.

Procedure. Standards of 0.010, 0.025, 0.050, and 0.100 mg/L of lead are prepared in distilled water previously acidified to pH 2.5 ($\pm$0.3) with HNO_3. Water samples are also acidified to pH 2.5 ($\pm$0.3) upon receipt. All glassware and plastics are washed with dilute HNO_3 and distilled water prior to use.

Transfer a 10-mL aliquot of each drinking water sample by pipette into a specimen tube. Then add 1 mL of ammonium tetramethylenedithiocarbamate solution followed by 5 mL of 4-methylpentan-2-one. Shake the mixture vigorously for 1 min and allow to stand for 3 min to permit the two phases to separate.

Transfer a 50 μL volume of the organic phase by pipette into the graphite furnace and run the sample using the miniflow program given in Table 8.10.

8.2.10 Determination of Be in Natural Waters by Furnace AAS*

An extraction–preconcentration is used to obtain beryllium in a form and at a level determinable in natural waters by furnace AAS. The extraction using acetylacetone is accomplished with carbon tetrachloride. Five hundred milliters of water can be extracted into 10 mL. A back extraction into 1:1 hydrochloric acid is done. This is followed by a reextraction into toluene using acetylacetone. The extraction step must be repeated in each case to get quantitative results.

Equipment. A S-112 (USSR) double-beam, atomic absorption spectrophotometer equipped with a graphite furnace and a deuterium lamp background corrector was used. The furnace was flushed with argon. The light source was an

*Procedure from Pilipenko and Samchuck (22), with permission of Plenum. Copyright by Plenum Publishing Corp.

LSP-1 hollow-cathode beryllium lamp. Operating conditions were as follows: FEU-106-1 photomultiplier potential of 1.2 kV and a lamp current of 25 mA.

Reagents. cp Hydrochloric and nitric acids (CP grade) and EDTA were used. The solvents were used without suppelementary purifications. The standard beryllium solution, 0.1 mg/mL, was prepared from 99.99% beryllium metal.

Procedure. To a 1-L Erlenmeyer flask add 500 mL of the previously filtered water sample, 20 mL of 1:1 H_2SO_4, and 10 mL of 10% ammonium persulfate. Boil the sample for 10 min, cool, and place in a separatory funnel. Adjust the pH to 7–9 with ammonium hydroxide and treat the solution with 10 mL of 0.1 *M* EDTA, 5 mL of acetylacetone, and 10 mL of CCl_4. Extract the mixture by shaking for 5 min. Allow the organic phase to separate and repeat the extraction with the same reagents. Back extract the beryllium from the CCl_4 by the addition of 10 mL of 1:1 HCl. Neutralize the back extract with ammonium hydroxide to a pH of 5–8. Add 0.3 mL of acetylacetone and 2.5 mL of toluene and shake the mixture to extract the beryllium. Repeat the extraction. Using a micropipette, introduce 50 μL of the extract into the graphite furnace. Use the following thermal program: dry for 0.5 to 1 min at $T = 90°C$, heat for 10 sec at $T = 700–800°C$, and atomize for 5 sec at $T = 2650°C$.

Make five measurements on each extract. The standard solutions used for plotting the curve should contain the following beryllium concentrations in the extract in micrograms per liter: 0.01, 0.03, 0.05, 0.1, and 0.2.

REFERENCES

1. H. M. Bowen, *Trace Elements in Biochemistry.* Academic Press, London, 1966.
2. R. E. Thiers, *Methods Biochem. Anal.* **5,** 273, (1975).
3. D. D. Nygaard, *Anal. Chem.* **51,** 881 (1979).
4. D. Eastwood, M. S. Hendrickad, and G. Sogliero, *Spectrochim. Acta* **35B,** 421 (1980).
5. J. R. Garbino and H. E. Taylor, *Appl. Spectrosc.* **34,** 584 (1980).
6. D. E. Nixon, V. A. Fassel, and R. N. Kinsely, *Anal. Chem.* **46,** 210 (1974).
7. A. M. Gunn, D. L. Millard, and G. F. Kirkbright, *Analyst* **103,** 1066 (1979).
8. M. Ihnat, A. D. Gordon, J. D. Gaynor, S. S. Berman, A. Desaulniers, M. Stoeppler, and P. Valenta, *Internat J. Envir. Anal. Chem.* **8,** 259 (1980).
9. P. D. Goulden and D. H. J. Anthony, *Anal. Chem.* **54,** 1681 (1982).
10. P. D. Goulden, D. H. J. Anthony, and K. D. Austen, *Anal. Chem.* **53,** 2027 (1981).
11. A. Sugimae, *Anal. Chim. Acta* **121,** 331 (1980).
12. J. D. Kinrade and J. C. Van Loon, *Anal. Chem.* **46,** 1896 (1974).
13. M. M. Kingston, I. L. Barnes, T. J. Brady, T. C. Rains, and M. A. Champ, *Anal. Chem.* **50,** 2064 (1978).

14. J. A. Dean and T. C. Rains (R. Maurodineanu, Ed.), NBS Special Publication 492, 1977.
15. R. G. Smith and H. L. Windon, *Anal. Chim. Acta* **113,** 39 (1980).
16. C. C. Patterson and D. M. Settle, *Accuracy in Trace Analysis, Sampling, Sample Handling and Analysis,* NBS, Special Publ. 427, 1976, p. 321.
17. V. B. Stein, E. Canelli, and A. H. Richards, *Internat. J. Envir. Anal. Chem.* **8,** 99 (1980).
18. U.S. EPA, *Methods for Chemical Analysis of Water and Wastes,* Vol. 87, 1974, p. 170.
19. R. E. Sturgeon, S. S. Berman, A. Desaulniers, and D. S. Russell, *Talanta* **27,** 85 (1980).
20. T. B. Webster, *Water Pollut. Control* **511** (1980).
21. R. M. Mitcham, *Analyst* **105,** 43 (1980).
22. A. T. Philipenko and A. I. Samchuk, *Zhur. Anal. Khim.* **37,** 614 (1982).

CHAPTER 9

AIR SAMPLES

9.1 INTRODUCTION

Metalic air pollutants occur in a variety of forms. Mainly they are present as particulate or as fumes, with particulate being the most common form. Table 9.1 lists metal species encountered in workroom air and threshold limit values (1).

It is common practice to analyze air particulate for trace metals. Coarse material (>5 μ) is less of a problem for human health because such particles are trapped in the nasal passages. Particles smaller than 2 μ, on the other hand, penetrate to the deeper recesses of the lungs and thus materials comprising these particles can be more easily ingested. The major sources of trace metal air pollutants are shown in Table 9.2, together with some of the effects of the metal on human health (2).

A variety of methods have been used to trap particulate. The most common are absorption in a liquid impinger, electrostatic precipitation, and filtration using high volume (Hi Vol) filtering apparatus. Filtration is favored by most workers. The filters are either glass fiber or cellulose organic-based membranes. Table 9.3 lists the concentration of impurities commonly found in these filters (2). It is thus evident that blanks must always be run to avoid contamination problems from the filter medium.

A variety of techniques have been proposed for preparing the sample for the determinative step. These fall into the two main categories of dry ashing followed by acid decomposition and wet ashing.

Dry ashing at elevated temperature may result in loss of some of the more volatile metals such as cadmium, lead, zinc, arsenic, selenium, and mercury. Table 9.4 indicates the magnitude of this loss (2). Table 9.4 also gives a similar compilation of data for a low-temperature oxygen plasma ash. In this case there is little loss for any of the elements. The disadvantage of oxygen plasma ashing is that it is very time consuming.

Wet ashing is the most widely accepted approach to the preparation of air filters for analysis. A number of acid mixtures have been used. If decomposition of the siliceous matter is essential, then hydrofluoric acid must be a component of the mixture. Many workers use $HClO_4$–HNO_3 in mixtures because they de-

Table 9.1 Threshold Limit Values—WorkRoom Air[a]

	Adopted Values (TWA)[b]	
	(ppm)	(mg/m^3)
Arsenic and compounds (as As)		(0.5)
Arsine	0.05	0.2
Beryllium		0.002
Cadmium, dust and salts (as Cd)		0.05
Calcium arsenate (as As)		1
Chromic acid and chromates (as Cr)		0.05, A2[b]
Chromium, soluble chromic, chromous salts (as Cr)		0.5
Cobalt metal, dust and fume		(0.1)
Copper fume		0.2
Lead, inorganic, fumes, and dusts (as Pb)		0.15
Lead arsenate (as Pb)		0.15
Lead chromate (as Cr)		0.05, A2[c]
Mercury (alkyl compounds), skin (as Hg)	0.001	0.01
Mercury (all forms except alkyl) (as Hg)		0.05
Nickel carbonyl	0.05	0.35
Nickel metal		1
Nickel, soluble compounds (as Ni)		0.1
Selenium compounds (as Se)		0.2
Selenium hexafluoride (as Se)	0.05	0.4
Tellurium		0.1
Tellurium hexafluoride (as Te)	0.02	0.2
Tin, inorganic compounds, except SnH_4 and SnO_2 (as Sn)		2
Tin, organic compounds (as Sn), skin		0.1
Uranium (natural), soluble and insoluble compounds (as U)		0.2
Vanadium (V_2O_5) (as VC dust)		0.5
Vanadium (V_20_5) fume		0.05
Zinc chloride fume		1
Zinc chromate (as Cr)		0.05, A2[c]
Zinc Oxide fume		5

[a]From (1).

[b]TWA: Time weighted average.

[c]A2: Suspected human carcinogenic potential.

Table 9.2 Major Sources of Air Pollution of Trace Metals and Their Effects on Human Health[a]

Element	Sources of Pollution	Effects on Human Health
As	Smelters processing arsenical ores; insecticides; herbicides	Dermatitis, bronchitis, and skin cancer
Be	Be–Cu alloys; rocket fuels and coals	Chemical pneumonitis, beryllіosis, chemical ulcer, and carcinogenesis
Cd	Electroplating; alloys, solders, pigments, and chemicals	Pulmonary emphysema, hypertension, kidney damage, and carcinogenesis
Cr	Chrome-plating, chrome-tanning, pigments, and chrome–alloy industry	Perforation of nasal septum, chronic catarrh, emphysema, and carcinogenesis
Fe	Iron and steel industry; coal, fuel oil, and incineration	Siderosis and pneumoconiosis
Mn	Iron and steel industry; fuel oil, incineration, and coal; dry cell batteries	Chronic manganese poisoning and/or manganese pneumonia
Ni	Steel and nickel-alloy industry; nickel-plating; asbestos; coal, fuel oil, and incineration	Dermatitis, respiratory disorders, and carcinogenesis
Pb	Automobile exhaust, coal, incineration; pigments	Lead poisoning
V	Coal and fuel oil	Cardiovascular disease and carcinogenesis
Zn	Smelting, refining, zinc galvanizing	Dermatitis, hypertension, and arteriosclerotic heart disease

[a]Reprinted with permission from (2), Hwang, *Anal. Chem.* **44**, 20A. Copyright 1972, American Chemical Society.

stroy organic matter very effectively. In some instances it is permissible to use HNO_3–HCl mixtures. In this case I have found that extraction efficiency is between 70 and 100% for the trace metals, depending on the element and the source of the particulate.

9.2 PROCEDURES

9.2.1 Determination of Metals in Particulate by ICPAES*

Thirteen metals can be determined by this procedure. These are listed in Table 9.5, which also shows the spectral lines employed. The sample is dissolved in a

*Procedure from Brezezinska and Van Loon (3).

Table 9.3 Impurity Elements in Filter Materials ($\mu g/cm^2$)[a]

Element	Glass Fiber	Organic Membrane
As	0.08	—
Be	0.04	0.0003
Bi	—	<0.001
Cd	—	0.005
Co	—	0.00002
Cr	0.08	0.002
Cu	0.02	0.006
Fe	4	0.03
Mn	0.4	0.01
Mo	—	0.0001
Ni	<0.08	0.001
Pb	0.8	0.008
Sb	0.03	0.0001
Si	7000	0.1
Sn	0.05	0.001
Ti	0.8	2
V	0.03	0.0001
Zn	160	0.002

[a]Reprinted with permission from (2), Hwang, *Anal. Chem.* **44,** 20A. Copyright 1972, American Chemical Society.

closed Teflon tube that is placed inside a pressure cooker. The filter is first digested in the closed tube in a nitric and perchloric acid mixture. After evaporation to fumes of perchloric acid, hydrofluoric acid is added and the sample digested open to the air. The procedure can be used for both membrane and glass fiber filters.

Equipment. An Applied Research Laboratories Model 34,000 was employed. The wavelengths and operating parameters used are given in Tables 9.5 and 9.6.

A commercially available pressure cooker suitable for home cooking was employed. Teflon bombs were made at the University of Toronto from Teflon availabale from Canplas Industries, Toronto, Canada.

Figure 9.1a and b are photographs of the Teflon digestion vessel used. These are cleaned using a concentrated HNO_3 leach, followed by $KMnO_4$ treatment, which is followed by HCl cleaning to remove residual MnO_2.

Reagents. Reagent grade (Baker Analyzed) chemicals were found to be pure enough for the trace elements covered in the proposed procedures. Stock 1000

Table 9.4 Effects of Ashing Methods on Metals Recovery[a]

Metal	Recovery (%)	
	Low-Temp. Ashing	Muffle Furnace (550°C)
Pb	101	46
Sn	95	87
Cu	98	92
Cd	92	53
Sb	99	46
Ba	97	99
Mn	99	107
Ni	97	99
Mo	98	116
Zn	96	39
Ti	95	92
Co	96	97
Cr	112	100

[a]Reprinted with permission from (2), Hwang, *Anal. Chem.* **44,** 20A. Copyright 1972, American Chemical Society.

Table 9.5 Spectral Lines Used

Element	Line (nm)	Order
Al	308.22	2
Ag	328.07	2
Cd	226.50	3
Co	228.62	3
Cr	267.72	2
Cu	324.75	2
Fe	259.94	2
Mn	257.61	3
Mo	202.03	3
Ni	231.60	3
Pb	220.35	3
Sb	217.59	3
Zn	213.86	3

Table 9.6 ICPES Operating Parameters

Plasma and Nebulizer/Premix Chamber

Argon flow rates (L/min)

Coolant gas, 12

Plasma gas, 0.8

Carrier gas, 1.0

Nebulizer

Concentric pneumatic with automatic tip wash

Sample uptake rate, 2.5 ml/min

Viewing height of plasma above coil is 15 mm

Radio-Frequency Generator

Frequency, 27.12 MHz

Operating power, 1.2 kW

Reflected power, <3W

Instrument

An ARL Model ICPQ 34,000

The grating spectrometer has 32 fixed channels and can be evacuated.

A wavelength scanning accessory consisting of a moving primary source was used.

A variable channel comprising a 0.25 m Spex Monochromator was also available.

Optics

Laser-ruled tripartite concave Al coated on SiO_2 blank, 1080 lines/mm

Metal–dielectric–metal narrow-band-pass filters for order sorting

Blaze angle, 600 nm (1st order)

Resolving power, 43,200 (1st order)

Reciprocal liner dispersion 0.926 nm/mm (1st order)

Four orders are used.

Primary slit, 20 μm

Secondary slits, 35 or 50 μm

ppm standard solutions were prepared from metals or metal salts. Perchloric–nitric acid (1 : 1 mixture) was added to stock and working solutions to give a 6% final acid content.

Procedure. Place the filter containing 1–20 mg of particulate matter into a Teflon vessel, add 2 mL concentrated HNO_3 and 0.5 mL concentrated $HClO_4$. Keep in the closed vessel at room temperature for about 1 h. Heat in the oven at 100°C for 1 h or in the pressure cooker for 0.5 h. Evaporate the sample to white fumes on a hot plate at 250°C. After cooling add 1 mL of concentrated HCl for Nuclepore or Milipore filters and 5 mL for glass fiber filters. Evaporate the sample to near dryness. Dilute the residue with 10 mL of 6% v/vk 1 : 1

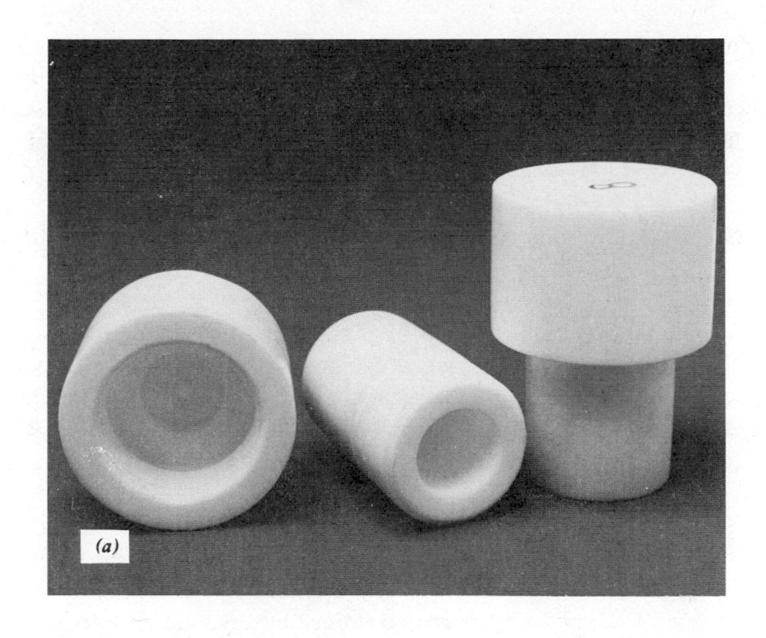

Figure 9.1 (a) Teflon decomposition vessels. (b) Household pressure cooker with teflon vessels.

mixture of HNO_3 and $HClO_4$ and heat on a hot plate for 5 min. If necessary, filter the sample through a Whatman 540 filter. Transfer the sample to a 10 mL volumetric flask and adjust the volume with the same acid mixture. Analyze 3 blank filters together with the samples using the instrument parameters given above.

9.2.2 Determination of As, Cd, Cu, Pb, and Zn in Air Particulate by Flame AAS*

This procedure is applicable to glass fiber air filters. Half of the filter is cut into small pieces. These are first attacked using a mixture of sodium chlorate and nitric acid. Subsequently hydrochloric acid and 0.4% EDTA solution are added to the decomposition mixture. The procedure was evaluated using NBS SRM 1648 Urban Air Particulate Matter and the results agreed well with the accepted values.

Equipment. A Varian-Techtron Model AA6 atomic absorption spectrometer with a hydrogen arc background corrector was used for all measurements. Cadmium, copper, lead, and zinc were determined by using an air–acetylene flame and the optimum parameters specified by the manufacturer. Varian hollow-cathode lamps were used and the monochromator was tuned to the most sensitive resonance wavelength for each lamp. A modified Varian Model 65 vapor generation accessory was used for arsenic determination in a nitrogen–hydrogen–entrained air flame. A Westinghouse electrodeless discharge lamp, run at 7 W, was the source for arsenic determination.

Signal readings for cadmium, copper, lead, and zinc were taken in the absorbance mode and concentrations were determined from six-point standard calibration graphs. Arsenic was determined in the absorbance peak height mode.

Reagents. Stock standard solutions (100 mg/L) were prepared from high-purity metals; the arsenic stock was made up from analytical reagent grade arasenic(III) oxide. Calibration standards were prepared each day by diluting the stock with a solution that was 20% v/v in HCl and 0.4% w/v in disodium EDTA. Distilled deionized water was used for the preparation of all soloutions. Analytical reagent grade acids and solids were used for all dissolution processes.

Procedure. Cut half of a glass fiber filter (original dimensions 20 × 25 cm) into small pieces and place in a 150 mL squat beaker. Add 0.5 g solid potassium chlorate followed by 100 mL of 20% v/v HNO_3. Cover the beaker with a watchglass and simmer the contents for 15 min on a hot plate. Then move the beaker

*Procedure from Byrne (4).

to the edge of the hot plate after it has cooled somewhat and evaporate the liquid to dryness overnight. Cool the beaker and contents and add 20.0 mL concentrated HCl. Warm the beaker to about 80°C and add 80.0 mL of the 0.4% EDTA solution. Allow the mixture to stand for 30 min and then stir well with a glass rod. Elemental determinations are made as described above. Acid, reagent, and filter paper blanks must be carried through the complete procedure.

9.2.3 Determination of Cd, Cu, Fe, Pb, and Zn in Aerosols by AAS*

This procedure is for the determination of metals of air particulate trapped on No. 41 Whatman filter papers. Presumably the procedure (involving $HClO_4$) would be suitable for other filter materials. No procedural step exists for the decomposition of siliceous material. In my view such a step may be necessary in some instances. It would be easy to insert a hydrofluoric acid (about 5 mL) step and do the decomposition in a Teflon dish.

Zinc is determined by flame atomic absorption. The other four elements are run using a furnace atomizer with atomic absorption. The validity of the procedure was checked using standard samples. These included Kodak TEG-50 (a gelatin spiked sample), NBS Coal Fly Ash, and 10 air filter samples analyzed independently by atomic absorption. Agreement was found to be excellent in all cases. The NBS fly ash was digested with hydrofluoric as well as perchloric and nitric acids.

Equipment. A Perkin-Elmer 503 atomic absorption spectrometer was used in combination with a Perkin-Elmer HGA-74 graphite furnace atomizer. Absorbances were read as peak height form a Hitachi-Perkin-Elmer 56 strip-chart recorder and from the built-in peak-read device on the spectrometer. The furnace was flushed with argon at a flow rate of 300 mL/min internally and 900 mL/min externally. The gas flow could be changed during the atomization stage to increase sensitivity by means of the stopped (Gas-Stop), reduced (Mini-Flow), and continuous flow settings. Solutions were injected with Eppendorf micropipettes with disposable polypropylene tips. Deuterium background correction was done with a Perkin-Elmer 303-0875 system.

The Perkin-Elmer 503 spectrometer was also used for flame analyses with a 10 cm single-slot burner with acetylene as fuel and compressed air as oxidant.

The wavelengths used in nanometers were Cd, 228.8; Cu, 324.8; Fe, 248.3 and 372.0; Pb, 283.3 and Zn, 214.0 and 307.6.

Reagents. Standard stock solutions of the elements investigated were prepared from Merck Titrisol (1000 ppm) solutions. Suprapur grade acids were used for

*Procedure from Geladi and Adams (5).

Table 9.7 Furnace Program Parameters Chosen for Routine Analysis

Parameter	Cd	Cu	Fe	Pb	Zn
Ashing temperature (°C)[a]	400	800	1200	500	600
Atomization temperature (°C)[b]	2300	2600	2500	2500	24000

[a]20 s ashing time.
[b]10 s atomization time.

destruction of the filter samples to minimize blanks. All dilutions were made with doubly distilled water from quartz apparatus. Acidified (~ 0.1 *M*) dilute solutions could be stored for several weeks without losses. Neutral solutions showed considerable losses by adsorption onto the glass vessels. Dilute solutions were therefore remade regularly.

Procedure. Treat about 10 cm^2 of aerosol-loaded filter paper with 7.5 mL of boiling concentrated HNO_3, then add 5 mL $HClO_4$ and heat until the solution is clear and colorless. Boil off the excess acid and redissolve the dry material in 2.5 mL HNO_3 of water by gentle heating. Cool the solution and dilute to 50 mL in a graduated flask. Known amounts of standard and filter paper blanks must be taken through the whole procedure. The volumes of acid added for heavily loaded filters can be adapted to the amount of material to be decomposed. The final volume can be reduced when very low concentrations are expected.

Dry the samples at 100°C and use a drying time (in seconds) that is equivalent to twice the injected volume in microliters. Clean-up should be done at 2700°C until disappearance of memory effects, usually after 10 s for Cd, Pb, and Zn and 15–20 s for Cu and Fe. The recommended furnace parameters are given in Table 9.7.

9.2.4 Determination of Metals in Air Particulate Matter by AAS*

The following metals on glass fiber filters can be determined by the procedure: iron, copper, nickel, cadmium, lead, vanadium, chromium, and manganese. A dry ashing at 500°C is employed to remove the organic matter prior to the wet chemical digestion. I suggest that the temperature not rise above 450°C if possible. The acid mixture consists of a 1 : 1 : 1 combination of hydrofluoric acid, nitric acid, and water. Standards are made to roughly match the matrix of the samples. Blank filters should also be run through the procedure. The procedure was not tested on standard reference materials because of the lack of such materials at the time of the procedural development.

*Procedure from Vijan et al. (6).

Equipment. A Perkin-Elmer Model 403 atomic absorption with air-acetylene and nitrous oxide-nacetylele capability was used. An aluminum block 26 × 6 × 11 cm was drilled with 40 holes to accommodate the 18 × 150 mm glass test tubes to a depth of 3 cm. Glass test tubes were marked at the 15 mL level. Teflon dishes of 100 mL capacity were used. Glass fiber filters 203 × 253 mm were employed (Mines Safety Appliance Co. Ltd.).

Reagent. A stock matrix solution is prepared by dissolving 126.7 g $Al(NO_3)_3 \cdot 9H_2O$, 88.6 g $(Ca(NO_3)_2 \cdot 4H_2O$, 4.75 g MgO, and 41.0 g $NaNO_3$ in 1000 mL of 3% HNO_3 solution. The working standard solutions of copper, nickel, cadmium, chromium, vanadium, and lead are prepared in 1 : 10 dilution and iron and zinc in 1 : 100 dilution of the stock matrix solution.

A composite 1000 ppm stock solution of copper, nickel, cadmium, manganese, vanadium, and chromium is prepared by dissolving 1.000 g of each metal in a minimum quantity of concentrated HCl. The solutions are transferred to a 1 L volumetric flask and made to volume with demineralized, distilled water.

A composite 1000 ppm iron and zinc stock solution is similarly prepared by dissolving the metals in hydrochloric acid.

A 1000 ppm stock solution of lead is prepared separately by dissolving the high-purity metal in a minimum quantity of nitric acid and making the volume to 1 L with demineralized distilled water. Analytical reagent grade soluble salts of metals, preferably nitrates, may also be used in place of high-purity metals.

The recommended concentrations in micrograms per ml of working standards are: for copper, nickel, cadmium, chromium, manganese, vanadium, and lead, 0.0, 0.2, 0.5, 1.0, 2.0, 3.0, 5.0; zinc, 0.0, 0.5, 1.0, 2.0, 3.0, 5.0, 10.0; iron, 0.0, 20.0, 30.0, 40.0, 60.0, 80.0, 100.0; lead, 0.0, 7.0, 10.0, 15.0, 20.0, 40.0, 60.0, 80.0, 100.0.

Procedure. Cut a 19 × 253 mm sample strip and place on an asbestos board. Ash at 500°C in a muffle furnace for 2 h. Transfer the strip to a 100 mL Teflon dish and add 15 mL of 1 : 1 : 1 HF, HNO_3, and water. Take to a bone-dry stage by slow evaporation. Add 5 mL concentrated HNO_3 and take to dryness once again. Wet the contents with 1 mL concentrated HNO_3, add 20 mL of water, simmer on the hot plate until the volume is reduced to about 10 mL, and then transfer quantitatively to the calibrated test tube. Place 40 prepared sample solutions in the aluminum block and heat for 2 h on the hot plate. Bring the volume up to the 15 mL mark at room temperature with demineralized, distilled water. Seal the test tube with parafilm and mix. Make a 1 : 10 dilution for Fe and Zn and analyze using matrix simuliated standard solutions to calibrate the instrument.

Samples are run on the atomic absorption unit using the manufacturer's recommended settings. Vanadium is done using a nitrous oxide–acetylene flame. The others are analyzed with an air–acetylene flame.

9.2.5 Determination of Pb in Air Particulate by Furnace AAS*

A study weas done (8) to investigate lead determinations by using the U.S. EPA lead reference method. Several laboratories participated. Samples of air filters were extracted using nitric acid at elevated temperature, nitric acid at room temperature with ultrasonic agitation, or a combination of hydrochloric and nitric acids. Little difference among the results of all three extraction methods was recorded.

Sample solutions were analyzed by atomic absorption (flame and furnace), anodic stripping voltametry, and optical emission spectrometry. The results from these techniques were similar except in precision. Coefficients of variation were as follows: ICPAES, 2%; furnace AAS, 6%; ASV, 10%; and spark excited AES, 15%.

Equipment. A description of the atomic absorption and experimental conditions used are given in Table 9.8.

Centrifuge tubes: polypropylene tube with screw tops of polypropylene, 50 mL volume (Nalgene 3119-0050).

Pipette: automatic dispensing, with accuracy of setting 0.1 mL or better and repeatability of 20 μL (Grumman Automatic Dispensing Pipette, Model ADP-30DT, or equivalent).

Ultrasonic bath: Heat Systems, Ultrasonic, Inc., Model B92H or equivalent. Cleaning power, 425 w; heating power, 400 w.

Before use, all labware should be scrupulously cleaned. The following procedure is recommended. Wash with laboratory detergent or ultrasonicate for 30 min with laboratory detergent, rinse, soak a minimum of 4 h in 20% v/v HNO_3, rinse three times with distilled, deionized water, and oven dry.

Reagent. Hydrochloric acid: ACS reagent grade, concentrated (11.7 *N*) (Fisher A-144 or equivalent).

Nitric acid: redistilled spectrographic grade (16 *N*). For preparing samples use Calvert Chemical spectrographic or equivalent).

Distilled deionized water: distilled building water supply passed through a deionizer system (Millipore Corporation Milli-Q or equivalent).

Procedure. In a 1 L volumetric flask, combine in order, and mix well the following: 500 mL distilled, deionized water, 64.6 mL concentrated (16 *M*)

*Procedure from Walling and Harper (7).

Table 9.8 Analytical Conditions of Techniques Used for Furnace Atomic

Spectrophotometer	Perkin-Elmer Model 403 atomic absorption spectrometer
Analytical line	283.3 nm
Atomizer	Perkin-Elmer Model 2100 heated atomizer
Dry	100°C for 30 sec
Char	500°C for 20 sec
Atomize	2500°C for 10 sec
Injection	Perkin-Elmer Automatic Sampler 1
Sample matrix	0.225 M HNO_3
Lamp	Lead EDL at 9 W
Slit width	0.7 nm
Purge gas	Argon at 50 units and 20 psi
Atomization tube	Graphite
Cooling water	1.5 L/min
Aliquot	20

redistilled spectrographic grade HNO_3, and 182.0 mL of ACS grade concentrated HCl. Cool and dilute to 1 L with distilled deionized water.

Cut a 1 × 8 in. strip from the folded particulate-bearing filter using a template and a pizza cutter. Using vinyl gloves or plastic forceps, accordion fold or tightly roll the filter strip and place it on its edge in a 50.0 mL screw top polypropylene centrifuge tube. Add 12.0 mL of extracting acid using a preset calibrated automatic dispensing pipette or a regular pipette. (The acid should cover the strip completely.) Cap the tube loosely to prevent pressure build-up during the ultrasonication step. Place tube in a sample rack. (*Note:* The sequence of adding the filter strip and acid to the centrifuge tube may be reversed, if more convenient, without affecting the results.) Identify the sample by affixing a sample identification label to the centrifuge tube with tape.

Put enough water, at a temperature of 100°C, in a clean ultrasonic bath so that the water level is slightly above the acid level of the centrifuge tubes in the rack. This water depth is well below the centrifuge tube caps and therefore will not cause contamination of samples during ultrasonication. Set the loosely capped centrifuge tubes containing the samples and the extracting acid upright in the rack in the ultrasonic bath. It is important for the vials to be upright in the rack during ultrasonication to prevent possible sample contamination. Ultrasonicate the samples 50 min. Remove the centrifuge tubes containing the samples from the ultrasonic bath and blot the tubes dry.

Uncap the centrifuge tube and add 28.0 mL of distilled deionized water to the tube using a calibrated automatic dispensing pipette or a regular pipette. Cap the centrifuge tube tightly and shake well.

Place the tube in the centrifuge and centrifuge 20 min at 2500 rpm. Decant the clear solution from the centrifuge tube into an acid-cleaned 30 mL polypropylene bottle (bearing sample ID label), taking care not to disturb the solids in the bottom of the tube. The sample is now ready for analysis. It has a 0.31 M HNO_3 and a 0.67 M HCl matrix and a volume of 40 mL. The solutions can be run by furnace atomic absorption using the conditions given in Table 9.8.

9.2.6 Determination of Total Hg Vapor in Air by AAS*

This method is for determining total mercury vapor. The mercury associated with particulate matter is filtered out. Air samples are passed through a cryogenic trap at up to 6 L/min and the mercury is trapped on glass spheres. The sample is then vaporized into a gold foil collector. Subsequently the gold foil is heated and the evolved mercury passed into a cold vapor absorption tube. Levels of mercury in air at levels below nanograms per cubic meter can be determined in this way.

Equipment. A Perkin-Elmer 305B atomic absorption spectrophotometer equipped with a Perkin-Elmer mercury hollow-cathode lamp and an Omni Scribe recorder (Houston Instrument) was used with the following conditions: wavelength, 253.7 nm; slit setting, 4; lamp current, 13 mA. An open-ended glass tube, 20 cm long and 1.5 cm i.d. with a 5 mm i.d. inlet tube fused in the middle was used as a flow-through cell. The cell was mounted on the burner head and aligned in the usual manner to let maximum light from the hollow-cathode lamp reach the detector. A Perkin-Elmer deuterium arc background corrector was also used throughout the work.

Reagents. To prepare gold-coated asbestos, about 3 g of asbestos (Bio-Rad Acid Washed Asbestos for column chromatography) was placed in a platinum

*Procedure from Oguma and Van Loon (9).

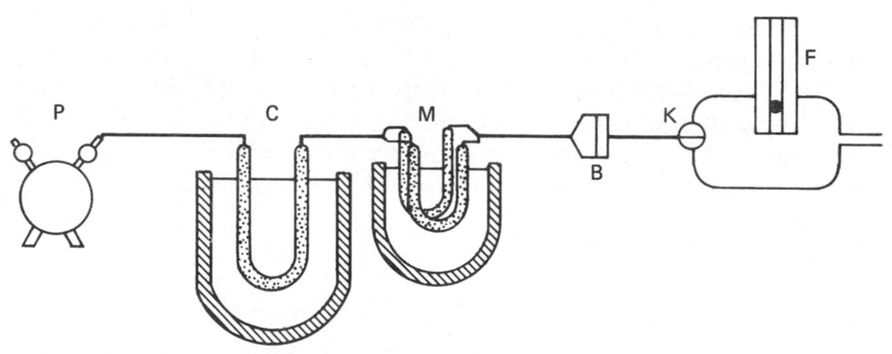

Figure 9.2 Collection apparatus for mercury in air (see text for component identification).

dish and ignited for 50 min. After being cooled to room temperature, the asbestos was coated with gold according to the procedure reported by Kunert et al. (10). All other reagents used were of reagent grade.

Procedure. Figure 9.2 shows the arrangement for the cryogenic trap technique. The air was pumped through the 0.45 μm Millipore filter (B), the moisture traps (M), and the collection tube (C), in this order, at a flow rate of 2 L/min. The flow rate was checked by passing air through the flow meter (F) by turning the three-way stopcork (K) at 10 min intervals. After an appropriate period of sample collection, the collection tube was disconnected from the system and installed in the measurement system (see Fig. 9.3).

The collection tube (C) was warmed up to room temperature and then heated at100–115°C for 5 min by means of a heating tape. During warm-up and heating of the collection tube, nitrogen gas was passed through the collection tube, the pyrolyzer (N), the cooling coil (G_1), and the amalgamation tube H, in this order, at a flow rate of 0.5 L/min. Then stopcocks (K) and (L) were closed and the

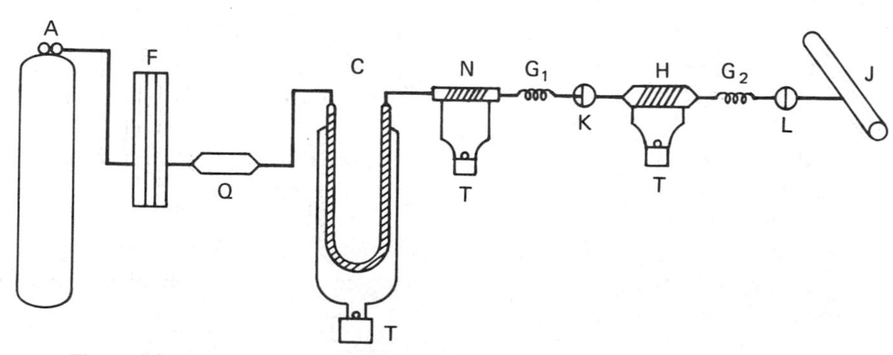

Figure 9.3 Mercury Determination System (see text for component identification).

amalgamation tube was heated to about 500°C for 30 sec by a heating coil. The mercury vapor thus liberated was swept into the flow cell (3) by nitrogen gas (1 L/min). The recorded peak heights were used for the quantificative determination of mercury. Standardization was achieved by injecting 0.25–1.5 mL samples of air saturated with elemental mercury vapor at a known temperature into the analytical system for a septum inserted between the cooling coil (G_1) and stopcock (K).

9.2.7 Determination of Be in Air Particulate by AAS*

The sample is collected on cellulose or glass fiber filters. A mixture of hydrofluoric and nitric acids is used for decomposition. Fuming in sulfuric acid is not done, presumably to avoid loss of beryllium as the insoluble oxide. Glass fiber filters must be prewashed with distilled water. Blank variations between batches of filters is a problem and hence blanks must be run with each sample set. It is possible to use either flame or furnace atomization.

Equipment. Glass filters, commercially available material in size 203 × 254 mm, are exhaustively washed prior to use. Cellulose filters, ashless, acid-washed, analytical grade, 203 × 254 mm in size are used. An atomic absorption spectrometer with meter, recorder or digital readout, and monochromator with wavelength dial reading to 0.1 nm are employed. The 234.8 nm beryllium line is used. Polyethylene bottles with screw cap (30 mL capacity) are used for storage of test samples. A polypropylene Buchner funnel is made with a sintered false bottom and vacuum connection, 216 × 267 × 85 mm in free depth. A Massman design graphite furnace should be used with a strip-chart recorder. Automatic Eppendorf pipettes with capacities 10–100 μL are used.

Cellulose filters may be used without further treatment. Glass fiber filters are purified by placing a group of 100 filters in the special Buchner funnel and extracting repeatedly with distilled water.

Mount the filter in a conventional Hi Vol or other sampler head. Draw air through the filter at a flow rate between 1.13 and 1.60 m^3/min for an appropriate period, such as 24 h. The resistance to flow offered by the cellulose filters is much greater than for glass, and an appreciably smaller total volume of air will be taken with cellulose. Calculate and record the total volume of air sample in cubic meters as the product of mean flow rate and time.

Reagent. Water should be distilled at least twice from glass or quartz. Hydrofluoric acid, 49%, reagent grade, in polyethylene containers is employed. Nitric acid is 71% reagent grade.

*Procedure from Zdrojewski et al. (11).

For the standard solution of beryllium, dissolve 22.757 g of beryllium nitrate tetrahydrate in water and make up to 1 L. In 1 mL of this stock solution there is 1000 μg of beryllium.

Procedure. Cut pieces from the exposed surface of a filter using a circular metal punch. The cutting edge of the punch is carefully wiped with lens tissue between each use to prevent carryover of contamination from one sample to another. Place one or more such disks in a Teflon beaker. Initiate the dropwise addition of 1 mL HF. Gently warm the contents of the beaker at low heat until the HF is almost completely evaporated. At this point, add 1–2 mL of HNO_3, and continue to heat gently until a few drops of HNO_3 are left. Add about 10 mL of water, bring nearly to boiling, and filter through a Whatman No. 41 filter paper into a glass beaker. Transfer to a 25 mL volumetric flask. Rinse down the Teflon beaker with another 10 mL of water, warm, and filter into the same beaker. Transfer to a 25 mL volumetric flask and make up test sample to the mark. Mix the contents of the volumetric flask thoroughly after adjustment to volume. Transfer contents of volumetric flask to a polyethylene storage bottle. The test sample is now raedy for analysis. This procedure applies to the digestion of glass fiber filters. Cellulose filters may be digested or extracted with HNO_3.

Flame (Nitrous Oxide–Acetylene) Analysis. Aspirate calibrating solutions over the concentration range 0–4 μg/mL with a blank solution between each standard. Similarly, aspirate sample solutions. Each solution should be run for 30 sec.

Furnace Analysis. With an automatic pipette with polyethylene tip, place identical microliter volumes of the dilute standard beryllium solutions and a distilled water blank in the furnace. Measure and record the response of each test portion following the predetermined measuring cycle. Cover the range 0.2–10 ng of beryllium in steps of 0.2 ng (0.01–0.05 μg/mL for a 20 μL sample). The recommended fuirnace parameters are drying, 20 sec, 100°C; charring, 20 sec, 1100°C ; atomizing, 20 sec, 2400°C. Samples are run in the same manner.

Blanks of both reagents and reagents plus filters are run.

REFERENCES

1. Threshold Limits Values for Chemical Substances in Working Room Air, American Conference of Industrial Hygenists in 1977.
2. J. Y. Hwang, *Anal. Chem.* **44,** 20A (1972).
3. A. Brezezinska and J. C. Van Loon, internal laboratory procedure, 1983.

4. R. E. Byrne, *Anal. Chim. Acta* **153,** 313 (1983).
5. P. Geladi and F. Adams, *Anal. Chim. Acta* **96,** 229 (1978).
6. P. N. Vijan, J. A. Pimenta, and A. C. Rayner, *Air Quality Laboratory Procedure*, Ontario Ministry of the Environment, Toronto, Ontario, 1978.
7. J. F. Walling and S. Harper, personal communication.
8. S. J. Long, J. C. Suggs, and J. F. Walling, *J. Air Pollution Contr. Assoc.* **29,** 28 (1979).
9. K. Oguma and J. C. Van Loon, internal laboratory procedure, 1980.
10. I. Kunert, J. Komarek, and L. Sommer, *Anal. Chim. Acta* **106,** 285 (1979).
11. A. Zdrojewski, L. Bubois, and N. Quickert, *Sci. Total Environ.* **6,** 165 (1976).

CHAPTER

10

SOILS AND RELATED SAMPLES

10.1 INTRODUCTION

Soils differ from most other samples covered in this book because of their relatively high silicate contents. The main components of soils are particles of primary rock minerals, for example, silica and silicates, carbonates, sulfides and oxides, secondary minerals such as clays that are produced by weathering, and organic matter both living and dead. The grain size of soils varies greatly, and within a single soil sample the particles may range from millimeter size to <300 mesh. In addition, soils are commonly inhomogeneous with depth. For example, in temperate wooded areas the top layer of a soil is typically composed of relatively large fragments of leaves and needles. Underlying this material would be a layer of decomposing organic substances, followed by layers that become progressively poorer in organic matter and richer in silica and/or carbonaceus material and secondary minerals.

For most environmental and biological purposes it is not desirable to determine the trace metals that are bound very firmly in the primary and secondary mineral fractions. That is, it is the trace metals that can be relatively easily mobilized that are of interest. This is because it is the more loosely bound fraction of the trace metals that can most readily interact with the biosphere.

The trace metals are associated in differing amounts with the various size fractions in soils. In general the smaller the grain size the higher the trace metal content. Thus sample preparation usually includes a sieving step in which some segregation on a grain size basis is made. (*Caution*: Sieving can result in severe contamination problems for some metals.) Various workers have recommended separation of differing size fractions for trace metal soil analysis. Generally there is agreement that anything above 80 mesh should be rejected.

In addition to soils, several other kinds of samples are closely related and are of interest in biological and environmental trace metal analysis. The most important of these are sediments found in oceans, lakes, and streams. The procedures given below for the trace metal analysis of soils are generally applicable to these samples.

This chapter also includes procedures for the trace metal analysis of fertilizers because fertilizers, particularly those derived from sewage sludge, are now commonly used in landscaping and agricultural applications. Sewage sludges, because they are relatively high in metals (i.e., up to a few percent for some urban sewage sludges on a dry weight basis), are important sources of metals in the soils where they occur.

10.1.1 Decomposition of Soils

A decision must first be made as to which fractions of the soil are to be decomposed or extracted because this will determine what decomposition strategy is chosen.

If total decomposition is desired, then fusions or acid attack using mixtures containing hydrofluoric acid can be used. Fusions are usually avoided because they result in solutions of very high salt content. These salts can cause serious interference in the determinative step.

If a fusion is to be used, lithium metaborate or alkali metal hydroxides mixed with sodium peroxide can be employed effectively. The presence of sodium peroxide assures oxidizing conditions and hence minimizes loss of volatile hydride-forming metals.

Hydrofluoric acid ensures decomposition of the silicate fraction of a soil and is important in total decomposition. However, to attack sulfides and organic matter it is necessary to employ oxidizing acids as well. For this purpose perchloric acid–nitric acid or nitric acid alone can be added with the hydrofluoric acid.

When it is desirable to dissolve precipitated, adsorbed, organically bound carbonate and sulfide-bound or loosely silicate-bound metals, acid mixtures containing oxidizing acids are used. Thus perchloric acid–nitric acid combinations are perhaps the most commonly used mixtures for attacking soil samples for biological and environmental purposes. Also often used but generally not as effective for this purpose is aqua regia—a 3 : 1 mixture of hydrochloric and nitric acid.

It is apparent that in most cases strong acid attack of soils, such as that listed above, releases much larger quantitites of metals than would be available to plants and other organisms present in the soils. Thus strong acid attack generally yields little information on the bioavailability of trace metals. At this time of writing a good deal of research is being done to find extraction mixtures that will yield good data on the bioavailability of trace metals. As yet there is no widespread agreement in this field. Some of the reagents that are being tested for this purpose include weak mineral acid mixtures, acetic acid mixtures, and solutions of chelating agents such as ethylenediaminetetraacetic acid.

10.1.2 Determination

Inductively coupled plasma emission spectrometry is finding wider and wider application in soil analysis. However, when the best detection limits are required, furnace atomic absorption spectrometry must still be employed.

Sample solutions prepared from soils are often high in elements such as silicon, iron, aluminum, calcium, magnesium, sodium, and potassium and thus it may be necessary to correct for background and spectral interference problems in inductively coupled plasma emission spectrometry. Likewise in furnace atomic absorption spectrometry, high levels of these and other elements may result in interference problems, which should be expected unless it is proven otherwise. To avoid interferences from these and other elements it may be possible to use solvent extraction or ion exchange procedures.

Table 10.1 lists the concentrations of selected trace elements in soils (1). These values will be helpful in deciding what levels to expect in analyses. The

Table 10.1 Levels of Selected Trace Elements in Soils[a]

Element	Mean	Range
	(ppm dry soil)	
Ag	0.1	0.01–5
As	6	0.1–40
B	10	2–100
Be	6	0.1–40
Cd	0.06	0.01–0.7
Co	8	1–40
Cr	100	5–3000
Cu	20	2–100
Fe	38000	7000–550000
Hg	0.03	0.01–0.3
Li	30	7–200
Mn	850	100–4000
Mo	2	0.2–5
Ni	40	10–1000
Pb	10	2–200
Se	0.2	0.01–2
Sn	10	2–200
V	100	20–500

[a]From Bowen (1).

values can also be employed to indicate whether elevated levels are being obtained. Care is necessary in interpreting the latter because slight increase, even up to three or four fold, does not always mean the value is elevated.

10.2 PROCEDURES

10.2.1 Determination of Trace Elements in Soils, Sediments, and Particulate by ICPAES*

The following elements can be determined by the procedure: calcium, aluminum, nickel, manganese, zinc, lead, copper, cobalt, chromium, and phosphorus. A total digestion with hydrofluoric acid is used in this procedure, but if only a strong mineral acid attack is desired, the hydrofluoric acid can be omitted. Otherwise, the procedure should be followed exactly as presented. Digestion is first done in a closed Teflon tube placed inside a pressure cooker. Subsequently the top of the Teflon tube can be removed and the acids evaporated as required. The procedure was tested with NBS SRM River Sediment and the results obtained were in good agreement with the accepted values. Background and spectral interference corrections were not found necessary for most of the elements for the materials analyzed, but when samples of soil or sediment contain metals at lower levels than in the NBS River Sediment the need for corrections should be reevaluated. Cadmium and lead values were corrected for the presence of aluminum, iron, calcium, and manganese.

Equipment. An Applied Research Laboratories Model 34,000 was employed. Wavelengths and operating parameters used are given in Tables 10.2 and 10.3. A commercially available pressure cooker suitable for home cooking was employed. Teflon bombs were made at the University of Toronto from Teflon available from Canplas Industries, Toronto, Canada.

Figure 9.1 is a photograph of the Teflon digestion vessels used. These are cleaned using a concentrated HNO_3 leach, followed by $KMnO_4$ treatment, which in turn is followed by HCl cleaning to remove residual MnO_2.

Reagents. Reagent grade (Baker Analyzed) chemicals were found to be pure enough for the trace elements covered in the proposed procedures. Stock 1000 ppm standard solutions were prepared from metals or metal salts. Perchoric-nitric acids (1 : 1 mixture) was added to stock and working solutions to give a 6% final acid content.

*Procedure from Brzezinska et al. (2).

Table 10.2 Spectral Lines Used

Elements	Lines (nm)	Order
Al	308.22	2
Ag	328.07	2
As	189.04	3
Ca	317.93	2
Cd	226.50	3
Co	228.62	3
Cr	267.72	2
Cu	324.75	2
Fe	259.94	2
Mg	279.08	2
Mn	257.61	3
Mo	202.03	3
Ni	231.60	3
P	178.28	4
Pb	220.35	3
Sb	217.59	3
Se	196.03	3
Zn	213.86	3

Procedures.

Soils and Sediments.

(a) Weigh 0.4–0.6 g of well-powdered sediment into a Teflon vessel. Add 3 mL concentrated HNO_3 and 1 mL concentrated $HClO_4$. Close the vessel and keep at room temperature for about 2 h. Heat in the oven at 100°C for 1 h or in a pressure cooker about 0.5 h. Evaporate the sample to fumes on a hot plate. Cool and add 10 mL of HF. Close the tube and heat in the oven or in the pressure cooker as above. After cooling, evaporate the solution to near dryness. Dilute the residue with 10 mL of 6% 1 : 1 mixture of HNO_3 and $HClO_4$ and heat on hot plate for about 10 min. Filter the sample through a Whatman 540 filter into a 25 mL volumetric flask and adjust the volume with the same acid mixture. When the silica content is higher (e.g., sands), the volume of HF should be increased. Run three blanks with the samples by the same procedure.

(b) Weigh 0.1–0.2 g of sediment into a Teflon vessel. Add 2 mL concentrated HNO_3, 0.5 mL concentrated $HClO_4$, and 4 mL HF. Keep the samples in closed vessels at room temperature for about 1 h. Heat the sample in the oven at 140°C for 2 h or in the pressure cooker for 1 h. Evaporate the sample to 2

Table 10.3 ICPES Operating Parameters

Plasma and nebulizer–premix chamber
Argon flow rates (L/min)
Coolant gas, 12
Plasma gas, 0.8
Carrier gas, 1.0
Nebulizer
Concentric pneumatic with automatic tip wash
Sample uptake rate, 2.5 mL/min
Viewing height of plasma above coil is 15 mm
Radio-frequency generator
Frequency, 27.12 MHz
Operating power, 1.2 kW
Reflected power, < 3W
Instrument
An ARL Model ICPQ 34,000.
The grating spectrometer has 32 fixed channels and can be evacuated.
A wavelength scanning accessory consisting of a moving primary source was used.
A variable channel comprising an 0.25 m Spex Monochromator was also available.
Optics
Laser-ruled tripartite concave Al coated on SiO_2 blank
1080 lines/mm
Metal–dielectric–metal narrow-band-pass filters for order sorting
Blaze angle 600 nm (1st order)
Resolving power 43,000 (1st order)
Reciprocal linear dispersion 0.926 nm/mm (1st order)
Four orders are used
Primary slit, 20 μm
Secondary slits, 35 or 50 μm

mL on a hot plate at 250°C. After cooling add 3 mL of concentrated HNO_3. Place 1 g of boric acid in a 100 mL volumetric flask and then add about 50 mL of distilled water. Transfer the sample to the flask and shake the flask to dissolve the boric acid (1–2 h). Adjust to volume with distilled water. Run three blanks with the samples by the same procedure.

Method (a) is much more time consuming than method (b) and is subject to contamination problems. Method (b) should be used when applicable. It is only when very low levels are encountered (e.g., nickel in these samples) that the greater dilution that is required in method (b) invalidates its use.

Particulate on Filters. Place the filter containing 1–20 mg of suspended matter in a Teflon vessel, add 2 mL concentrated HNO_3 and 0.5 mL concentrated $HClO_4$. Keep in a closed vessel at room temperature for about 1 h. Heat in the oven at 100°C for 1 h or in the pressure cooker for 0.5 h. Evaporate the sample to white fumes on a hot plate at 250°C. After cooling, add 1 mL concentrated HF for Nuclepore or Milipore filters and 5 mL for glass fiber filters. Evaporate the sample to near dryness. Dilute the residue with 10 mL of 6% v/v 1:1 mixture of NHO_3 and $HClO_4$ and heat on a hot plate for 5 min. If necessary, filter the sample through a Whatman a 540 filter. Transfer the sample to a 10 mL volumetric flask and adjust the volume with the same acid mixture. Analyze three blank filters together with the samples.

10.2.2 Determination of Mo, Co, and B in Soil Extracts by ICPAES*

Molybdenum, cobalt, and boron are important trace elements in soils. Generally atomic absorption sensitivities are poor for boron and molybdenum and the inductively coupled plasma possesses characteristics (particularly high temperature) that makes it especially suited for determining these elements.

Sample sizes up to 50 g can be handled by the procedure, so that the low levels of boron, molybdenum, and cobalt usually found in soils can be easily determined. The ICP detection limits for the wavelengths chosen are 0.01, 0.05, and 0.05 ppm for molybdenum, cobalt, and boron, respectively.

Interference problems were encountered. The most sensitive molybdenum line (379.8 nm) that was used in this work is interfered with to some extent by the iron line at 379.8 nm. Aluminum has an effect on the molybdenum intensity (enhancement). The presence of 1000 ppm of aluminum was found to be a good interference suppressant (e.g., 2000 ppm calcium had no effect on the signal of 10 ppm molybdenum).

Only a partial extraction of the elements is done. Amonium acetate, acetic acid, and boiling water were used for molybdenum, cobalt, and boron, respectively.

Equipment. The ICP system used in this work utilized a Radyne 2.5 kW free-running valve oscillator operating at 36 MHz and a Techtron grating scanning monochromator. The plasma torch was a demountable Greenfield type, and a dual-tube aerosol chamber similar to the one used by Scott et al. (4) combined with a pneumatic nebulizer was employed for sample introduction. Details of the instrumental system are included in Table 10.4.

*Reprinted with permission from (3), Manzoori, *Talanta* **27**, 682. Copyright 1980, Pergamon Press, Ltd.

Table 10.4 Instrumentation

Plasma power supply	SC15; operating frequency 36 MHz; power output 0–2.5 kW, continuously variable; load coil $2\frac{1}{2}$ turns, 6.3 o.d., d. of coil 32 mm
Plasma torch	Demountable silica torch with brass base; coolant tube, 25 mm bore, 1.5 mm wall; plasma tube, 22 mm bore, 1 mm wall; injector tube, 11 mm o.d., 9 mm i.d., and 2 mm orifice diameter
Nebulizer	Pneumatic nebulizer and spray chamber uptake rate 1.5 mL/min at argon flow rate of 2 L/min, efficiency 7%
Spectrometer	Techtron AA4 grating monochromator
Slit width	25 μm
Readout	Signal from PMT after amplification was displayed on a Servoscribe chart-recorder, model RE 511-20

Reagents. All chemicals were analytical reagent grade. The Mo and Co stock solutions (1000 ppm) were prepared by dissolving molybdate and cobalt chloride in distilled water. Boric acid was used to prepare boron stock solutions (1000 ppm).

Procedure. The standard extraction methods employed at the Macaulay Institute were used to extract Mo, Co, and B from soil samples.

Molybdenum Extraction. Shake a 50 g sample of soil overnight with 800 mL of neutral 1 *M* ammonium acetate. Filter the whole extract through a 18.5 cm Whatman No. 540 filter paper, transferring as much of the soil as possible to the paper to minimize the passage of clay particles. Wash the paper several times with distilled water, evaporate the filtrate to dryness, partially on a hot plate and then to completion on a steam bath. Transfer the residue to a 100 mL standard flask and make up to the mark. No oxidation with HNO_3 is applied, because organic matter at the concentration levels generally occurring in the soil extracts did not cause any disturbance in the plasma. However, some extracted solutions contained solid particles that made the nebulization process unstable. In these cases a single-step oxidation with HNO_3 is used in order to obtain clear solutions (1 mL of concentrated HNO_3 is added to 10 mL of solution and the mixture boiled to low volume and then diluted to 10 mL again).

Cobalt Extraction. Shake a 20 g sample of air-dried soil overnight on an end-over-end shaker with 800 mL of 0.5% acetic acid solution. Filter the whole

Table 10.5 Plasma Operating Conditions

Net forward radio-frequency power	1200 W
Spectrometer slit	25 μm (entrance)
Argon coolant-gas flow rate	14 L/min
Argon plasma-gas flowrate	5.0 L/min
Injector gas flow rate	1.6 L/min
Height of observation	25 mm above work coil

extract in the same manner as the ammonium acetate extracts and evaporate to dryness. Take up the residue in 100 mL 0.1 *M* HCl.

Boron Extraction. Boil a 50 g sample of air-dried soil under reflux with 100 mL of water for 10 min. After partial cooling, filter the extract through an 18.5-mm Whatman No. 3 filter paper into a conical beaker and transfer a 40 mL aliquot of the filtrate to a silica beaker. Take the solution to dryness and oxidize the residue twice with 10 mL of 6% H_2O_2 solution. Dilute the residue to 50 mL. Run the samples on the ICP using the conditions outlined in Table 10.5.

10.2.3 Determination of Cd, Co, Cu, Ni, and Pb in Soil Extracts by Solvent Extraction and Furnace AAS*

Procedures for the selective extraction of trace metals from a particular fraction of a soil are becoming more and more important to agricultural scientists. I treat such procedures with some degree of skepticism because past experience suggest that it is not possible in most cases to restrict the extraction to a particular soil fraction. However, in spite of these problems agricultural scientists find these procedures helpful.

The following procedure gives a determination of trace metals obtained after using the selective extractants acetic acid, oxalic acid, and pyrophosphate. Commonly the levels of cadmium, cobalt, copper, nickel, and lead are so low in these extracts that furnace atomic absorption is necessary. However, direct determination of these elements by furnace atomic absorption of metals in common selective extractants such as acetic acid, pyrophosphate, and oxalic acid is not possible because of interference problems. Thus a dithizone extraction into chloroform is used to recover the desired metals. Determinations of the metals are then done using the chloroform phase. Recoveries of 92–108% of the metals added to the selective extractants were obtained.

Equipment. All analyses were done with a Perkin-Elmer 306 spectrophotometer and a HGA-74 graphite furnace.

*Procedure from Iu et al. (5).

To optimize the reproducibility, the sample was injected into the furnace by holding the syringe in the vertical position. The standard error of the mean based on six replicate injections of the sample adopting this method was less than 1.5%. An autosampling device would be preferable.

Reagents. Where possible, analytical grade reagents were used; deionized water was used for all solutions unless otherwise specified.

DITHIZONE IN CHLOROFORM (0.1 AND 0.05% W/V). Appropriate amounts of dithizone were dissolved in 500 mL of chloroform. This reagent should be stored in a refrigerator in a dark bottle and made up fresh after 3–4 days.

AMMONIUM CITRATE. Citric acid (50 g) was dissolved in not more than 30 mL of water, and 60 mL of ammonia liquor was added. The solution was left until cool and all crystals of citric acid had dissolved; it was then made up to 100 mL with water.

HYDROXYLAMMONIUM CHLORIDE. Hydroxylammonium chloride (20 g) was dissolved in 30 mL of ammonia liquor; it was then made up to 100 mL with water.

Procedure. Pipette a 10 mL sample into a 100 mL separating funnel, followed by 5 mL of ammonium citrate solution and 2 mL of hydroxylammonium chloride solution. In the case of oxalic acid extracts, add 10 mL of water to prevent the formation of a precipitate. Mix the contents of the separating funnel well and adjust the pH to between 9 and 10 with (1 + 4) HCl or (1 + 1) ammonia solution. Add dithizone in chloroform (5 mL of 0.1%). Shake the funnel and release the pressure before shaking on a reciprocating shaker for 5 min. When the two phases separate, collect the organic phase in a 10 mL volumetric flask. Adjust the pH of the aqueous phase to between 8 and 9 and carry out a second extraction with 5 mL of 0.05% w/v dithizone in chloroform. Repeat the shaking procedure as before and add the organic phase to the 10 mL volumetric flask. If necessary, fill the flask to the mark with chloroform. This organic phase is suitable for injecting directly into the graphite furnace. Use the manufacturer's recommended instrument parameters.

10.2.4 Determination of Ag, Bi, Cd, Cu, Pb, and Zn in Soils and Stream Sediments by Solvent Extraction and Flame AAS*

Tricaprylymethylammonium chloride is used to extract the metals into a 15% Aliquat 336–MIBK solvent in the presence of ascorbic acid and potassium iodide

*Reprinted in part with permission from (6), Viets, *Anal. Chem.* **50,** 1097. Copyright 1978, American Chemical Society.

for determination by flame atomic absorption. This approach eliminates interferences that occur from the use of potassium chlorate–hydrochloric acid digestion mixture. The acid mixture is effective in removing silver, besmuth, cadmium, copper, and zinc from all but the silicate-bound phase in soils and stream sediments. If a total digestion is required, hydrofluoric acid modification of the procedure can be used. This is also given. Detection limits in the solvent are 0.02 μg/ml for silver, cadmium, copper, and zinc and 0.2 μg/ml for bismuth and lead. An assessment of accuracy was made by comparing results obtained by this method with other published results for the same samples. Acceptable agreement was obtained.

Equipment. All atomic absorption measurements were made with a Perkin-Elmer 103 atomic absorption spectrometer and many measurements were verified with a background-corrected Perkin-Elmer 360 instrument. A single-slot 10 cm burner was used with an air–acetylene flame. Operating conditions were those recommended by the instrument manufacturer. The burner positions, flame parameters, and aspiration rates were optimized for maximum absorbance and linear response while aspirating known standards. Conventional hollow-cathode lamps were used for all metals determined.

Reagents. All chemicals were reagent grade. The aqueous reagent solutions were prepared with distilled, deionized water. Aliquat 336 was obtained from General Mills, Chemical Division, Minneapolis, MN.

Individual 1000 μg/ml standards of Ag, Bi, Cd, Cu, Pb, and Zn were prepared in 10 *N* HCl using spectrographically pure metals or metal oxide obtained from Johnson Matthey Chemicals. A combined standard containing 1000 μg/ml containing each of the six metals was also prepared in 10 *N* HCl. From the 1000 μg/ml standards, 100, 10, and 1.0 μg/ml standards were prepared in 10 *N* HCl.

Procedure. Accurately weigh 1.00 g of 80 mesh sample into a 16 $\times$ 150 mm disposable test tube containing a clean boiling chip. Add 1 g of $KClO_3$ using a plastic scoop and mix thoroughly. Carefully add four 1 mL portions of concentrated HCl, allowing ample time for the reaction to subside between additions. Mix the contents of the tube after the first and last additions of acid.

Certain samples containing large amounts of carbonate or organic materials may require a powder funnel to contain the foaming solution. Rinse the residue in the funnel into the tube with subsequent reagent additions. After 30 min, place the tube in a heating block and bring to a moderate boil for 2 min or until all yellow fumes have evolved.

Add 8 mL of a 20% w/v ascorbic acid, 10% w/v potassium iodide solution to the cooled tube and mix completely. Accurately deliver 5.0 mL of 10% Aliquat 336–MIBK solution to the tube, cap it with a polyethylene stopper, and shake for 5 min. Allow the organic phase to separate, centrifuge if necessary, and aspirate the organic phase to determine metal contents by atomic absorption spectrometry.

If a total digestion is required, a HF digestion may be employed as follows. A 1.00 g sample is weighed into a 100 mL Teflon beaker and is taken to dryness with 15 mL HF over low heat. The residue is again taken to dryness after the addition of 10 mL of a 5% w/v $AlCl_3$ solution in 6 *N* HCl. This residue is treated in the Teflon beaker with $KClO_3$ and HCl as above and washed into a 16 × 150 mm test tube with subsequent reagent additions and extracted.

The organic standards are prepared by adding the appropriate amounts of the 10 *N* HCl combined standards to 25 × 200 mm screw-capped test tubes. The standards are treated as normal samples except that all reagents are increased fourfold. A 10% Aliquat 336–MIBK solution shaken over 4 *N* HCl may be used as a blank solution, but reagent blanks should be utilized to monitor possible contamination.

10.2.5 Determination of Cd and Pb in Organic and Silica-Rich Sediments by Flame AAS*

Two procedures for dissolving organic and silica-rich sediments are given. Both require a mixture of nitric, perchloric, and hydrofluoric acids. One procedure involves three steps in an open Teflon beaker and the other is performed in a closed Teflon bomb. A higher concentration of perchloric acid is employed compared to with competitive approaches allowing a temperature of 150°C to be reached in the bomb. Furthermore, with the bomb approach boric acid is employed to complex excess fluoride after the decomposition.

Both decomposition approaches were tested using NBS SRM 1645 River Sediment and flame atomic absorption. The results are in good agreement with the accepted values.

Equipment. A Perkin-Elmer Model 303 atomic absorption spectrometer was used with an air–acetylene burner and a Perkin-Elmer Model 54 strip-chart recorded. Electrodeless discharge lamps (Perkin-Elmer EDL) were used. Instrument settings were those recommended by the manufacturer. For bomb digestions, a Parr Model 4745 acid digestion bomb was used with a 25 mL Teflon cup.

Reagents. Deionized, glass-distilled water was used in the preparation of all solutions. The acids were Fisher ACS reagent grade. Standard metal stock solutions were prepared from certified atomic absorption solutions (Alfa Products). Standard working solutions were prepared for the bomb-digested samples

*Procedure from Hsu and Locke (7).

by dilution with 1% v/v HNO_3. The working standards were prepared to contain 8.0% v/v of concentrated HNO_3, 4.0% v/v of 60% $HClO_4$, 8.0% v/v of 49% HF, and 5.0% w/v boric acid to match the samples. Blanks were prepared similarly. The NBS SRM 1645 River Sediment was used to evaluate the accuracy and precision of the methods; it was dried in a desiccator over magnesium perchlorate before weighing.

Procedure.

Open-Beaker Digestion Method. To 0.2 g or less of sediment, accurately weighed in a 100 mL Teflon beaker with Teflon cover, add 5 mL concentrated HNO_3 and boil the mixture gently for 30 min on a hot plate. Cool the beaker and add 2 mL of 60% $HClO_4$, 5 mL of concentrated HNO_3, and 5 mL of 49% HF. Heat the beaker until the mixture has again evaporated nearly to dryness. Wash the cover and the wall of the beaker with 5 mL of water, and heat the solution to the evolution of dense white fumes. Cool the beaker and add 10 mL of 5% HNO_3 to dissolve the salts. Transfer the solution to a 50 mL volumetric flask and dilute with water washings from the digestion beaker. Aspirate directly into the spectrometer.

Bomb Digestion Method. To an accurately weighed sediment sample (not more than 0.2 g) in the Teflon cup add 4.0 mL of concentrated HNO_3, 2.0 mL of 60% $HClO_4$, and 4.0 mL of 49% HF. Seal the bomb and heat for approximately 3 h at 150°C in a drying oven. Cool the bomb and transfer the contents of the Teflon cup quantitatively to a 125 mL polyethylene bottle containing 2.5 g of boric acid in about 15 mL of water. After shaking to dissolve salts, transfer the solution to a 50 mL volumetric flask and dilute with water washings of the bottle. If necessary, solutions are stored in a refrigerator in 100 mL polyethylene bottles until ready for analysis. Aspirate samples directly into the spectrometer.

10.2.6 Determination of Metals in Sewage Sludge by AAS*

Urban sewage sludges are a significant source of metal in the environment. Contamination of air (incineration), water (effluent from sewage treatment plants), and soil (land disposal or use as fertilizer) can occur. The following procedures give the determination of Cd, Cr, Cu, Fe, Hg, Mn, Ni, Pb, V, and Zn in sewage sludge. Samples for the analysis of all elements except mercury are decomposed with a 1:3 v/v mixture of HNO_3 and HCl. Tests were carried out to see what fraction of total metals is extracted by this rapid procedure. In most cases the HCl–HNO_3 dissolved over 90% of the metals which is certainly adequate for most purposes. Sewage sludge samples to be analyzed for mercury

*Procedure from Van Loon et al. (8).

Table 10.6 Atomic Absorption Parameters

Element	Analytical line (nm)	Background line (nm)[a]	Oxidant	Slit (μm)
Al	309.2	nn	N_2O	80
V	218.4	319.6	N_2O	80
Cr	357.9	(366)[b]	Air	40
Cd	228.8	226.4	Air	80
Pb	283.3	282.0	Air	80
Ni	232.0	231.6	Air	40
Zn	213.9	(212)	Air	80
Cu	324.7	296.1	Air	80
Mn	279.5	281.7	Air	40
Fe	248.3	nn	Air	40

[a]nn = not needed; or use deuterium arc.

[b]Line found suitable for our lamp; accurate wavelength not known.

are decomposed with nitric acid, sulfuric acid, and potassium permanganate. Procedures were tested by analysis of NBS standards and standard sewage sludges. Good agreement with accepted values was obtained.

Equipment. Operating parameters for the Instrumentation Laboratories Model 153 atomic absorption unit are given in Table 10.6. In this work background correction was done by the nonabsorbing line method, but deuterium arc or Zeeman-effect-based background correction is preferred.

Reagents. All acids and chemicals were reagent grade, Fisher Certified. Blanks were run at all times to eliminate problems from contamination.

Standard aqueous metal solutions (1000 ppm metal) were prepared from reagent grade metals or metal salts. Working solutions were prepared by diluting these by an appropriate amount, being careful to maintain an acidity (HNO_3) of approximately 5% by volume.

Standard sewage sludges are available from Canadian Center for Inland Waters, Burlington, Ontario.

Procedures.

Procedure for Cd, Cr, Pb, Ni, Zn, Cu, Mn, Fe, Al, and V.

SAMPLING. No firm guidelines on sampling are possible. The worker must be cautioned that failure to investigate sampling constraints fully may invalidate the data.

Weigh 0.5 g of a representative sample into an acid-washed (aqua regia) 100 mL beaker. Add 9 mL HCl and 3 mL HNO_3. Heat on medium heat for 1 h. (During this time samples evaporate to near dryness.) If any is dry add 1 mL HNO_3. Dilute to 10 mL with water and filter (Whatman #42) into a 25 mL flask. Wash the filter well with water and dilute to volume.

For sludge and sludge-based fertilizer samples the following dilutions are commonly required. Dilute 2 mL of stock solution to 25, 100, and 250 mL. Dilute 2 mL of the preceding 25 ml dilution to 250 mL. Make the acid content 5% with HNO_3 in each case.

Run the solutions on the atomic absorption instrument using the conditions in Table 10.6. (Note the need for background correction.)

Standard working solutions are run at the same time to obtain calibration curves. In addition, a reference standard, either an internal laboratory standard or a standard sewage sludge, is run with each batch.

Procedure for Hg. Weigh a 0.1–10 g sample, depending on Hg content, into a BOD bottle supplied with the Coleman MAS 50. Add 15 mL of 7 *N* HNO_3, 15 mL of 1 H_2SO_4 and 1 g of $KMnO_4$ crystals as required to keep a persistent purple color in the solution. Heat at less than 75°C for 15 min. Cool to room temperature and add 10 mL of 0.1% hydroxylamine hydrochloride followed by 10 mL of 10% $SnCl_2 \cdot 2H_2O$ solution and run on the Coleman MAS 50 analyzer. This procedure is modified from that recommended by Hatch and Ott (9) for other sample types.

10.2.7 Determination of Metals in Fertilizers by ICPAES*

This procedure is for the determination of boron, calcium, magnesium, manganese, phosphorus, potassium, and zinc in fertilizers. The instrument was calibrated using two standard solutions and a blank (water). High levels of potassium and phosphorus in fertilizers can affect the results. The method was tested on 17 Magruder fertilizer standards. Most of the values obtained were in satisfactory agreement with the accepted values, although iron results tended to be slightly lower in some cases.

Equipment. Equipment consisted of a Jarrell-Ash Model 750 ICAP emission spectrograph (Fisher Scientific, Jarrell-Ash Division, 590 Lincoln St., Waltham, MA) with 50 μm entrance and exit slits, and 1180 rulings/mm grating, blazed at 2500 Å in the first order. Sample was atomized by cross-flow nebulization with an argon flow rate for aerosol transport of 2 L/min, giving 2.0 mL/min sample uptake rate. Argon coolant flow rate was 16 L/min. Forward power frequency to the coupling unit was 1 kW, with reflected power <1 w. Observation

*Procedure from Jones (10).

Table 10.7 Spectral Lines Employed

Element	Spectral Line (nm)
B	249.7
Ca	422.6
Cu	327.4
Fe	259.9
K	766.4
Mg	279.5
Mn	257.6
P	214.9 × 2
Zn	213.8

height above the load coil was 13 mm. Integration time was 10 sec following a 5 sec preburn. The elements reported and their spectral lines are listed in Table 10.7.

Reagents. Two solution standards are prepared from 1000 ppm atomic absorption standards by diluting appropriate aliquots with water to give final concentrations as shown in Table 10.8. The spectrograph is calibrated with standards 2 and 3, with water serving as the blank, using a two-point calibration technique. This routine procedure is used for elemental analysis of water; therefore no special standard preparation was used for elemental analysis of fertilizer digests. These standards also contained other elements (Al, Ba, Cd, Co, Na, Ni, Pb, Si, and Sr) at known concentrations.

Table 10.8 Standard Solutions

	Concentration (ppm)	
Element	Standard 2	Standard 3
B	10	0
Ca	10	0
Cu	0	10
Fe	0	10
K	17.5	0
Mg	10	0
Mn	0	10
P	10	0
Zn	0	10

Procedure. Weigh a 1.0 g fertilizer sample and transfer to a 200 mL beaker. Add 10 mL 38% HCl and 5 mL 69% HNO_3 and heat the sample to HNO_3 fumes and then to HCl fumes. Cool the sample, transfer quantitatively to a 1 L volumetric flask, and dilute to volume with water.

Transfer an aliquot of the prepared fertilizer digest to a sample cup and analyze by ICAP emission spectroscopy. Conditions used by Jones (11) are employed.

REFERENCES

1. H. M. Bowen, *Trace Elements in Biochemistry,* Academic Press, London, 1966.
2. A. Brzezinska, A. Balicki, and J. C. Van Loon, internal laboratory procedure, 1983.
3. J. I. Manzoori, *Talanta* **27,** 682 (1980).
4. R. H. Scott, V. A. Fassel, R. N. Kinseley, and D. E. Nixon, *Anal. Chem.* **46,** 75 (1974).
5. K. L. Iu, I. D. Pulford, and H. J. Duncan, *Anal. Chim. Acta* **106,** 3A (1979).
6. J G. Viets, *Anal. Chem.* **50,** 1097 (1978).
7. C. G. Hsu and D. C. Locke, *Anal. Chim. Acta* **153,** 313 (1983).
8. J. C. Van Loon, J. Lichwa, D. Rutton, and J. Kirade, *Water, Air and Soil Pollution,* **2,** 473, (1973).
9. W. R. Hatch and W. L. Ott, *Anal. Chem.* **40,** 2085 (1968).
10. J. B. Jones, Jr., *Assoc. Off. Anal. Chem.* **65,** 781 (1982).
11. J. B. Jones, Jr. *Com. Soil Sci. Plant Anal.* **8,** 349 (1977).

CHAPTER

11

DETERMINATION OF METAL SPECIES

11.1 INTRODUCTION

In the fields of biological, clinical, agricultural, geological, and environmental sciences, it is often crucially important to know the chemical forms as well as the total amount of an element in a sample because the chemical form determines the mode of interaction of the element in biological systems. To this date, although there are abundant total element data, there is in general very little information on the forms of most elements at trace levels in complex samples from these fields. It is important, therefore, to accelerate the development of methods for elemental speciation so that problems in these subject areas may be more readily solved.

When compounds are relatively stable, separation using the various techniques of chromatography is possible. If the element-specific techniques of atomic absorption, fluorescence, and emission spectroscopy are used as chromatography detectors rather than (or in addition to) the conventional, nonselective detectors, then the chromatograms will be much easier to interpret. It is expected that rapid advances in the methodology for trace elemental speciation will be made using element-specific detectors with chromatography.

The use of element-specific detectors with chromatography was pioneered by McCormack et al. (1) (atomic emission) and Kolb et al. (2) (atomic absorption) in 1965 and 1966. Also during these years, a series of papers by Bache and Lisk (3) describing microwave plasma atomic emission detectors for gas chromatography appeared. However, little practical application of research using these approaches was evident until the mid-1970s.

11.1.1 Choice of Atomizer and of Atomic Spectrometry Technique

The atomizer is the point of interface of chromatography with the techniques of atomic spectrometry. Table 11.1 summarizes the atomizer choices for gas and liquid chromatography that have appeared in the literature to date.

Unlike the situation with interfacing gas and liquid chromatography with mass spectrometry, interfaces with techniques of atomic spectrometry are usually relatively simple. The atomizer and type of atomic spectrometry detector that

Table 11.1 Atomizers for Chromatography Interfaces

	Gas Chromatography	Liquid Chromatography
Atomic emission spectrometry	Flame MIP DCP ICP Nitrogen afterglow	Flame DCP ICP
Atomic absorption spectrometry	Flame Graphite furnace Carbon rod Cold vapor generation Quartz tube	Flame Graphite furnace Carbon rod
Atomic fluorescence spectrometry	Flame Carbon cup Quartz furnace	Flame Carbon cup

are chosen depend on a number of factors; the most important are (1) availability of equipment; (2) detection limit required; (3) ease of making the interface; and (4) the requirement of single or multielement detection.

The microwave-induced plasma (MIP), particularly the argon or helium atmospheric-pressure MIP using the Beenakker cavity (4) can be highly recommended for gas chromatographic detection. Unlike other types of microwave plasmas, this one is very stable over extended periods of operation. Detection limits for elements such as mercury are among the best achievable (5). Solvents will cause instability or even plasma extinction and must be vented away from the plasma. Microwave plasmas are not recommended with liquid chromatography.

The direct current plasmas (DCPs) and inductively coupled plasmas (ICPs) are excellent choices for use with liquid chromatography. They are more stable with respect to solvent introduction than are microwave plasmas. Although these plasmas have been used for gas chromatography, generally detection limits are poorer than for other competitive approaches. The ICPs and DCPs are particularly recommended when refractory elements are to be determined in liquid samples. Furthermore, these plasmas give the least species-dependent response of all atomizers.

Flame atomizers, particularly air–acetylene and nitrous oxide–acetylene, find wide application in chromatographic detection. Detection limits for most elements in the flame are poorer than with furnace atomizers. Furnace atomizers used with atomic absorption or atomic fluorescence give the best detection limits for most elements. They will be the atomizers of choice in many cases because it is frequently essential to have the best detection limits when working with real samples from the subject areas mentioned above. The most difficult atomizer interface is that of liquid chromatography with furnace-type atomizers. It is difficult to make a direct connection between a liquid stream and commercial furnace atomizers because of the necessary stepwise nature of furnace heating cycles. Of the available furnace-type atomizers, the Perkin-Elmer or similar-type tube furnace for absorption and the Varian-Techtron carbon cup or similar type for fluorescence give the best detection limits and are recommended in most situations. A L'vov-type platform should, of course, be used when liquid samples are to be analyzed. Other resistively heated furnaces (e.g., quartz tubes wound with Nichrome wire) can be recommended for gas chromatography detection. For work with the hydride-forming elements (both direct gas chromatographic detection and liquid chromatographic work where hydride generation is involved), the quartz tube atomizers give best detection limits.

Multielement (simultaneous in most cases) detection may be necessary in some applications. The available choices are atomic emission spectrometry and nondispersive atomic fluorescence spectrometry. The experience with the latter technique is not extensive because, until recently, commercial instrumentation was not available. Generally therefore, ICP and DCP emission spectrometry for liquids and gases or microwave plasma emission spectrometry for gases can be recommended.

It is of interest to compare the relative detection capabilities of atomizers with different techniques of atomic spectrometry for an actual analysis. Radzuik et al. (6) and Coe et al. (7) compared relative detectable amounts for organolead and organomanganese compounds, respectively, using flame and furnace atomizers with atomic absorption and atomic fluorescence spectrometry detectors. Table 11.2 summarizes the results obtained.

Haraguchi et al. (8) compared the detection limits, linear dynamic ranges, and reproducibilities for inorganic, monomethyl, and dimethyl arsenic compounds (see Table 11.3) using MIP, ICP, and AAS with hydride generation. Because each arsenic hydride has a different boiling point (AsH_3, $-55°C$; $MeAsH_2$, $2°C$; and Me_2AsH, $36°C$), these compounds were evolved sequentially from the trap by heating.

Species dependency of the detector signal is also an important consideration. Small variations in signal for different compounds of the same element containing similar amounts of analyte are also noted in the data of Haraguchi et al. (8). In general ICP, DCP, and flame atomizers show little or no variation in signal levels in such cases. Furnaces with AAS may give significant signal variations dependent on species identified. Thus calibration must be made with this problem in mind. Vickery et al. (9) in a liquid chromatographic study of lead compounds using a furnace atomizer did a matrix modification prior to atomization to minimize this problem.

Table 11.2 Comparison of Detectable Amounts for Tetraalkyllead and an Organomanganese Carbonyl Separated by Gas Chromatography Using Different Detectors

	Relative Detectable Amounts[a]			
	AAS		AFS	
Atomizer	Pb	Mn	Pb	Mn
Flame				
Flame	50	40	10	
Quartz-tube-in-flame	3	20		
Furnace				
HGA 2100[b]	1	1		
CRA 63 tube[c]	2			
CRA 63 cup[c]	5			
Quartz furnace[c]	3	1	3	

[a] Relative to the best value, which is 1.

[b] HGA, heated graphite atomizer from Perkin-Elmer Co.

[c] CRA, carbon rod atomizer from Varian-Techtron Co.

Table 11.3 Comparison of Detection Limits, Dynamic Ranges, and Reproducibilities Obtained by ICP, MIP, and AAS[a]

	Compounds	ICP	MIP	AAS
Detection limit (ng)	Inorg. As[b]	0.09	0.5	1
	MeAs	0.3	0.2	2
	Me_2As	1	2	4
Dynamic range	Inorg. As[b]	10^4	10^3	10^2
	MeAs[c]	10^4	10^4	10^2
	Me_2As	10^3	10^3	10^2
Reproducibility (%)	Inorg. As	3.7	9.0	3.5
	MeAs	4.0	8.4	4.1
	Me_2As[d]	5.2	10.1	5.2

[a] These data were obtained by hydride-generation-liquid-nitrogen-trap-atomic-spectrometric detection methods.

[b] Inorg. As is inorganic arsenic.

[c] MeAs is monomethyl arsenic.

[d] Me_2As is dimethyl arsenic.

11.1.2 Recommended Gas Chromatography Interfaces

Interfacing gas chromatography with atomic spectrometry atomizers is relatively straightforward because the gas flow from the chromatograph is usually compatible with flame, plasma, and furnace operation. In the case of microwave plasmas, a helium or argon atmospheric pressure plasma using a Beenakker cavity is recommended.

The DCP, ICP, and flame atomizers may be directly interfaced with gas chromatography. In the case of plasmas, an argon or helium carrier gas should be used in the chromatograph. The chromatograph carrier gas can then be used as part of the plasma sample carrier gas flow. In interfacing flames the chromatographic carrier gas may be used as part of the auxiliary gas stream. A better approach with flames is to use a quartz tube held in the flame. The gas from the chromatograph is then passed into the hot quartz tube.

Problems due to atomization of organometallic compounds in hot transfer lines near the atomizer may occur with furnace-type atomizers (e.g., for tetraalkylleads the gas flow from the chromatograph is injected into the furnace through the injection port with a temperature-resistant tantalum tube (6)). The magnitude of premature atomization was found to vary with the composition of the tube:

$$SiO_2, Al_2O_3 > \text{stainless steel, carbon} > Ta > \text{Teflon}$$

This problem was obviated by DeJonghe et al. (10) by injecting the gas chromatographic effuent flow directly into the internal gas flow of the device. Although this approach appears to be preferable, I know of no data to show how detectable amounts compare using the two methods.

Ebdon et al. (11) described four atomizers for interfacing gas chromatography with atomic absorption spectrometry. They stress the optimizations necessary during the development of such cells. Their main goal was to increase atom residence times in the flame. Simplex optimization was employed. The four atomizers are:

1. An air–propane burner with a glass-lined interface tube that directs gas chromatography effluent along the slot of the burner. The air–propane flame was employed because it has a four times slower burning velocity than air—acetylene, thus increasing atom residence times.
2. A ceramic tube with a hole at the midpoint in the bottom of the tube held over the burner. This consisted of a ceramic tube held over an air–acetylene burner with a glass-lined interface tube extending horizontally to the base of the flame but not attached to the ceramic tube.
3. A ceramic tube held over the burner as in (2), but with the glass-lined

interface tube passing up vertically through the burner slot and positioned just below the hole in the atomizer tube.

4. To prevent exhaust gas from the flame from passing through the tube and thus reducing atom residence time, a ceramic tube device was designed in which atomization and tube heating were separated. The air–acetylene flame was used to heat the tube, but a H_2 diffusion flame burning on the end of the glass-lined interface tube directed into the ceramic tube was the atomizer.

Detection limits for tetraethyllead species were 2000, 75, 71, and 17 pg, respectively, for designs (1)–(4).

11.1.3 Recommended Liquid Chromatography Interfaces

Interfacing of liquid chromatography with ICPs and DCPs is relatively easy. The sample uptake rate of these plasmas is about 1–2 mL/min, which is generally comparable with the flow rate of liquid from a high performance liquid chromatograph. Uden et al. (12), using a dc argon plasma, found that a conventional nebulization system is satisfactory with reverse phase and ion exchange solvents. However, a special impact nebulizer interface was necessary to deal with the hydrocarbon and halocarbon solvents used in normal phase liquid chromatography.

Burner–nebulizer flow rates are usually 3–6 mL/min. Thus in interfacing flame atomizers with high performance liquid chromatography it is necessary to use a method that takes account of the slower flow rate (1–2 mL/min) of the liquid chromatograph. Although it is possible to use an auxiliary liquid flow as suggested by several workers, this approach is not recommended because of the dilution involved. A better method was suggested by Slavin and Schmidt (13). Liquid from the chromatographic column is allowed to form a drop on the end of the column (about 100 μL when it fell off). The drop is then allowed to fall into a Teflon microsampling cup attached directly to the nebulizer. The 100 μL drop size is sufficient to produce a steady-state signal with a Perkin-Elmer 603-type burner–nebulizer system. Thus no loss in sensitivity is obtained with this method.

Interfacing of liquid chromatography with furnace atomizers cannot be direct because of the stepwise nature of the heating cycles of commercial furnaces. The following two interfacing strategies proposed by Brinckman et al. (14) can be recommended:

1. Pulse mode (periodic stream sampling): The effluent is passed through a modified single sampling cup of an automatic sampler. Liquid enters the bottom of the cup and overflows continuously out of an overflow tube attached near the top.

2. Survey mode (sequential stream analysis): The automatic sampler of the atomic absorption unit is used as a fraction collector and each fraction is analyzed in turn by the atomic absorption instrument in the conventional way.

In using these methods, a loss in sensitivity will occur because the exact top of the peak for a constituent will be missed. To overcome this problem Vickrey et al. (9) proposed a peak storage method. For this approach an ultraviolet detector is used to identify when a desired component is coming out of the chromatograph and the solution containing all of this constituent is stored in a length of thin diameter tubing. Each consecutive fraction of sample in the tube is later analyzed.

When hydride elements in species in a liquid sample are to be determined, the following proposal by Ricci et al. (15) is highly recommended. These workers proposed the interfacing of ion chromatography with an atomic absorption detection for the determination of arsenite, arsenate, dimethylarsenate, and *p*-aminophenylarsenate by continuous evolution of the gaseous arsines produced *after* the chromatographic separation. This is an excellent approach in that it combines the high sensitivity of the hydride generation technique with the superior separation characteristics of the ion chromatograph. It also avoids the criticism of the frequently used approach of gas chromatography for separating the evolved hydrides of arsenic compounds from solution prior to their separation. (The latter method may result in altering the original solution composition.)

A great deal of research is still needed on optimizing the interface between liquid chromatography and atomic spectrometry detectors. No systematic studies of this type have yet been done.

The effect of variations in the location of the spray chamber on the ICPAES signal was investigated by Whaley et al. (16) when ICPAES is used as a detector for liquid chromatography. The two extremes in placement of the spray chamber in LC/ICPAES are near the base of the torch (the conventional position) and near the outlet of the LC column. With the former, the longer distance of liquid flow transport results in peak broadening and distortion. When the spray chamber is near the outlet of the LC column, and thus the spray from the nebulizer is transported a long distance, a signal loss occurs. In both situations the peak area decreases with increasing flow rate over about 1.5 mL/min. However, this parameter was least affected when the spray chamber was near the outlet of the LC column.

A number of researchers have recently done studies involving improvement of the interface between the column and the inductively coupled plasma. Jinno and Tsuchida (17) placed a micro high performance liquid chromatographic column directly in the line between the pump and the input of the nebulizer. Various metal diethyldithiocarbonates were separated using a Jasco SC-01 (ODS silica, 5 *M*) Teflon column of 0.5 mm i.d. × 22 cm.

Research into developing a spray chamber for use in interfacing a size-exclusion liquid chromatograph with an inductively coupled plasma when highly volatile organic solvents are used is reported by Hauser and Taylor (18). A cooling-jacketed spray chamber was found to eliminate carbon build-up and undesirable abnormally high nebulization efficiencies when using toluene. Detection limits compared with those obtained with conventional spray chambers were ten to 100 times better.

Work is presently underway in my laboratory employing a graphite furnace to volatilize microliter samples of LC effluent into an ICP or MIP. Early results for copper, chromium, lead, and cadmium show a sensitivity enhancement up to ten times as compared to conventional interfacing approaches.

11.1.4 Methodology

Element-specific detectors can be used with most column chromatograph techniques. Sometimes it will be important to make slight changes in the procedures to accommodate better the special characteristics of the atomizer being used. For example, it is best if possible to choose a solvent with good burning characteristics when using a flame. However, in most cases no change in chromatographic methodology is necessary.

In developing a procedure for operation of the atomic spectrometry detector, care must be taken to investigate interferences characteristic of the technique used. Because atomic spectrometry detectors are element specific, interference problems are less severe than with conventional detectors. Also because a separation is occurring, offending constituents may be resolved in time from the analyte species, resulting in the absence of a usual interference.

Plasma emission spectrometry suffers severely from atomic spectral interferences and thus a thorough investigation of potential problems should be made. Because nonspecific background-type interferences are common with all atomic spectrometry techniques, they must be investigated. Solute–analyte interactions cause greater problems with atomizers used in atomic absorption and fluorescence. The very high temperature of the ICP and DCP makes these devices relatively free from this source of problem. Transport interferences occur in cases in which tubes of a small internal diameter must be employed.

Although element-specific detectors have a distinct advantage in most aspects of elemental speciation work, the following limitations must be taken into consideration:

1. Column decomposition products of the compounds of interest that do not contain the element(s) being determined will not be identified.

2. Important compounds in the sample that do not contain the element(s) of interest will not be detected.

When these problems must be avoided, a conventional detector may be used in series or parallel with the element specific detector. Series operation is preferable where applicable because the sensitivity will not be greatly affected.

Species dependency of the detector signal may be a problem in some instances. If so, standardization must be done with each constituent individually. Alternatively, a chemical treatment step can be performed to convert the compounds to a species independent form prior to atomization.

In some determinations, such as the gas chromatographic determination of substances in air, there is a need for a dynamic method for standardization. Although this is difficult in most instances, it is much superior to a static approach and should be employed whenever possible.

When organometallic compounds are being separated, it is important to remember that many of these are relatively unstable under conditions commonly employed in chromatography. It is important therefore to use nonreactive equipment, for example, glass or Teflon-coated columns and transfer lines. Injection ports must not contain exposed metal. As low a temperature as possible should be used.

Table 11.4 lists speciation procedures of environmental and biological interest.

11.2 PROCEDURES

11.2.1 Water Samples

*11.2.1.1 Determination of Se(IV) and Se(VI) in Ground Waters by Ion Chromatography AAS**

Organic materials in ground waters interfere with the determination of Se(IV) and Se(VI). Oxidative destruction of these substances results in a change of the Se(IV) to Se(VI) ratio. In this method ion chromatography is used to separate the two selenium species from organics in ground waters. Hydride generation and atomic absorption detection are then used. Both Se(IV) and Se(VI) can be determined by the procedure. Analysis of selenium-spiked ground water shows that 96 to 102% recoveries can be expected using Amberlite XAD-8 resin.

*Reprinted in part with permission from (30), Roden and Tallman, *Anal. Chem.* **54**, 307. Copyright 1982, American Chemical Society.

Table 11.4 Speciation Applications of Environmental and Biological Interest

Element	Samples	Species	Chromatography	Atomizer Detector	Reference
Pb	Waters	Dialkyllead, trialkyllead, tetraalkyllead, Pb (II)	Gas (after butylation)	Quartz tube AAS	19
Pb	Waters	Trialkyllead	Gas (after butylation)	He–MIP	20
Pb	Review	—	Gas	Flame and furnace	21
Hg	Air	Hg, $HgCl_2$, CH_3HgCl, $(CH_3)_2Hg$	Absorption train	AFS	22
Hg	Breath, saliva, hair	Volatile	Activated carbon bid	Quartz T tube AAS	23 and 24
Se	Air	$(CH_3)_2Se$, $(CH_3)_2Se_2$	Gas	Graphite furnace AAS	25
As	Seawater	Dimethylarsinic acid	Cation exchange	Furnace AAS	26
Se	Distilled water, synthetic water, Texas river water	Selenite, selenate	Ion	Furnace AAS	27
As Ge Sb	EPA waters	Hydrides	Gas	ICPAES	28
Sn	Waters	CH_3SnCl_3, $(CH_3)_2SnCl_2$, $(CH_3)_3SnCl$, Sn(IV)	Gas (after butylation)	Quartz tube AAS	29

Se	Waters	Se(IV), Se(VI)	Ion	Hydride-to-graphite furnace AAS	30
Se	Review (environmental) waters	Many	Many	AAS, AES, AFS	31
As	Water	Monomethyl-arsenic acid, dimethyl-arsinic acid, As(III), As(V)	Ion exchange	Graphite furnace AAS	33
C, P, Hg, Br, Ci, S, F, N, O	Gases	—	Gas	He–MIP–AES (with 9-channel polychromator)	34
B, C	Biological	Steroid carboranes	Gas	He–MIP–AES	35
B	Human urine	Catechol derivatives	Gas	He–MIP–AES	36
Cu, Zn, Fe	Seawater	Several	HPLC adsorption	Flame AFS	36–39
Pb, Cu, Cd, Zn, Fe	Foods	Assimilable metals	None—enzyme extraction	DCP–AES	40
Ca, Mg	Natural waters	Organically bound	HPLC size exclusion	ICP–AES	41
Zn	Human	Albumin–bound Zn	HPLC affinity	Furnace AAS	42

Equipment. The hydride generation and trapping apparatus was the same as that described by Shaikh and Tallman (43) except that their liquid-nitrogen trap was replaced by a smaller trap 26 cm long made from 5 mm i.d. Pyrex tubing. The inside surface area of this trap was increased by making numerous indentations in the tube by intense localized heating and reduced internal pressure. This smaller trap permitted substitution of a styrofoam cup of warm tap water for the Variac and Nichrome coil heater used by Shaikh and Tallman (43). The hydride generator was interfaced to a Perkin-Elmer Model 603 HGA-2100 graphite furnace atomic absorption spectrometer as described previously (43). Pyrolytically coated graphite furnace tubes were used, and the light source was a Perkin-Elmer selenium electrodeless discharge lamp. All data were collected with background correction. The instrument settings were as follows: slit width, 3 mm; wavelength, 196 nm; atomization temperature, 2650°C; atomization time, 12 sec; helium purge gas flow rates, 100–125 mL/min when sparging the hydride generator and 375–425 mL/min when sweeping hydride from the cold trap to the furnace. The HGA-2100 furnace controller was set to go directly to atomization upon initiation of the temperature program and was interfaced to the Model 603 microprocessor to simultaneously initiate the read function of the spectrometer. The peak-height mode of absorbance measurement was used throughout this study. The absorbance–time profiles were also recorded on a Heath Model EU-205-11 strip-chart recorder.

The ion exchange column was made of Pyrex glass tubing 28 cm in length with a 1.3 cm id. The column was slurry-packed to a depth of 14 cm with XAD-8 resin (Amberlite). Column flow was controlled with a Teflon stopcock.

Reagents. All chemicals were analytical reagent grade and were used without further purification. Solutions were prepared in distilled, deionized water (four-cartridge Milli-Q system, Millipore, Inc.).

Procedure. For those samples subjected to XAD-8 column chromatography, prepare the column before each sample by passing 60 mL of pH 12–13 KOH through the column followed by 60 mL of pH 1.6–1.8 HCl. After column preparation, measure out a 25 mL aliquot of sample and acidify to pH 1.6–1.8 with HCl. Place the aliquot on the column and follow by sufficient pH 1.6–1.8 HCl eluent to collect 50 mL of column effluent. Split the 50 mL of effluent into two 25 mL volumes, with each brought to 50 mL with deionized water. Add to one portion 50 mL of concentrated HCl and follow by boiling for 8 min. This portion is then analyzed for total Se. Add to the second portion 50 mL of concentrated HCl immediately before analysis. The results of this portion yield only the Se(IV). The Se(VI) can then be determined by the difference.

Place samples subjected to either of the pretreatments described above (100 mL final volume) in the hydride generator. The generator is mounted, helium sparging is begun, the trap is immersed in liquid nitrogen (1 min cool down), and 10 mL of 10% $NaBH_4$ is injected over a period of 1 to 1.5 min. Eight minutes after the initial injection of $NaBH_4$, stop the sparging and direct the helium flow to bypass the generator. Initiate the graphite furnace temperature program and immediately remove the liquid nitrogen from the trap and replace by a styrofoam cup containing warm tap water (50–60°C). The hydride is rapidly vaporized and swept into the furnace. The atomization peak is recorded on the strip-chart recorder and peak height is indicated by the spectrometer's microprocessor.

*11.2.1.2 Determination of Species of Arsenic in Water by Selective Volatilization AAS**

The following arsenic species can be determined in water: arsenate, arsenite, monomethyl, dimethyl, and trimethyl arsine, monomethylarsonic acid, and dimethylarsinic acid, and trimethylarsine oxide. Detection limits for the compounds are in the nanogram per liter range. First the arsines are bubbled out of the water using a stream of helium. Next the other compounds are converted to their corresponding arsines by borohydride reduction. In both cases the arsines are frozen out in a liquid-nitrogen trap. Separation occurs by selective volatilization by slowly heating the trap. Each compound is atomized in a flame-heated quartz tube.

Equipment. The apparatus for the volatilization and trapping of the arsines (Fig. 11.1) is constructed from Pyrex glass, with Teflon stopcocks and tubing, and with nylon Swagelok connectors. The sample trap consists of a 6 mm o.d. Pyrex U-tube of ~ 15 cm length, filled with silane-treated glass wool. The interior parts of the six-way valve that interfaces the volatilization system with the gas chromatograph are made of Teflon and stainless steel.

The atomic absorption detection system consists of a Varian AA5 with a hollow-cathode arsenic lamp; the standard burner head is replaced by a 9 mm i.d. quartz burner cuvette (Fig. 11.2) modified after a design by Chau, Wong, and Goulden (45).

Reagents Arsenate and arsenite standards were prepared from Baker Analyzed NaH_2AsO_4 and "Primary Standard" arsenic trioxide. Monomethylarsonic acid (99.9% pure) and dimethylarsinic acid (99.87% pure) standards were obtained from Ansul Corp., Weslaco, TX. Monomethylarsine and dimethylarsine standards were prepared by reduction from the corresponding acids and purified

*Reprinted in part with permission from (44), Andrae, *Anal. Chem.* **49**, 820. Copyright 1977, American Chemical Society.

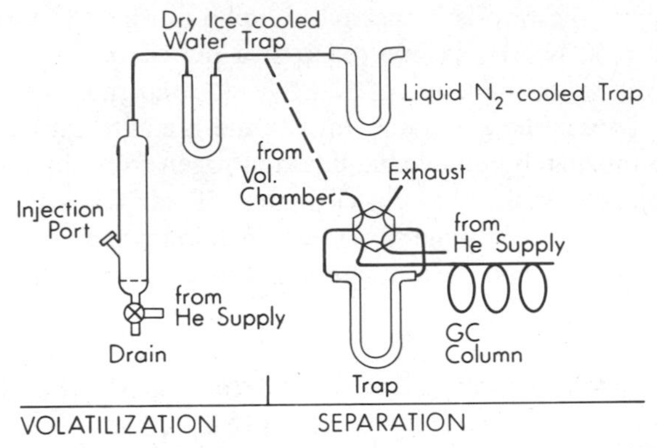

Figure 11.1 Apparatus for the volatilization, trapping, and separation of the arsines (44).

by fractional distillation. Trimethylarsine was obtained from Ventron Corp., Danvers, MA. Stock solutions with 1000 ppm As were prepared for each of the acids and diluted daily before use. Arsine standards were prepared in oxygen-free methanol and/or toluene in sealed containers immediately before use.

Sodium borohydride was obtained from Ventron. From blank runs this reagent did not contain detectable amounts of As(III) or organoarsenicals. The As(V) content varied between 1 and 30 ppb from lot to lot. Baker Analyzed Reagent HCl and Sigma Reagent Grade Tris and Tris-HCl were used for the buffers. The Tris buffer solution is 2.5 *N* in Tris and 2.475 *N* in HCl, giving a pH of 6.2 after dilution to 0.05 *N*.

Procedure

Isolation of the Arsenic Species. Introduce the 1–50 mL sample into the gas stripper with a hypodermic syringe through the injection port. If an air-free transfer is necessary, the septum is replaced by a fitting, that connects to a four-

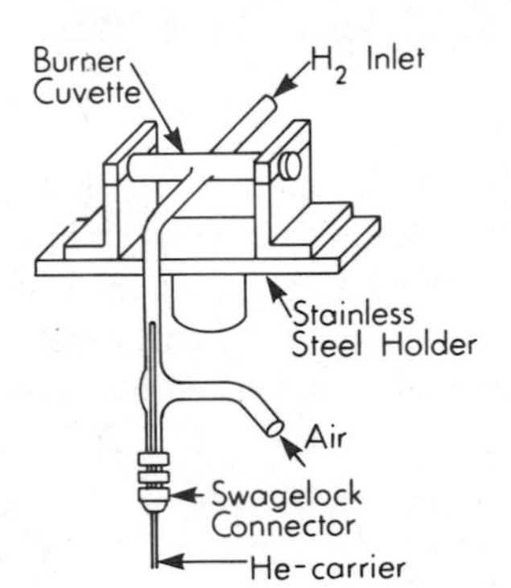

Figure 11.2 Quartz cuvette burner head (44).

way valve with a sampling loop. The system is first flushed with helium, and then the sample is injected by turning the valve.

Strip out any volatile arsines in the sample by bubbling a helium stream through the sample. Then add 1 mL of the Tris buffer solution for each 50 mL sample, giving an initial pH of about 6. Inject into this solution 1.2 mL of 4% $NaBH_4$ solution while continuously stripping with He. After about 6–10 min, the As(III) is converted to arsine and stripped from the solution. The pH at the end of this period is about 8. Then add 2 mL of 6 *N* HCl, which brings the pH to about 1. The addition of three aliquots of 2 mL of 4% $NaBH_4$ solution during 10 min reduces As(V), monomethylarsonic acid, dimethylarsinic acid, and trimethylarsine oxide to the corresponding arsines, which are swept out of the solution by the He stream and the evolved hydrogen.

Drying of the Gas Stream. It is found necessary to dry the helium gas stream coming from the reaction vessel because water clogs the gas trap. Various chemical drying agents were tried (such as $CaCl_2$, K_2CO_3, $Mg(ClO_4)_2$, silica gel, and Drierite. These materials either failed to remove the water adequately or irreversibly trapped the arsines. Effective drying is obtained with a 28 cm long Pyrem U-tube, 7-mm id, immersed in a dry ice–isopropyl alcohol bath.

Trapping and Separation of the Arsines. For direct detection by the atomic absorption technique, the stripping gas stream can be used to carry the sample into the burner. The arsines are isolated by immersing the gas trap in liquid nitrogen and released by slowly warming it up to room temperature. This slow warming results in a sequential separation of the arsines on the basis of their boiling points.

*11.2.1.3 Determination of Dialkyllead, Trialkyllead, Tetraalkyllead, and Pb(II) Ions in Water by Gas Chromatography AAS**

Sodium diethyldithiocarbamate is used to chelate R_2Pb^{2+}, R_3Pb^+ (R is Me and Et) and Pb^{2+}, which are then extracted together with R_4Pb into benzene. The benzene phase is carefully evaporated to a small volume in the presence of butyl Grignard reagent. The butylated compounds are then separated by gas chromatography using a temperature program of 80 to 200°C at 5°C/min. Water samples can be analyzed and the detection limit for the compounds (as lead) is 0.1 μg/L based on using a 1 L sample.

Equipment. The details of the gas chromatography AAS system have been described previously (45). The sample extract was introduced directly to the chromatographic column by a syringe. The chromatographic column was made

*Procedure from Chau et al. (19).

of glass, 1.8 m long × 6 mm diameter, packed with 10% OV-1 on 80–100 mesh Chromosorb W with carrier gas (N_2) flow rate of 65 mL/min. The injection port and transfer line were at 150 and 160°C, respectively; the column was programmed from 80 to 200°C at a rate of 5°C min. The 217.0 nm line from a lead electrodeless discharge lamp operated at 10 W was used. An electrically heated quartz furnance at 900°C with hydrogen flowing at 85 mL/min was employed and a deuterium lamp was used for background correction. Peak areas were integrated with an Autolab Minigrator (Spectra-Physics, CA).

Reagents The lead compounds used were dimethyllead chloride and diethyllead chloride (both from Associated Octel., S. Wirrel, Great Britain); tetramethyllead, tetraethyllead trimethylead acetate, and triethyllead acetate (from Alfa Chemicals, Danver, MA); and mixed tetraalkyllead compounds (Ethyl Corp., Ferndale, MI. The commercial trialkyllead salts are reasonably pure, whereas the dialkyllead compounds generally contained trialkyllead and other lead(II) impurities. A convenient procedure based on the reaction of trialkyllead and iodine monochloride can be used to synthesize dialkyllead if pure compounds are not available (46). Trialkyllead compounds can also be synthesized if required by the reaction of iodine and tetraalkyllead (47).

Standard dialkyllead solutions (100 μg Pb/mL) were prepared by first dissolving 0.0149 and 0.0162 g of dimethyllead chloride and diethyllead chloride, respectively, in small amounts of methanol and diluting to 100 mL with distilled water. Standard trialkyllead solutions (100 μg Pb/mL) were prepared by dissolving 0.0152 g of trimethyllead acetate and 0.017 g of triethyllead acetate in 100 mL of water.

Sample Storage. All the ionic alkyllead compounds slowly degrade in the presence of light. However, lake water samples enriched with dimethyllead chloride and trimethyllead acetate at the 100 μg/L level are stable over a period of at least one month in the laboratory when stored in the dark and refrigerated. There is no need to add any preservative to the sample, but storage in a cold dark room is recommended. Alternatively, the samples can be extracted, butylated, and dried over anhydrous sodium sulfate.

Procedure. To 1 L of water add 50 mL of aqueous 0.5 *M* sodium diethyldithiocarbamate (NaDDTC), 50 g of sodium chloride, and 50 ml of benzene, and shake the mixture for 30 min. Evaporate the benzene phase carefully in a rotary evaporator to 4.5 mL in a 10 mL centrifuge tube to which 0.5 mL of butyl Grignard reagent is added. Gently shake the mixture for 10 min and wash with 5 mL of 0.5 *M* H_2SO_4 to destroy the excess of Grignard reagent. Pipette about 2–3 mL of the organic phase into a small vial and dry with anhydrous sodium sulfate. Appropriate amounts (5–10 μL) are injected into the gas chromatography AAS.

11.2.2 Clinical Samples

*11.2.2.1 Determination of Protein-Bound Zinc and Copper Species in Blood by Gel Filtration and AAS**

Copper and zinc proteins can be separated by gel filtration using Sephadex G-100 by the following procedure. The determination of Copper and Zinc is by furnace atomic absorption spectrometry. Proteins can be measured by immunonephelometry. Advantages are that only a 1 mL sample is required, the method is semiautomated, and no sample pretreatment is needed. It is necessary to purify the Tris-HCl buffer using passage through Chelex-100. The detection limits for Zinc and Copper are reported to be 0.31 and 1.3 μg/L, respectively using 10 μL of zinc and 20 μL of copper aliquots.

Experimental

Equipment. The columns used (2.6 × 100 cm) were supplied by Pharmacia Fine Chemicals Ltd. The eluent from the column was collected in polystyrene vials with an automatic fraction collector (LKB Model 7000 UltraRac).

For the protein measurements and identification a Technicon Auto-Analyzer II specific protein analyzer, consisting of a pump, cartridge with mixing coils, nephelometer, and a recorder was used.

The fractions were analyzed for copper and zinc by using a Perkin-Elmer Model 272 spectrophotometer equipped with a Perkin-Elmer HGA–500 heated graphite atomizer and a Perkin-Elmer Model AS-1 autosampler. The atomic absorption signals were recorded with a Perkin-Elmer Model 56 strip-chart recorder. Total serum copper and zinc were determined with a Perkin-Elmer Model 403 spectrophotometer with the burner and acessories for use with an air–acetylene flame.

A Wang 600 programmable calculator was used for calibration and sample concentration calculations.

Reagents. Columns were prepared from Chelex-100 (100–200 mesh; Bio-Rad Laboratories) and Sephadex G-100 (40–120 μm; Pharmacia Fine Chemicals). The buffer was prepared from tris(hydroxymethyl)aminomethane hydrochloride and tris(hydroxymethyl)aminomethane.

The atomic absorption spectrometry standards were individual 100 μg/mL stock solutions (British Drug House Ltd.) containing zinc and copper nitrates. Intermediate stock solutions (10 μg/mL) were prepared daily for subsequent preparation of the working standards.

*Procedure from Gardiner et al. (48).

Purification of Reagents and Columns

PREPARATION OF GEL. To remove endogenous trace metal contamination the following procedure is used. Measure approximately 400 mL 0.02 *M* HCl into a 1 L vacuum flask and sprinkle 25 g of dry Sephadex G-100 beads into the solution. Allow the beads to swell for 24 h at room temperature. Decant the supernatant liquid and add a further 400 mL 0.02 *M* HCl to the gel. Repeat this procedure twice more. Subsequently wash the already swollen gel three times with deionized water, leave overnight, and then deaerate. At this point extra care must be taken to avoid contaminating the already cleaned gel.

REMOVAL OF METAL CONTAMINATION IN THE TRIS-HCl BUFFER. Copper and zinc contamination in the Tris-HCl used in preparing the buffer can be removed by passing the solution through a Chelex-100 column; 2 g of resin in its sodium form is mixed with deionized water. Pour the mixture into a glass column (1 × 30 cm). Equilibrate the column by passing through it 100 mL 1 *M* HCl followed by 200 mL of deionized water. A solution of 0.1 *M* Tris-HCl prepared by dissolving 14 g in 1 L of deionized water is passed through the column. Monitor the pH of the eluent; at pH 2.7 the eluent is collected for the subsequent preparation of the buffer. Dilute 500 mL of the purified buffer to 1 L and bring the pH to 7.4 by adding Tris-base.

PACKING THE COLUMN WITH SEPHADEX G-100. Before packing, the column must be decontaminated by cleaning with 3 *M* HNO_3. The acid is left in the column and tubing for 3 days and washed out thoroughly with deionized water before use. Pour the prepared gel into the column and pack as prescribed by the manufacturer. Monitor the pH of the eluent until it is the same as that of the starting buffer. Operate the column at 20 ml/h.

Procedures

Chromatography of the Serum Sample. Apply a 1 mL portion of serum to the column and, after eluting 100 mL of buffer, forty 3.2 mL fractions are collected in polystyrene vials prerinsed with deionized water and air dried.

Determination of Zinc and Copper in the Fractions. The Zn and Cu standards used in the analysis of the column fractions are prepared in 0.05 *M* Tris-HCl buffer. For the Cu determination, standards containing 0–80 μg/L Cu are prepared by measuring 0, 10, 20, 40, 80, 120, and 160 μL of the 10 μg/mL intermediate stock solution into separate polystyrene vials and making up the volumes to 20 mL with buffer. Standards (0–40 μg/L Zn) for the determination of Zn are prepared by transferring 0, 10, 20, 40, 60, and 80 μL of the 10 μg/mL

TABLE 11.5 Instrumental Conditions for the Determination of Copper and Zinc in the Gel Chromatography Column Fractions[a]

	Cu	Zn			Cu	Zn
Wavelength (nm)	324.8	213.8	Drying:	Temp. (°C)	100	100
				Ramp time (sec)	20	20
				Hold time (sec)	30	30
Lamp current (mA)	13.0	15.0	Ashing:	Temp. (°C)	700	500
				Ramp time (sec)	10	10
				Hold time (sec)	10	10
Spectral bandwidth (nm)	0.7	0.7	Atomization	Temp. (°C)	2700[b]	2500[c]
				Ramp time (sec)	1	5
				Hold Time (sec)	5	5
Sample size (μL)	20	10	Cleaning:	Temp. (°C)	2700	2700
				Ramp time (sec)	2	2
				Hold time (sec)	3	3

[a] Background correction used, scale expansion ×2.
[b] No gas flow through the tube during atomization.
[c] 100 mL min gas flow through the tube during atomization.

solution into vials and making up the volumes to 20 mL. Instrumental conditions for the determination of both elements are given in Table 11.5.

The amounts of analyte found in each fraction were added to give the total metal content. The total obtained was corrected for the blank, which was calculated by taking the mean of the amount of analyte in three fractions after the last protein had left the column. The mean blank value was then multiplied by the total number of fractions collected and the figure obtained was regarded as the overall blank.

Protein Identification and Measurement. A modification of the automated procedure for specific proteins (49) was used.

*11.2.2.2 Separation and Determination of Fe Proteins in Human Liver by Liquid Chromatography AAS**

This procedure is applicable to needle biopsy samples. Furnance AAS is employed in the determinative step. Total Fe, transferrin, ferritin-Fe, hemoprotein-Fe, and hemosiderin-Fe may be determined after chromatographic separation in a citrate buffer at pH 4.95 using carboxymethyl cellulose.

*Procedure from Selden and Peters (50).

Equipment. A Perkin-Elmer 360 dual-beam atomic absorption spectrophotometer with a deuterium background corrector was used in conjunction with a heated graphite furnance (HGA-76A), an automated sampler (AS-I), and a Model 165 chart recorder. The instrument settings were: single-element hollow-cathode iron lamps, current 30 mA; wavelength, 248.3 mn; and slit width, 0.2 nm. The furnace settings were: stage I, 0–100°C at 3.8°C/sec, then maintained for 150 sec; stage II, 100–1150°C at 19.1°C/sec, then maintained for 20 sec; and stage III, 2000°C for 7 sec with gas stop. A 10 mV chart recorder was employed. All glassware and plastics were washed in 6 *N* HCl and rinsed in distilled, deionized water before use.

Reagents Analar sodium citrate, citric acid, and imidazole, and Aristar sucrose and sodium hydroxide were obtained from BDH Chemicals, Poole, Dorset. Ferrichloride solution (BDH) was used as the standard for atomic absorption spectrophotometry. Human hemoglobin was purchased from Sigma Chemical Co. Carboxymethyl cellulose CM 32 was obtained from Whatman. Carboxymethyl cellulose was cycled as described by the manufacturer and equilibrated with 6 m*M* sodium citrate–4 m*M* citric acid buffer, pH 4.95. It was suspended in this buffer and poured into Pasteur pipettes plugged with glass wool to form a 1.5 cm column.

Procedure. Human liver is obtained by needle biopsy either percutaneously from patients being investigated for pyrexia of unknown origin or from laparotomy from patients undergoing surgery for cholelithiasis or peptic ulceration. A portion of the tissue is processed for routine histological examination. The remainder of the tissue is collected in ice cold isotonic 0.25 *M* sucrose containing 3 m*M* imidazole-HCl buffer, pH 7.2. The liver tissue is disrupted by 15 strokes of a loose-fitting "A" pestle in a Dounce homogenizer (Knotes Glass Co., Vineland, NJ) and the homogenate stored at −20°C.

Prepare an aliquot of liver homogenate (10–200 μL) for chromatography by sonication with 25 μL of 0.1 *M* sodium acetate buffer, pH 4.5, containing 0.5% Triton-X 100. The pH of the final mixture should be 4.9. Layer the sample onto the column and elute the iron proteins successively as follows: 4 mL of equilibration buffer (transferrin fraction); 2 mL of 50 m*M* sodium citrate–14 m*M* citric acid, pH 5.5 (ferritin); 2 mL of 42 m*M* sodium citrate–8 m*M* citric acid–0.3 m*M* NaCl pH 5.7 (hemoprotein); and 2 mL of 0.1 *M* NaOH (hemosiderin). Use a flow rate of 0.3 mL/min and perform the chromatographic separation at room temperature. Several columns are run simultaneously. The 20 μL samples of column eluates, together with blanks and standards, are analyzed for iron in duplicate by atomic absorption spectrophotometry. Adjust the chart recorder to 10 mV and use a chart speed of 60 mm/min with automated recording of the charring and atomization steps. The total iron in liver homogenates is estimated in the same manner with suitable dilutions of iron-free water. Prepare a standard curve over the range 0–400 ng/mL.

11.2.3 Air Samples

*11.2.3.1 Determination of Tetraalkyllead Compounds in Air Using Gas Chromatography AAS**

In this procedure street air samples are stripped of the tetraalkyllead compound in a short piece of chromatographic column held at −72°C. The individual compounds [$(CH_3)_4Pb$, $(CH_3)_3C_2H_5Pb$, $(CH_3)_2(C_2H_5)_2Pb$, $(C_2H_5)_3CH_3Pb$, and $(C_2H_5)_4Pb$] are then separated by a gas chromatograph equipped with a furnace atomic absorption detector. The detection limit for all the compounds is 0.5 ng/m^3 using a 20 L sample.

Equipment. The Perkin-Elmer 603 atomic absorption spectrophotometer used was equipped with deuterium background corrector, an HGA-2100 graphite furnance, and a Perkin-Elmer Model 56 recorder. The radiation source was a Perkin-Elmer electrodeless discharge lamp operated at 10 W. A Pye gas chromatograph (Series 104) was interfaced to the graphite furnace with a tantalum connector machined from a 6.4 mm diameter rod (Ventron Corp.). The glass chromatographic column (150 cm long × 0.6 cm o.d.) was packed with 3% OV-101 on 80–100 mesh Chromosorb W. The effluent was transferred to the furnace by Teflon-lined aluminum tubing (3 mm o.d.) heated electrically to 80°C.

Reagents. Tetramethyllead and tetraethyllead were obtained from Ventron Corp. Trimethylethyllead, dimethyldiethyllead, and methyltriethyllead were provided by the Ethyl Corp. (Ferndale, MI). Freshly diluted lead standards were prepared daily in analytical grade methanol or benzene. The carrier gas for gas chromatography was purified dry nitrogen. The graphite furnace was flushed with purified argon. Analytical grade HNO_3 was used for the leaching of lead from the filters.

Procedures. The absorption tubes for air samples were U-shaped Teflon-lined aluminium tubes (30 cm long × 3 mm o.d.) packed with 3% OV-101 on 80–100 mesh Chromosorb W. Moisture was condensed from air by using glass U tubes at −15°C. The air was sampled with a peristaltic pump, Model LG-100 (Little Giant Pump Co., Oklahoma City).

Atmospheric particulates were collected on 37 mm diameter Whatman No. 40 ash-free cellulose filter paper. The diameter of the exposed filter area was 20

*Procedure from Radzuik et al. (54).

mm. The air intake was inside a 2 L vessel placed upside down so that only particules smaller than 10 μ were sampled. The filter holder was connected to a Sargent-Welch Duo-Seal vacuum pump, Model 1400 B (Skokie, Il) calibrated with a flow meter at the start and end of the sampling period. Connections were made using new Tygon or Teflon tubing.

The atomic absorption spectrometer should be operated with a deuterium arc background correction at the 283.3 nm lead resonance line with a spectral bandwidth of 0.7 nm. The graphite furnace is heated continously for 20-min periods at 1500°C by using the charring stage at the longest time setting. For optimal gas flow from the gas chromatograph, the quartz windows are removed from the furnace assemblies. The graphite furnace is operated with an internal flow of 40 mL/min. Because of the high carrier gas flow, the sensitivity is unaffected by the internal gas flow. These optimal operating conditions were arrived at while using a gas chromatograph oven temperature of 150°C and a carrier gas flow rate of 140 mL/min.

Air and exhaust were sampled directly for organic lead compounds with four parallel traps maintained at −72°C in a dry ice–methanol bath. Water was precondensed in U tubes at −15°C (sodium chloride in crushed ice). The flow through the traps averaged 70 mL/min over periods of 30 min (exhaust) to 18 h.

Keep each trap in dry ice until it can be attached to a 4-way valve installed between the carrier gas inlet and the injection port of the gas chromatograph. Immerse the trap in boiling water. After 1 min, introduce the volatilized fraction of the sample to the gas chromatographic column by diverting the carrier gas through the 4-way valve. Program the gas chromatographic oven from 50 to 200°C with a heating rate of 40°C/min. Add mixed alkyllead standards to blank traps that are run under identical conditions for calibration.

*11.2.3.2 Determination of Dimethylmercury and Methylmercury Chloride in Air by Gas Chromatography AAS**

The following procedure is suitable for the determination of dimethylmercury and methylmercury chloride either in building air or urban street air. Detection limits are 2 and 5 ng for dimethylmercury and methylmercury chloride, respectively. A Silyl-8-coated Tenax column is employed for the gas chromatographic separation with a thermal program of 150 to 180°C at 20°C/min. A pyrolyzer unit is installed in the effluent line to the cold vapor atomic absorption cell tc decompose the mercury compounds. It is operated at 900°C. Because there must be no metal parts in contact with the mercury compounds in the gas flow,

*Reprinted with permission from (55), Brzezinska et al., *Spectrochim. Acta* **38*B***, 1339. Copyright 1983, Pergamon Press, Ltd.

the gas chromatographic column and transfer line system are constructed of Teflon and glass.

Equipment

Chromatography. A Fisher 2400 gas chromatograph was employed. It was modified to contain no metal surfaces in contact with the gas stream. A Teflon insert was placed in the injector. Glass columns 1 or 2 m long, 5 mm o.d., and 2 mm i.d. were employed. The packing was Tenax or mixed Tenax-DEGS. A Tenax coating using 250 μL of Silyl-8 was also studied. The recommended chromatographic operating conditions are gas (argon) flow rate, 100 mL/min; injector temperature, 200°C; column temperature program, 150–180°C at 20°C/min; and a Teflon tube transfer line heated to 140–180°C.

Atomic Absorption Detector. A Perkin-Elmer 305B atomic absorption unit equipped with a deuterium arc background corrector was used. The radiation source was a mercury hollow-cathode lamp. An Omniscribe (Houston Instruments) single-pen chart recorder was employed at a chart speed of 0.5 cm/min. The wavelength and band pass were 253.6 nm and 0.7 nm, respectively.

A quartz T-tube atomizer was employed. This consisted of a quartz capillary (5 nm o.d. and 1 mm i.d.) pyrolyzer (at 900°C) terminated by an open-ended, unheated 15 cm long × 1–5 cm i.d. quartz tube absorption cell (Fig. 11.3). The addition of CuO to the pyrolyzer as recommended by others was found to have little useful effect in this work.

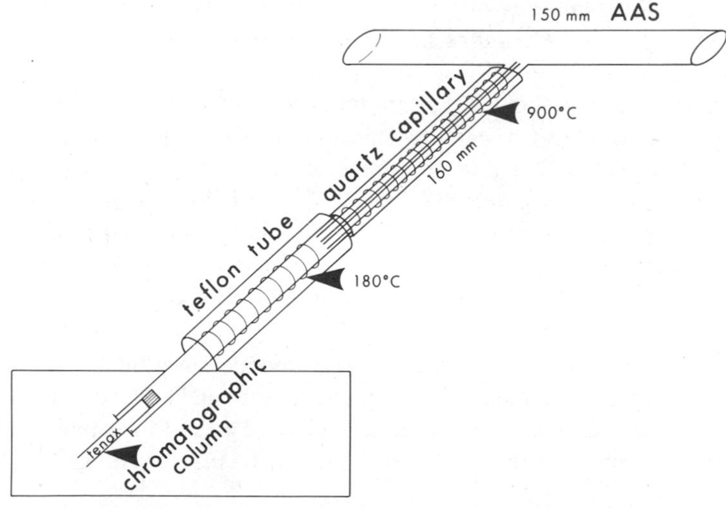

Figure 11.3 Pyrolizer–atomizer system for mercury compounds.

Reagents

DIMETHYLMERCURY (ALFA INORGANICS). A stock solution was prepared by weighing 0.05 mL into a 50 mL volumetric flask and diluting it with benzene. This was kept in a refrigerator at about 4°C. Working solutions were prepared fresh daily by dilution in benzene.

METHYLMERCURY CHLORIDE (ALFA INORGANICS). A sample weighing 100 mg was dissolved in 50 mL of benzene. The stock solution was kept in a refrigerator at about 4°C. Working solutions were prepared fresh daily in benzene.

Procedure. Collect air samples on a short piece of glass chromatographic column, 14 × 0.8 cm packed with 20% Tenax and 80% 1 mm glass chips. Pump air at 2 L/min through a Millipore membrane filter of 0.45 μm pore size and then through the trap. After the air sample had been collected on the short column, connect the latter to an argon supply and elute the mercury species by heating the column to about 200°C for 0.5 h. Use an argon flow rate of 10 cm/min. Collect the mercury compounds in 0.1 mL of benzene in a microimpinger. Cool the impinger solution in an ice–salt water bath (temperature about 6°C). Inject appropriate aliquots of benzene solutions (10 or 20 μL) containing standards or sample into the gas chromatograph. Use a temperature program of 150–180°C at 20°C/min. Use a 1 m column with a carrier gas flow rate of 100 mL/min. Retention times for dimethylmercury and methylmercury chloride are 0.3 and 1.8 min, respectively.

*11.2.3.3 Determination of Cr(III) and Cr(VI) in Welding Fumes by Ion Exchange Chromatography AAS**

When stainless steel is welded, chromium is evolved in the fumes produced. It is well known that Cr(VI) is carcinogenic and that Cr(III) is relatively harmless. Obviously it is not sufficient to determine total chromium in welding fumes if their hazard to human health is to be assessed. The following procedure is given for the selective determination of Cr(VI) and Cr(III) after water extraction of fumes and separation by ion exchange. The proposed procedure was tested by analyzing various mixtures of known concentrations of Cr(III) and Cr(VI).

Equipment. A Perkin-Elmer Model 603 atomic absorption spectrometer equipped with a deuterium arc background corrector and a Perkin-Elmer Model 56 recorder was used with an air–acetylene flame. Perkin-Elmer Intensitron hollow-cathode lamps were used at the recommended current settings.

*Procedure from Naranjit et al. (52).

The pH measurements were done with a Fisher Accumet Model 144 pH meter. The undissolved welding fumes were characterized with a Jeol Model JSM-35 scanning electron microscope equipped with a PGT-100 energy-dispersive microanalyzer.

Reagents. All reagents used were analytical grade. Anga 316 anion-exchange resin and Dowex 50W-8X cation-exchange resin were employed for the separation. Buffer solutions with pH values ranging from 3 to 5 were prepared (53). Single-element or multielement secondary standard solutions were prepared from 100 ppm stock solutions and acidified to a pH of about 2.

Procedures

Sampling. A fume box with a capacity of 0.3 m^3 was used for collecting welding fumes from manual metal arc welding on stainless steel (18.5% Cr, 9% Ni) with two types of electrodes: AROSTA 316 L (16–18% Cr, 10–14% Ni) and ESAB OK 67.52 (18% Cr, 9% Ni). The fumes were collected on membrane filters (Acropore AN 1200) by pumping at a rate of 20 L/min.

Determination of Total Metals. For the determination of total metals, each sample of the welding fumes was acid digested as follows. A 4.5 mL portion of a mixture of aqua regia and hydrofluoric acid (8 + 1) is added to 100 mg of the sample. Heat the solution at 100°C in a Teflon decomposition vessel for 30 min. Add 3 mL boric acid solution and reheat the solution for 15 min more. Cool the solution and dilute to 100 mL with water.

Procedure for Cr(III) and CR(VI) in Aqueous Extracts of Welding Fumes. Extract the welding fumes (powder or sample filter) with an aqueous buffer solution of pH 3 to 5. Filter the extracts through membrane filters (0.45 μm pore size). Pass portions of the extracts through both an anion and a cation exchange resin and rinse with distilled water. Dilute the effluents to a known volume. Analyze for chromium by atomic absorption spectrometry by using either flame or electrothermal atomization with appropriate standardization.

11.2.4 Botanical and Zoological Samples

*11.2.4.1 Determination of Se Alkanes in Gases Given off by Astragalus Racemosus Using Gas Chromatography AAS**

In this procedure a homemade very simple and inexpensive (less than $200) gas chromatograph is used. However, any commercial unit would be preferable.

*Procedure from Radziuk and Van Loon (51).

Astragalus racemosus is a selenium accumulator. Gases containing dimethyl selenide and dimethyl diselenide are transpired and can be separated using gas chromatography. The gas chromatographic column was packed with 60–80 mesh Chromosorb W coated with Silicone Oil DC 550. Retention time of dimethyl selenide and dimethyl diselenide are 271 and 4368 sec, respectively, using a thermal program of room temperature of 135°C at a rate of 4°C/min. A resistively heated quartz tube atomizer held at 900°C with a H_2 diffusion flame burning at its end was employed.

Equipment. A Perkin-Elmer 305B atomic absorption unit with deuterium arc background corrector was used. The hollow-cathode lamp was purchased from Westinghouse.

The quartz T-tube furnace is shown in Figs. 11.4 and 11.5. All tubing is quartz. The section containing the column is 20 cm long × 4.5 cm i.d., narrowing to a neck 8 cm long × 8 mm i.d. that connects to the atomizer. The atomizer is 10 cm long × 2 cm i.d. The gas inlet tubes in the neck and on the male part of the ground quartz joint are 6 mm i.d. The column and atomizer sections are wound with Chromel C wire, 20 gauge, with a resistance of 0.65 Ω/ft. Asbestos insulation is used to cover the windings to a 1 cm thickness. The windings can be powered by Variacs. A temperature of 900 to 1000°C was used in the atomizer.

Hydrogen gas is injected into the atomizer using the gas inlet quartz tube closest to the atomizer. This tube runs through the neck to a point 5 mm inside the atomizer. The hydrogen flow was 1 L/min. The nitrogen inlet tube opens directly into the neck. Nitrogen flow was 6 L/min.

The columns were constructed by Chromatographic Specialties, Brockville, Ontario, in a helix-wound form (Fig. 11.4). The smallest diameter that could be wound without causing flattening and hence obstruction of the flow was 4 cm. The column was constructed from a 122 cm length of 3 mm aluminum tubing. A length of empty aluminum tube was left on both ends of the column. The outlet was long enough to reach to within 5 mm of the inside of the atomizer.

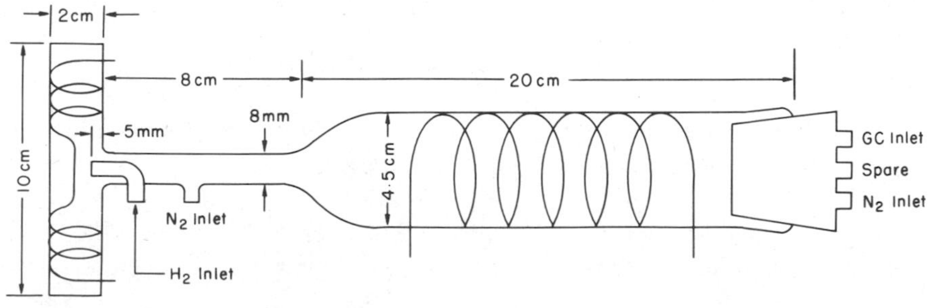

Figure 11.4 Diagram of home made quartz T-tube GC-atomizer system.

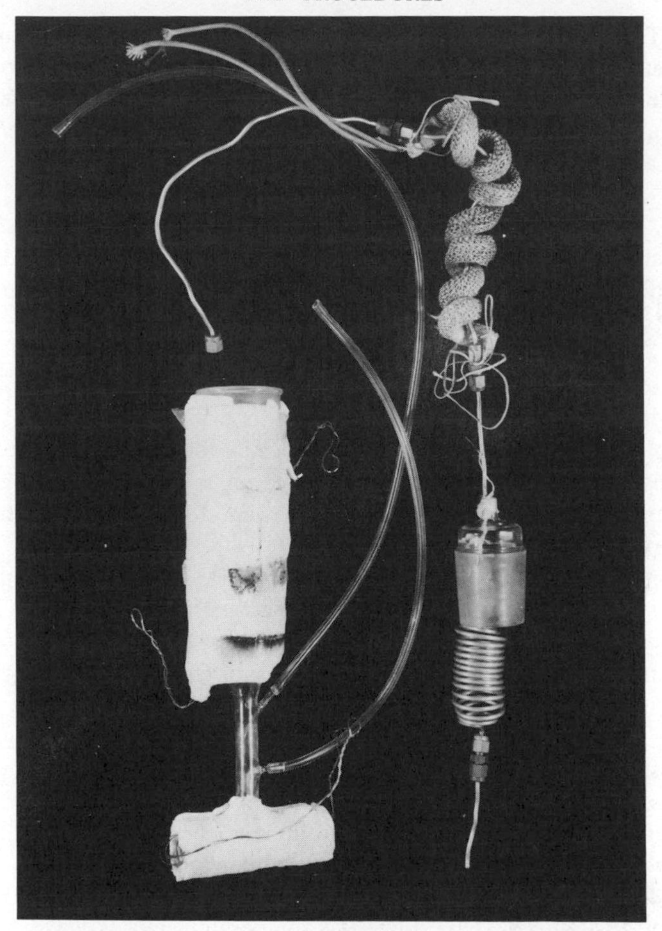

Figure 11.5 Photograph of quartz T-tube GC-atomizer system showing gas chromatographic column.

A 20% five-ring polymetaphenyl ether on 60–80 mesh Chromasorb W column was found to be suitable for selenium work. This column is stable to 250°C.

The preabsorption column was 35 cm long and similar in other dimensions to the main column. Silicone Oil DC 550 on 60–80 mesh Chromosorb W was used as a packing material. After the collection period was over, the precolumn, wrapped in heating tape, was connected to the main column using Swagelock fittings. Columns must be cured for several hours at 200°C with the carrier gas on.

A Sargent Model DSRG double-pen recorder was used. One channel was connected to a calibrated platinum–rhodium thermocouple and a cold reference junction. Thus temperature can be plotted if desired. The other channel monitored the atomic absorption signal. The T-tube atomizer is attached to the AAS burner. While monitoring the signal from the hollow-cathode lamp, the atomizer is moved up and down and in and out with the burner adjustment screws until minimum absorbance is noted.

Reagents. Reagents were obtained from Alfa Products, Beverley, MA. The selenium alkanes, dimethyl selenide, diethyl selenide, and dimethyl diselenide, were shipped as liquids sealed in ampules. Working standards of the selenium compounds were prepared by diluting to an appropriate level in 95% ethanol.

Procedure. In a manner similar to previous workers, volatile selenium compounds were released from plants and standards using heat.

Place plants or standard solutions in a U-tube that is connected to the precolumns by plastic tubing. Place the precolumn in dry ice and connect to a vacuum pump. Apply low heat to the U tube while drawing air through at a slow rate (15 mL/min) using the vacuum pump.

Connect the precolumn to the main column while maintaining the precolumn in dry ice. Connect a nitrogen cylinder to the other end of the precolumn and move the dry ice to the main column. Wind heating tape around the precolumn and elute the selenium compounds onto the main column using a N_2 flow rate of 15 mL/min and an elution temperature of 175°C. Elution requires 10 min.

During this interval, turn on the atomizer and warm up for 20 min. The atomic absorption unit, deuterium arc, and recorder should also be stabilizing during this interval. The instrument parameters are those recommended by the manufacturer. It is crucial that the deuterium arc beam be properly aligned with the hollow-cathode beam. The N_2 flow to the atomizer is adjusted to 6 L/min.

Insert the column into the cool column chamber of the T tube. Initiate the H_2 flow to the atomizer at 1 L/min. If a temperature plot is desired, the thermocouple is inserted so that the tip rests on the column. Turn the N_2 column flow on at 15 mL/min and initiate the heat program to the column by adjusting the Variac to a reading that will give the desired heating rate (e.g., 4°C/min). After the desired maximum temperature has been reached, in this case 135°C, the procedure is discontinued.

*11.2.4.2 Determination of Methylmercury in Fish by Gas Chromatography AAS**

The following is a modification of the West*öö* approach (57). A cold vapor atomic absorption apparatus is used as the detector for gas chromatography.

Methylmercury is converted to the bromide and is then extracted into toluene.

*Procedure from Selden and Peters (50).

To remove fatty acids and amino acids that might poison the column a back extraction into an aqueous phase of the mercury compounds formed in reaction with thiosulfate is carried out.

An aliquot of the final aqueous phase is injected into the gas chromatograph. The effluent from the chromatograph is passed through a pyrolysis tube and into the cold vapor absorption cell.

Standardization is done by standard additions. To ensure that the standard additions are subject to the same conditions as the mercury in the fish, the mercury addition is done as close to the beginning of the procedure as possible. The absolute detection limit (not defined) of the method is 3.5 ng mercury.

Equipment. A schematic diagram of the apparatus is shown in Fig. 11.6. A Perkin-Elmer Model 800 gas chromatograph was used for all experiments. The following operating conditions were found satisfactory: column, 10% SP 2300 on 80–100 mesh Chromosorb W; oven temperature, 145°C; inlet temperature, 200°C; carrier gas, nitrogen at a pressure of 3.5 kPa cm^2 measured at the gas chromatography inlet; flow rate, 90 mL/min.

The furnace was made by winding a Nichrome resistance wire around a quartz tube 6 cm long, 4 mm o.d., 2 mm i.d. This unit was placed inside another quartz tube (7 cm long, 8 mm o.d., 6 mm i.d.). The Nichrome wire coil was connected to a 0–230 V Variac transformer. The circuit was equipped with a voltmeter and an ammeter. The furnace was operated at about 10 V and 2.3 A; the temperature was then about 620°C. From the end of the stainless steel tube, a PVC tube led to the inlet device for the cuvette. To homogenize the sample, a Bellco No. 1977 12 mL graduated tissue grinder was used.

The Perkin-Elmer Model 303 atomic absorption spectrometer was run using the 253.7 nm mercury line. Deuterium background correction was essential. The signals were recorded on a Perkin-Elmer 159 chart recorder.

Reagents. All reagents were of analytical reagent grade. The water was freshly distilled.

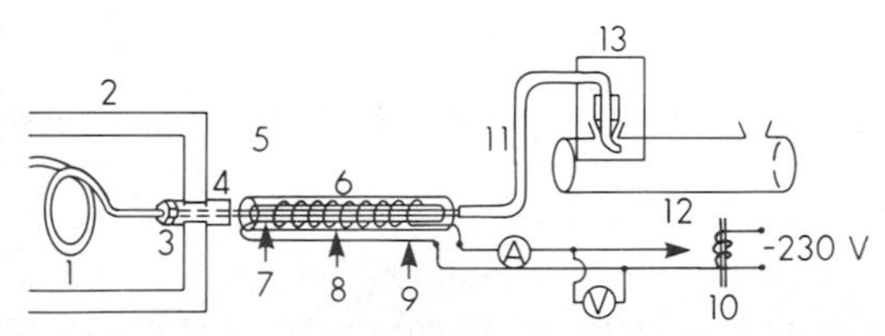

Figure 11.6 Apparatus for measuring organic mercury compounds in GC effluents (50): (1) column stainless steel; (2) gas chromatograph; (3) GC oven; (4) Teflon joint; (5) stainless steel tube; (6) electrical furnace; (7) inner quartz tube; (8) outer quartz tube; (9) resistance wire; (10) Variac transformer; (11) PVC tubing; (12) 10 cm quartz cuvette; and (13) inlet device for cuvette.

BROMINE REAGENT. Potassium bromide (360 g) was dissolved in 700 mL of water. Concentrated H_2SO_4 (110 mL) was added to 100 mL of water. After cooling to room temperature, the solutions were mixed and made up to 1 L with water.

POTASSIUM IODIDE (3*M*). The solution must be kept dark and cooled.

STANDARD ORGANIC MERCURY SOLUTION (20 ppm Hg per Compound). The following salts were weighed on a microbalance: 25.03 mg of methylmercury chloride (CH_3HgCl), 26.43 mg of ethylmercury chloride (C_2H_5HgCl), and 31.22 mg of phenylmercury chloride (C_6H_5HgCl). The salts were transferred to a 1 L volumetric flask, dissolved in benzene, and made up to volume with the same solvent.

Procedure. Transfer three weighed 0.5 g portions of frozen fish to three tissue grinders. Add 0.5 g of 1 *M* copper sulfate solution to each portion and 50 and 100 μL of the standard mercury solution to two of the samples; add 2 mL of bromine reagent and homogenize well. After homogenization, lift the rod carefully from the solution and rinse with water until the total volume of sample, reagents, and water is exactly 7 mL. Pipette 3.5 mL toluene into each solution, and shake the grinders for 2 min. Then centrifuge the mixture for 10 min at 2000 rpm.

Pipette a 3 ml portion of each toluene phase into one of three centrifuge tubes, with a separate pipette for each sample. Add to the residue in each grinder 3.5 mL of toluene and repeat the operation, with the 3.0 mL portion of each toluene phase transferred with the same pipettes as before to the appropriate centrifuge tube. Add to each tube 2.0 mL of 5×10^{-3} *M* sodium thiosulfate solution, shake the tubes 2 min, and centrifuge for 15 min.

Insert graduated pipettes (2 mL), equipped with Peleus balloons, through the organic phases into the lower aqueous phases, and take as much aqueous phase as possible (but equal volumes of 1.6–1.9 mL) from each tube, starting with the tube in which the cloudy layer between the phases was thickest. Transfer these volumes to three centrifuge tubes and add the same volumes of potassium iodide solution to each. From that moment the solutions must be kept in the dark. Add 1.0 mL benzene to each tube; shake the tubes vigorously for 2 min and centrifuge for 10 min. Transfer suitable aliquots (5–25 μL) of the solutions into the gas chromatograph and measure the signal on the atomic absorption instrument optimized at a wavelength of 253.7 nm with deuterium background correction and suitable scale expansion.

11.2.5 Determination of As Species by High-Pressure Liquid Chromatography AAS*

The following procedure is for the automated determination of arsenate, arsenite, methylarsonic acid, and dimethylarsinic acid. It has been applied to the determination of these species in soil extract. Arsenite, methylarsonic acid, and dimethylarsinic acid can be separated and determined using a strong anion–exchange column with an acetate buffer mobile phase. A strong cation–exchange resin can be employed to separate arsenite, dimethylarsinic acid, and methylarsenic acid using an ammonium acetate mobile phase. All four compounds can be separated and determined with a C_{18} reverse-phase column using methanol–water mixtures saturated with tetraheptylammonium nitrate. Graphite furnace atomic absorption is used as the detector in each case and signal peak areas are employed. I recommend that nickel be added to the sample in the furnace to allow a high temperature ash step to be used.

Equipment. Altex Scientific Model 110A single piston and Model 100 dual piston liquid chromatographic pumps were used to deliver solvents and analytes through SAX columns (25cm × 3.2mm i.d.) packed with 10 μ particles. The columns were purchased from Altex Scientific. The HPLC-GFAA system consisted of an Altex single-wavelength detector (254 nm) arranged in series to a Perkin-Elmer Model 360 atomic absorption spectrometer; it was equipped with a programmable graphite furnace and a Perkin-Elmer autosampler Model AS-l that periodically transferred 20 μL aliquots from a 50 μL flow-through sample cup to the graphite furnace for analysis. A detailed description of the system, its mode of operation, and its detection capabilities have been published (14). A Waters Assoc. Model 6000 A dual-piston pump was used to force the mobile phase through an Altex C_{18} reverse-phase column (25cm × 3.2mm i.d.) packed with 10 μ particles. The analytes were placed on the column through a Waters Assoc. Model UK6 sample injector. The detector system consisted of a Beckman Model 25 variable wavelength UV-VIS spectrophotometer in series with a Waters Assoc. differential refractometer Model 401 and an Altex slider injection eight-port valve. This pneumatically activated sampling valve delivered 40 μL of the column effluent through an injector into the graphite cuvette of a Hitachi-Zeeman GFAA Model 170-70. A description of the interface between the HPLC and the GFAA is available in the literature (59).

Reagents. Sodium arsenite ($NaAsO_2$), methylarsonic acid (MAA), disodium methylarsonate (DSMA), dimethylarsenic acid (DMAA), and sodium dimethy-

*Procedure from Brinkman et al. (58).

larsinate (SDMA) were obtained from Dr. E. Woolson, USDA Agricultural Research Service, Beltsville, MD. All other chemicals were purchased from commercial sources and were used as recieved. The concentrations of solutions, prepared with deionized water, were based on the individual chemical assays provided by the manufacturers.

Procedures.

HPLC-GFAA Operational Parameters. Before use, approximately 60 mL of the mobile phase is passed through the column. Depending on the desired separation, the solutions are chromatographed on SAX or SCX columns at flow rates ranging from 0.15 to 1.0 mL/min. Aqueous buffer solutions in the pH range 2.9–6.9 are employed with total concentrations of the anionic components of the buffers between 0.025 and 0.075 *M*. Twenty microliter volumes of the solutions containing the desired concentration of one or any combination of the arsenic compounds to be investigated are injected automatically into the graphite cuvettes every 45 sec. The samples are dried in the graphite cuvette for 15 sec at 150° and for 5 sec at 200° and the atomized at 2700° for 10 sec. Arsenic concentrations are measured at 193.7 nm. (The samples were not subjected to an ashing cycle to minimize volatilization. An addition of nickel would allow ashing up to 1500°.) During atomization the interrupt mode of operation is employed, assuring a longer residence time of the arsenic atoms in the AA optical path. For automatic background correction a deuterium lamp is used.

Extraction of Arsenic Compounds From Soils. Add a 25 mL volume of 1.0 *M* NH_4Cl to a 125 mL Erlenmeyer flask containing 0.5 g of soil. Shake the flask on a Burrel wrist-action shaker for 30 min. After the mixture has settled for 5 min, decant the supernatant into 15 mL culture tubes and centrifuge for 20 min at 1470*g*. In turn, ultracentrifuge two 1.5 mL samples, one from each supernatant (15,100*g* for 5 min), then analyze for total arsenic by atomic absorption spectrometry and arsenic compounds by HPLC-GFAA.

Shake the residual soil from the above extraction for 17 h with 25 mL of solution that is 0.5 *M* NH_4F and 0.1 *M* NaOH. Treat the supernatant as described above.

Separation of Arsenic Compounds on C_{18} Reverse-Phase Columns. Before analysis the columns are conditioned by pumping the mobile phase through the columns at 1 mL/min for 1 h. Solutions containing 20 ppm of arsenic of each compound to be investigated are employed for the separation and calibration experiments. The amounts of arsenic compound are varied by injecting volumns in the range 5 to 25 μL onto the column. Water–methanol mixtures 0.005 *M* with respect to tetrabutylammonium hydroxide (pH adjusted with phosphoric acid) or water–methanol mixtures saturated with tetraheptylammonium nitrate (0.002M) (pH adjusted with 0.2 *M* NaOH) serve as mobile phases. Flow rates of 0.5 and 1.0 mL/min are used. For the GFAA analyses automatically inject 40 μL samples of the column effluent into the graphite cuvette every 30 to 45 sec. Dry samples in the graphite cuvette at 75° for 10 sec, ash at 300° for 10 sec (addition of Ni will allow ashing up to 1500°), and atomize at 2500° for 7 sec. Adjust the carrier gas flow during the atomization of the sample to increase sensitivity. Arsenic is determined using the 193.7 nm line.

11.2.6 Determination of Forms of Pb in Fish, Vegetation, Sediment, and Water Samples by Gas Chromatography AAS*

These procedures are for the determination of total, organic-solvent-extractable, and volatile lead, and the individual tetraalkyllead compounds in fish, vegetaion, sediment, and water samples. Total lead is determined after acid digestion of the sample. Mixtures of HNO_3–$HClO_4$ and H_2SO_4 are used and the ratios depend on the type of sample analyzed. Organic-solvent-extractable lead is determined by shaking the sample with hexane. The volatile portion of lead is determined by heating the sample at 150°C and sweeping the gases into the atomic absorption atomizer. Individual tetraalkylleads [$(CH_3)_4Pb$, $(CH_3)_3C_2H_5Pb$, $(CH_3)_2(C_2H_5)_2$ Pb, $CH_3(C_2H_5)_3Pb$, and $(C_2H_5)_4Pb$] are volatilized from the sample as above and then trapped in a short section of chromatographic column. Gentle heating is used to volatilize the compounds into the gas chromatograph for separation and determination. It is important in determining tetraalkylleads to avoid metal transfer lines to prevent losses of the compounds on the line surfaces. Glass or Teflon-lined metal tubes can be used.

Detection limits for total lead and hexane-extractable lead determinations are given in Table 11.6.

Equipment. Atomic absorption equipment was supplied by Perkin-Elmer and consisted of a Model 603 spectrophotometer equipped with a deuterium arc background corrector, a Model 500 graphite furnace, and a Model AS-1 autosampler. An HGA 2100 furnace was modified for use as the atomizer for the gas chromatographic detection of lead compounds. Effluent from the chromatograph was passed through a heated (80°C) 3 mm o.d. stainless steel transfer tube from the end of the chromatographic column to a tantalum connector (54), friction fitted into the injection port of the graphite furnace element.

A Pye series 104 gas chromatograph and a glass chromatographic column 2.3 m long by 0.6 cm o.d. were used. The chromatographic column was packed

*Reprinted with permission from (60), Cruz et al., *Spectrochim. Acta* **35B,** 775. Copyright 1980, Pergamon Press, Ltd.

Table 11.6 Detection Limits[a]

Sample	Total Pb (ppb)	Hexane-Extractable Pb (ppb)
Fish	30	2
Vegetation	10	2
Sediment	50	2
Water	1	0.001

[a] Volatile Pb, 0.5–1.5 ng/g for all samples; tetraalkyllead, 0.5 ng/g for all samples.

with 3% OV-1 on 80–100 mesh Chromosorb W. Sample traps for tetraalkyllead compounds consisted of 20 cm pieces of Teflon-lined aluminum tubing packed similarly to the chromatographic column tubes. A methanol–dry ice mixture held in a beaker was used for cooling. Moisture was removed from the incoming air with a drying tube containing Drierite.

Figure 11.7 (top) shows the sample collection apparatus. A is a flow meter; B a glass U tube half-filled with Drierite; C a glass U tube containing the sample; D a glass U tube water trap; E a Teflon-lined aluminum sample trap packed with OV-1 on 80–100 mesh Chromobsorb W; and F leads to a vacuum pump. Connections are made with Swagelok fittings.

Figure 11.7 (bottom) is the gas-chromatograph–sample-oven–atomic-absorption-detector system: A connects to an N_2 cylinder; B is the sample oven containing the sample trap and a four-way valve; C is the gas chromatograph with

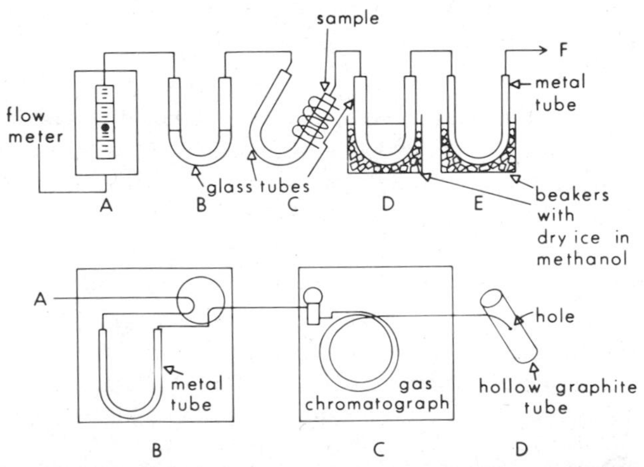

Figure 11.7 Top: Sample collection apparatus. Bottom: sample oven–gas-chromatograph–graphite-furnace–atomic-absorption-detector system. See text for description of components.

the tubing from the sample oven connected through a standard sample injection valve to the chromatographic column; and D is the graphite furnace tube that is interfaced to the transfer line by a tantalum connector. Other connections are made with Swagelok fittings.

Apparatus for volatile lead determinations was similar to that used for tetraalkyllead except that no gas chromatographic column was employed. Samples for analysis are placed in a glass U tube. All beakers and other glassware were thoroughly acid washed and distilled water rinsed just before use.

Reagents. Tetraalkyllead compounds were obtained from Ventron Corp., Danvers, MA and Ethyl Corp., Benton, MI. Freshly diluted working solution was prepared daily in reagent grade hexanes (Caledon Laboratories, Georgetown). These standards were used for the preparation of volatile lead (use trimethylethyllead only) and tetraalkyllead working solutions.

An inorganic lead stock solution was prepared by dissolving 1 g of Pb metal in a minimum quantity of HNO_3. The solution was diluted to 1000 mL with water and HNO_3 to give a final concentration of 1000 μg/mL Pb and 5% HNO_3.

National Bureau of Standards, SRM 1571 Orchard Leaves and SRM 1577 Bovine Liver were used as control samples.

Very pure acids such as Aristar (British Drug Houses) or Ultrex (J.T. Baker) must be used.

Procedures.

Total Lead. Weigh 1.0 g fish and vegetation and 0.25 g sediment into a 100 ml beaker. Add the appropriate volume of the acid mixture given in Table 11.7. Leave overnight on the hot plate. Next morning turn the hot plate on to medium heat and leave until the solution has evaporated to fumes of H_2SO_4 (fish and vegetation) or incipient dryness (sediment). Add another 5 mL of the acid mixture to the sediment and heat to incipient dryness again. Cool the samples and dilute with water to 10 mL in a volumetric flask. Allow any sediment to settle and pour into cups of the autosampler.

Table 11.7 Acid Mixtures

		Ratios of Acids		
	Volume (mL)	HNO_3:	$HClO_4$:	H_2SO_4
Fish	10	4	1	0.5
Vegetation	5	4	1	1
Sediment	10	4	1	0

Table 11.8 Graphite Furnace Thermal Program

Program	Temperature (°C)	Ramp Time (sec)	Hold Time (sec)
Dry	110	20	15
Char 1	450	20	10
Char 2	600	10	10
Atomize	2500	0	5
Clean-out	2700	0	3

Fish, vegetation, and the soil samples with low values are run by standard additions as follows. Pipette 20 μL of the sample, using the autosampler, into the furnace. Run the sample using conditions in Table 11.8. Repeat this procedure twice more, but add manually each time a suitable addition of a standard inorganic lead solution immediately after the autosampler injection. The second addition should be twice the volume of the addition to the first sample. Background correction must be employed in all cases.

Soil sample solutions that are higher than 1 μg/mL in concentration can be run by flame atomic absorption. Calibration is done by direct comparison using inorganic lead standards.

Water samples are placed in the sample cups of the autosampler. The samples are run by direct comparison with inorganic lead standard solutions. Background correction is essential. The graphite furnace is operated using the program in Table 11.8.

Hexane-Extractable Lead. Weigh 1 g of fish, vegetation, or soil into a screw-capped 20 × 120 mm test tube. Add 2 mL of hexane and place a piece of Teflon tape over the mount of the tube. Screw on the cap. Place the tube in an ultrasonic bath for 15 sec. Then manually shake the tube for 1 min and let settle overnight. Remove enough of the hexane to fill an autosampler cup two-thirds full.

For water samples add 200 mL of water to a separatory funnel together with 5 mL of hexane. Shake for 20 min. Let phases separate for 20 min. Remove the water and put enough hexane layer into the cup to fill two-thirds full. Run the samples by electrothermal atomic absorption. Calibration can be done by direct comparison using inorganic lead standards.

Volatile Lead. Weigh 2 g of fish, sediment, and vegetation or place 2 mL of water into a glass U tube. Fit a piece of Teflon tape over the ends of the tube before screwing on the cap. Assemble the sample collection apparatus as in top of Fig. 11.7. Wrap heating tape around the U tube. Use a procedure identical to that given below for tetraalkyllead except that no gas chromatograph is employed because the sample trap in the sample oven is directly connected to the transfer line. After the sample trap is heated isothermally, the evolved gases are swept directly into the atomizer.

Use trimethylethyllead for standardization. Inject an appropriate size aliquot into the spent sample tube. Run as described previously.

Tetraaklyllead. Preparation of sample. Weigh 1 g of homogenized fresh or frozen sample into one arm of a glass U tube. (Place water samples in the bottom of the tube.) Using the flat end of a 25 cm stainless steel spatula spread the solid samples thinly on the wall of the tube between the middle and the top. Cover each end of the U tube with Teflon tape (2 layers on the sample end and 1 on the other). Screw on the caps tightly. Samples may be stored in this way in the freezer until needed.

Collection of Tetraalkylleads. Assemble collection apparatus as shown at the top of Fig. 11.7. All the glass U tubes must have Teflon tape over their ends before capping. The glass U tube (D) and Teflon-lined aluminum sample trap (E) are immersed in a methanol–dry ice bath. Swagelok fittings are tightened gently with wrenches to ensure leak-free operation.

Wrap 60 cm of 36 W heating tape around the section of the sample tube containing the sample. The U tube containing the sample (tube C) must be inclined at about 30° so that water vapor is efficiently transferred to the water trap (tube D). Start thc pump and initiate the heating. The flow through the system is adjusted using flow meter (A) to be 70 mL/min. Heat the sample at 150°C for 20 min. Remove the methanol–dry ice bath from around drying tube (D) and place a coiled heating tape around this tube. Heat to 100°C and continue the collection for 10 min more. Check the flow meter from time to time to ensure that a rate 70 mL/min is maintained throughout the collection procedure. Remove the sample trap (E) from the collection apparatus and keep in the methanol–dry ice bath or a freezer until needed.

Gas Chromatographic Procedure. The apparatus is assembled as shown at the bottom of Fig. 11.7. All transfer lines between the gas chromatograph (C), sample oven (B), gas chromatograph (C), and graphite furnace (D) must be wrapped in 36 W heating tape and heated to 150°C.

With the four-way valve (oven B in Fig. 11.7, bottom) in the bypass position, place the sample trap in the sample oven and connect the Swagelock fittings. Adjust the N_2 carrier gas throug the system to 100 mL/min (reduces to 80 mL/min when sample trap is switched in). Heat the sample trap at 130–150°C for 10 min (no flow through the sample trap). Turn the four-way valve to start the flow through the sample tube. Immediately initiate atomic absorption furnace and gas chromatograph programs. Tables 11.9 and 11.10 list the atomic absorp-

Table 11.9 Atomic Absorption Parameters

Analytical line	283.3 nm
Slit	0.7 nm
Expansion	×5
Dry	Not used
Ash	450°C for 50 sec
Atomize	1600°C for 420 sec
Argon flow	40 (on HGA meter)

Table 11.10 Gas Chromatograph Parameters

Nitrogen flow rate	80 mL/min
Temperature program	Isothermal at 50°C for 2 min and then 15°C/min to 150°C

tion and the gas chromatograph parameters. Continue the procedure until 150°C has been reached in the gas chromatograph oven. Between runs the gas chromatograph oven must be cooled for 10 min with the door open and fan on high (or until the column temperature reaches 50°C).

Calibration is done by injecting a mixture of the 5 tetraalkylleads directly through the injection valve of the gas chromatograph. The apparatus is left with a spent sample trap in the oven. A chromatographic and atomic absorption procedure identical to that used for the samples is employed.

11.2.7 Determination of Cr(III)–Cr(VI) Using Direct Atomization of Cr from Chelating Resin by Graphite Furnace AAS*

Chelex-100 chelating resin is used to complex Cr(III) quantitatively from solution mixtures of Cr(III) and Cr(VI). Chromium(VI) passes through the resin bed. The Cr(VI) is then reduced to Cr(III) and the separation onto the Chelex-100 is repeated. In each case an aqueous suspension of the resin is prepared and injected directly into the furnace.

Equipment. A Nippon Jarrell-Ash FLA-100 carbon tube atomizer was used in conjunction with a Nippon Jarrell-Ash Model AA-8500 two-channel atomic absorption spectometer. A single-element chromium hollow-cathode lamp (Hamamatsu TV L-233) was employed as the line source to determine chromium at 357.9 nm, and the neon line at 352.0 nm from this lamp was used for mea-

*Procedure from Isozaki et al. (61).

suring background absorption. Peak areas were measured by using a Tokyo Kagaku RD-202 digital integrator. The numerical values of peak area and peak height were recorded with a Tokyo Kagaku RD-10-21 digital printer. The peak atomic absorption signals of chromium during atomization were stored in a Tokyo Kagaku TM-707 sampling memory, and the signals were recorded with a Hitachi 056 recorder after amplification. The resin suspension was injected into the atomizer with a micropipette (Eppendorf Model 4700) fitted with disposable plastic tips.

A 15 mL polystyrene centrifuge tube with a screw cap was used for treatment with the resin; a 5.0 mL mark was etched on the tube. A Sartorius suction bottle with a membrane filter (2.0 μm pore size) was used for separating the resin from the sample solution. Apparatus such as polyethylene beakers and centrifuge tubes were washed thoroughly with 10% HNO_3 and with water.

Reagents. All chemicals were of super-special grade. Fresh distilled, deionized water was used throughout.

Standard Cr(III) and Cr(VI) Solutions (1000 mg/L). Dissolve 0.500 g of 99.9% Cr metal in 10 mL of 6 *M* HCl and dilute to 500 mL with 0.1 *M* HCl. Dissolve 1.414 g of potassium dichromate in 500 mL of water. Prepare working solutions by appropriate dilution and store these solutions in glass bottles in the dark.

Chelating Resin. Chelex-100 (Bio-Rad Laboratories, Richmond, CA; Na form) prepared to a grain size below 400 mesh was used. After removing fine particles in the resin by decantation, add 500 mL of 2 *M* HCl to 50 g of the resin in a beaker. Stir the mixture for about 2 h with a magnetic stirrer and wash the resin with water by decantation. Add 500 mL of 2 *M* NH_4OH to the resin and stir the mixture for 2 h. Filter the resin through a 2.0 μm membrane filter, wash with water, and dry in vacuum for 6 h at 60°C. (The resin was used in the NH_4^+ form because in its H^+ form it tends to lose its chelating capacity.)

Buffer Solution (pH 4.0). Purify a 0.8 *M* sodium acetate solution by passing it through a column packed with Chelex-100 (Na form), and dilute to 0.4 *M*. Use a 0.2 *M* acetic acid–0.05 *M* sodium acetate solution as buffer.

Procedure. Transfer 250 mL of sample solution containing less than 0.5 μg of Cr(III) and Cr(VI) to a polyethylene beaker. Adjust the sample solution to pH 4 with ammonia solution or acetic acid. Add 5 mL of pH 4.0 buffer solution and 0.10 g of Chelex-100 and stir for 20 min with a magnetic stirrer. Separate the resin from the aqueous phase with the membrane filter, retain the filtrate (see below), transfer the resin with water to a centrifuge tube and centrifuge. Decant the supernatant liquid from the tube, taking care to leave the resin, and prepare

Table 11.11 Instrumental Operating Conditions[a]

Operating Steps	Current (A)	Time (sec)	Mode	Approximate final temp. (°C)
Drying	40	15	Ramp	300
Ashing	120	75	Ramp	1200
Atomization	270	10	Flash	2500

[a] Wavelength, 357.9 nm; lamp current, 10 mA; argon gas flow rate 3 l/min; damping, 1; sensitivity 1.

5.0 mL of resin suspension by adding water to the tube. After mixing thoroughly to form a uniform suspension, inject 10 μL (0.2 mg of resin) of the suspension into the carbon tube atomizer. Measure the atomic absorption of Cr (peak area) under the instrumental conditions shown in Table 11.11.

For the determination of Cr(VI), transfer the filtrate to a polyethylene beaker. Add 5 mL 3 *M* HCl and 5 mL 3% H_2O_2 to the filtrate and stir for 20 min. Adjust to about pH 4 with ammonia solution. Measure the peak area Cr absorbance as described above after treatment with Chelex-100.

11.2.8 Determination of Dialkyllead, Trialkyllead, Tetraalkyllead, and Pb(II) in Sediment and Biological Samples*

Biological samples are dissolved using a tissue solubilizer. The chemical forms of lead are not altered using this technique. Subsequently the lead species are complexed using sodium diethyldithiocarbamate and extracted. The lead compounds are butylated and then separated by gas chromatography.

Detection is by atomic absorption spectrometry utilizing a heated quartz tube atomizer. This procedure is similar to the one in Section 11.2.1.3 for water samples (19).

Equipment. The interfacing of the gas chromatograph–atomic absorption spectrometer (GC–AAS) system has been described (19). The glass chromatographic column was 1.8 m long, 6 mm diameter, containing 10% OV-1 on 80–100 mesh Chromosorb W. Nitrogen gas was used as carrier gas at 65 mL/min. The injection port and the transfer line were heated to 150 and 160°C, respectively. The column oven was programmed from 80 to 200°C at 50°C/min. A lead electrodeless discharge lamp was operated at 10 W and the 217 nm line was used. The atomization source was an open-ended quartz furnace electrically

*Reprinted in part with permission from (62), Chau, Wong, Bengert, and Dunn, *Anal. Chem.* **56,** 271. Copyright 1984, American Chemical Society.

heated to ~900°C with hydrogen flowing through it at 85 mL/min. Background correction by a deuterium lamp was employed. Signals in the form of chromatographic peaks were integrated with an Autolab Minigrator (Specto-Physics, CA).

Reagents. Standard trialkyllead solutions (100 μg/mL as lead) were prepared by dissolving 0.0152 g and 0.0171 g, respectively, of trimethyllead acetate and triethyllead acetate (Alfa Chemicals, Danvers, MA) in 10 mL of distilled water in amber volumetric flasks.

Standard dialkyllead solutions (100 μg/mL) were conveniently prepared by adding 3 drops of iodine monochloride solution to 3–4 mL of the above trialkyllead standards. The trialkyllead compounds are spontaneously converted to the corresponding dialkyllead species (46). The addition of a small amount of ICl did not significantly change the concentration of the trialkyllead standard solutions. Dialkyllead standards prepared in this manner should not be mixed with the trialkyllead standards in case the residual ICl should act on the trialkyllead. Commercially supplied dialkyllead compounds generally contained trialkyllead and Pb (II) impurities because of their instability on storage.

Iodine Monochloride. Potassium iodide (11 g) was added to 40 mL of water containing 44.5 mL of concentrated HCl. Potassium iodate (7.5g) was slowly added with stirring until the iodine so formed gradually redissolved to give a light brown solution. This solution can be kept for months if stored in an amber bottle at room temperature.

The following were used: tetramethylammonium hydroxide (TMAH) (Fisher Chemicals), 20% in water; 0.5 *M* sodium diethyldithiocarbamate (NaDDTC) (Baker); and *n*-butyl Grignard reagent, 1.9 *M* in tetrahydrofuran (Alfa, Danvers, MA).

Procedure. Determination of alkyllead in fish tissue consists of homogenizing fish samples a minimum of five times in a commercial meat grinder. Digest about 2 g of the homogenized paste in 5 mL of TMAH solution in a water bath at 60°C for 1–2 h until the tissue completely dissolves to a pale yellow solution. After cooling, neutralize the solution with 50% HCl to pH 6–8. Extract the mixture with 3 mL of benzene for 2 h in a mechanical shaker after adding 2 g of NaCl and 3 mL of NaDDTC. After centrifugation of the mixture, transfer a measured amount (1 mL) of the benzene to a glass-stoppered vial and butylate with 0.2 mL *n*-BuMgCl with occasional mixing for about 10 min. Wash the mixture with 2 mL of 1 *N* H_2SO_4 to destroy the excess Grignard reagent. Separate the organic layer in a capped vial and dry with anhydrous Na_2SO_4. Suitable aliquots (10–20 μL) are injected to the GC–AAS system for analysis.

Clams and fish intestine can be treated similarly. Macrophyte samples are shredded to thin pieces before the TMAH digestion; they generally dissolve much slower than fish. It was found convenient to leave the mixture in a capped tube in a water bath at 60°C overnight.

Determination of Alkyllead in Sediment. Extract 1–2 g dried or 5 g wet sediment samples in capped vials with 3 mL of benzene after addition of 10 mL of water, 6 g of sodium chloride, 1 g of potassium iodide, 2 g of sodium benzoate, 3 mL of NaDDTC, and 2 g of 20–40 mesh coarse glass beads for 2 h in a mechanical shaker. After centrifugation of the mixture take 1 mL aliquot of the benzene for butylation as described for fish samples.

Preparation of Standards. Add standard solutions containing 5 μg each of trimethyllead acetate and triethyllead acetate to the same volume digestion or extraction mixture and process side by side with the samples. The same volume is taken for butylation as for the samples. Peak area of the trimethyllead is taken for calculation of the other alkyllead species.

REFERENCES

1. A. J. McCormack, S. C. Tong, and W. D. Cooke, Anal. Chem. **37,** 1470 (1965).
2. B. Kolb, G. Kemmner, F. H. Schleser, and F. Wiedeking, *Z. Anal. Chem.* **221,** 166 (1966).
3. C. A. Bache and D. J. Lisk, *Anal. Chem.* **37,** 1477 (1965).
4. C. I. M. Beenakker, *Spectrochim. Acta* **32B,** 173 (1977).
5. K. Tanabe, K. Chiba, H. Haraguchi, and K. Kuwa, *Anal. Chem.* **53,** 1450 (1981).
6. B. Radziuk, Y. Thomassen, L. R. P. Butler, Y. K. Chau, and J. C. Van Loon, *Anal. Chim. Acta* **108,** 31 (1979).
7. M. Coe, R. Cruz and J. V. Van Loon, *Anal. Chim. Acta* **120,** 171 (1980).
8. H. Haraguchi, Y. Watanabe, Y. Nojiri, T. Hasegawa, and K. Fuwa, Ninth International Conference on Atomic Spectroscopy Tokyo, September 4–8, 1981, Abstr., p. 94.
9. T. M. Vickrey, H. E. Howell, and M. T. Paradise, *Anal. Chem.* **51,** 1880 (1980).
10. W. DeJonghe, D. Chakraborti, and F. Adams, *Anal. Chim. Acta* **115** 89 (1980).
11. L. Ebdon, R. W. Ward, and D. A. Leatherhard, *Analyst* **107,** 195 (1982).
12. P. C. Uden, B. D. Quimby, R. M. Barnes, and W. G. Elliot, *Anal. Chim. Acta* **101,** 99 (1978).
13. W. Slavin and G. J. Schmidt, *J. Chromatogr. Sci.* **17,** 610 (1979).
14. F. E. Brickman, W. R. Blair, K. L. Jewett, and W. P. Iverson, *J. Chromatogr. Sci.* **15,** 493 (1977).
15. G. R. Ricci, L. S. Shepard, G. Colovos, and N. E. Hester, *Anal. Chem.* **53,** 610 (1981).
16. B. S. Whaley, K. R. Snable, and R. F. Browner, *Anal. Chem.* **54,** 162 (1982).
17. K. Jinno and H. Tsuchida, *Anal. Let.* **15,** 427 (1982).

18. D. W. Hauser and L. T. Taylor, *Anal. Chem.* **53,** 1223 (1981).
19. Y. K. Chau, P. T. S. Wong, and O. Kramar, *Anal. Chim. Acta* **146,** 211 (1983).
20. S. A. Estes, P. C. Uden, and R. M. Barnes, *Anal. Chem.* **54,** 2402 (1982).
21. W. R. A. DeJonghe and Fl. C. Adams, *Talanta* **29,** 1057 (1982).
22. R. A. Jackson and W. C. Li, Pittsburgh Conference on Analytical Chemistry and Applied Spectroscopy 1983, Abst., p. 22.
23. J. W. Robinson and E. M. Skelly, *Spectrosc. Lett.* **14,** 519 (1981).
24. J. W. Robinson and E. M. Skelly *Spectrosc. Lett.* **15,** 631 (1982).
25. S. Jiang, W. DeJonghe, and F. Adams, *Anal. Chim. Acta* **136,** 183 (1982).
26. J. A. Persson and K. Irgum, *Anal. Chim. Acta* **138,** 111 (1982).
27. D. Chakraborti, D. C. J. Hillman, K. J. Irqolic, and R. A. Zingaro, *J. Chromatogr.* **249,** 81 (1982).
28. M. A. Eckhoff, J. P. McCarthy, and J. A. Caruso, *Anal. Chem.* **54,** 165 (1982).
29. Y. K. Chau, P. T. S. Wong, and G. A. Bengert, *Anal. Chem.* **54,** 246 (1982).
30. R. Roden and D. E. Tallman, *Anal. Chem.* **54,** 307 (1982).
31. R. Ropbberecht and R. van Grieken, *Talanta* **29,** 823 (1982).
32. G. E. Pacey and J. A. Ford, *Talanta.* **28,** 935 (1982).
33. M. Danczìger, R. Jackson, and M. Shats, Pittsburgh Conference on Analytical Chemistry and Applied Spectroscopy 1983, Abstr., p. 648.
34. I. S. Krull, S. W. Jordon, S. Kahl, and S. B. Smith, *J. Chromatogr. Sci.* **20,** 489 (1982).
35. S. W. Jorden, I. S. Krull, and S. B. Smith, *Anal. Let.* **15,** 1131 (1982).
36. D. J. Mackey, *J. Chromatogr.* **236,** 81 (1982).
37. D. J. Mackey, *J. Chromatogr.* **237,** 79 (1982).
38. D. J. Mackey, *J. Chromatogr.* **242,** 279 (1982).
39. D. J. Mackey, *Marine Chem. 11,* **169** (1982).
40. D. J. McWeeny, H. M. Crews, and J. A. Burrel, Pittsburgh Conference on Analytical Chemistry and Applied Spectroscopy, 1983, Abstr., p. 771.
41. W. S. Gardner, P. F. Landrun, and D. A. Yates, *Anal. Chem.* **54,** 1198 (1982).
42. J. W. Foote and H. T. Delves, *Analyst* **107,** 121 (1982).
43. A. U. Shaikh and D. E. Tallman, *Anal. Chim. Acta* **98** 251 (1978).
44. M. O. Andrae, *Anal. Chem.* **49,** 820 (1977).
45. Y. K. Chau, P. T. S. Wong, and P. D. Goulden, *Anal. Chem.* **49,** 2279 (1975).
46. F. G. Noden, *Lead in the Marine Environment.* M. Branica and Z. Konrad (Eds.), Pergamon Press, New York, 1980, p. 83.
47. L. Newman, J. F. Philip, and A. R. Jensen, *Ind. Eng. Chem. Anal. Ed.* **19,** 451 (1947).
48. P. E. Gardiner, J. M. OHaway, G. S. Fell, and R. R. Burns, *Anal. Chim. Acta* **124,** 281 (1981).
49. L. W. Killingsworth, C. F. Buffone, M. Sonawane, and C. Lunsford, *Clin. Chem.* **20,** 1548 (1974).
50. C. Selden and T. J. Peters, *Clin. Chim. Acta* **98,** 47 (1979).
51. B. Radziuk and J. C. Van Loon, *Sci. Total Environ.* **6,** 251 (1976).
52. D. Naranjit, Y. Thomassen, and J. C. Van Loon, *Anal. Chim. Acta.* **110,** 307 (1979).
53. *Handbook of Chemistry and Physics,* 49th ed., CRC Press, Cleveland, 1968, p. D79.

54. B. Radziuk, Y. Thomassen, and J. C. Van Loon, *Anal. Chim. Acta* **105,** 255 (1979).
55. A. Brzezinska, J. Van Loon, D. Williams, K. Oguwa, K. Fuwa, and I. H. Huraguchi, *Spectrochim. Acta* **38B,** 1339 (1983).
56. R. Bye and P. E. Paus, *Anal. Chim. Acta* **107,** 169 (1979).
57. G. Westöö, *Acta Chem. Scand.* **20,** 2131 (1966).
58. F. E. Brinckman, K. L. Jewett, W. P. Iverson, K. J. Irgolic, K. C. Ehrhardt, and R. A. Stockton, *J. Chromatogr.* **191,** 311 (1980).
59. R. A. Stockton and K. J. Irgolic, *Int. J. Environ. Anal. Chem.* **6,** 313 (1979).
60. R. B. Cruz, C. Lorouso, S. George, Y. Thomassen, J. P. Kinrade, L. R. P. Butler, J. Lye, and J. C. Van Loon, *Spectrochim. Acta* **35B,** 775 (1980).
61. A. Isozaki, K. Kumagai, and S. Utsumi, *Anal. Chim. Acta* **153,** 15 (1983).
62. Y. K. Chau, P. T. S. Wong, G. A. Bengert, and J. L. Dunn, *Anal. Chem.* **56,** 271 (1984).

CHAPTER

12

SOME IMPORTANT DEVELOPING AREAS IN TRACE METAL ANALYSIS FOR BIOLOGICAL AND ENVIRONMENTAL SAMPLES

12.1 DIRECT TRACE METAL ANALYSIS OF SOLIDS BY ATOMIC SPECTROMETRY

The direct elemental analysis of solids (methods not requiring chemical pretreatments) is an important priority in analytical methods development. In the case of the trace metal analysis techniques, many strategies for direct solid sample analysis have been proposed. However, with the exception of arc and spark AES, no approach to direct solid sample analysis for these spectrometric techniques has received widespread acceptance. In addition, most of the proposed methods for solid sample analysis are plagued by poor precision and difficulties with standardization.

Ideally, a method for direct trace element solid sample analysis should have the following attributes:

1. It should be applicable to a wide range of sample compositions.
2. The method should be relatively fast.
3. Standardization should be simple.
4. To avoid inhomogeneity problems, the method should be capable of handling fairly large samples.
5. Simultaneous multielement analysis is desirable.
6. Cost per analysis should be as low as possible.
7. Repeatability and accuracy should be suitable for the particular application.

Graphite furnace AAS has been widely publicized as a method for the direct analysis of solids. Langmyhr (1) has reviewed this material recently. L'vov (2), in his pioneering paper proposing the use of furnace atomization for AAS, suggested that such an atomizer would be useful for the analysis of solids.

However, real progress in this direction has been slow. Most early commercial furnace atomizers suffered from a variety of technical problems, the foremost of which was the failure to permit atomization under isothermal conditions. As a result interference problems were complex and great. Use of the L'vov platform together with the rapid heating rates of recent commercial furnaces has to a large extent obviated this problem. In addition, Zeeman background correction, now available with a few furnaces, allows compensation for the larger magnitude of nonspecific interferences that are commonly encountered during solid sample analysis.

In spite of this progress, a number of drawbacks to direct solid sample analysis by furnace AAS still exist. These include failure of even Zeeman background correction to compensate for nonspecific absorption in many instances; the necessity of using a small sample weight (i.e., less than 2 mg); the need for standards that closely approximate the chemical composition and the physical characteristics of the samples; and skeletal remains of samples (usually organic samples) that block the light beam.

Interesting variations on conventional electrothermal atomization have been proposed by L'vov and his co-workers (3, 4). These are the capsule-in-flame atomizer and the circular cavity furnace. In the former, atomization occurs in a flame, whereas in the latter the sample vapor diffuses through the wall of the inner tube into the optical beam.

Solids have been placed directly in the flame for AAS analysis. For example, the sample mixed with sodium chloride was deposited on an iron screw and placed in a flame (5). Rubidium was determined in rocks by this approach. Because of the relatively low temperatures of flames, such an approach is restricted to the most volatile elements.

Spark AES comes closest to the ideal but is deficient because it is applicable mainly to conducting samples. It is the commonest method used by industry for routine analysis of metals. Arc AES, although applicable to a wide range of sample compositions, gives poor repeatability and accuracy. Both of these methods require that the standards used closely approximate the chemical composition and physical characteristics of the samples.

Powdered samples rolled in filter paper were introduced into the flame for flame emission analysis (6). This method was improved and automated by Roach (7) and by Stewart and Harrison (8).

A furnace mounted below an arc was used by Preuss to inject sample into the atomizer (9). The aerosol thus produced was entrained in a stream of inert gas. More recently this approach has been expanded for the production of aerosols from solid samples with subsequent injection into flames and plasmas.

Other methods of injecting solids into the arc or spark include the forcing of a sample up through a hollow electrode using a piston (10), and the spreading out of the sample on a cellulose tape that is fed between the gap of the electrodes (11). Both of these approaches have been modified and subsequently used with plasma atomizers.

The ICPs and DCPs possess properties that make them potentially attractive devices for the direct trace metal analysis of solids. The main one is the high temperature attained in the zone of sample introduction. This may result in excellent breakdown of the sample, atomization, and excitation. In this regard ICPs are somewhat superior to DCPs. In addition, ICPs and DCPs are very stable, and, if the method for introduction of solids does not greatly disturb the plasma, good precision of measurements should be possible. Microwave plasmas on the other hand are fairly intolerant to the introduction of all but dilute gaseous samples.

12.1.1 Applications

Table 12.1 is a compilation of useful methods proposed for introducing biological and environmental solids directly into atomizers.

12.1.2 Standardization

In most cases, whether AAS, AFS, or AES, when solids are analyzed directly it is necessary to employ a standard that is similar in composition to the samples. This reflects the strong matrix dependency of the signal in most of the approaches. The techniques that might be expected to be least matrix dependent are those involving the inductively coupled plasma. It is often difficult to obtain suitable reference samples of a wide range of solid samples. Thus as a substitute it may be possible to make up synthetic standards from carefully weighed pure constituents that are subsequently combined into the desired form. In all situations, to minimize matrix dependency it is important to use signal integration.

12.1.3 Future Work

Atomic fluorescence is generally the least attractive technique for solid sample analysis because of the severe problems due to scatter of incident radiation by particles in the atomizer. Perhaps recent progress in using inductively coupled plasmas as atomizers for fluorescence will diminish this problem.

Inductively coupled plasma emission spectrometry is a particularly attractive method for solid sample analysis. Although equipment and research to date have concentrated on analysis of solutions, the extremely high temperature and relatively lengthy sample residence times of the inductively coupled plasma are highly conducive to efficient vaporization, atomization, and excitation. Thus the high matrix dependency noted in other atomizers, requiring the use of solid standards similar in composition to the samples, should to some extent be reduced with this plasma. Of course, the inherent multielement characteristics of AES are another plus for this approach.

Table 12.1 Direct Solid Analysis[a]

Sample Type	Method Used	Atomizer–Analyzer	Elements	Ref.
Solution residues	Graphite disk in torch	ICP–AES		12
Coal, leaves	Graphite rod (with cup) in center of torch	ICP–AES	As, Bi, Cu, Ga, In, Na, Pb, Ti, Zn	13
Seawater, organic tissue	Graphite T tube furnace	Graphite furnace–AAS	Pb, Hg	14
Sulfide ores, rocks, biological samples, air	Electrothermal aerosol generation (graphite furnace)	flame–AAS–AFS	As, Se, Hg, Pb, Fe, Mn, Cd, Zn	15, 16
Residues	Chloride generator	Flame–AAS; Microwave plasma AES	30 Elements	17
Air particulate	CO_2 laser volatilizer	ICP–AES	Many heavy metals	18
Solution residues	Electrothermal aerosal generation (Ta filament)	ICP–AES	As, Sb, Se, Te, Hg, Cd, P, Pb, Be, Ti, Mn, Sn, Ag, B, Ba, Bi	19
Solution residues	Electrothermal aerosal generation (graphite rod)	ICP–AES	Ag, As, Au, Be, Cd, Ga, Hg, In, Li, Mn, Ps, Pb, Re, Sb, Ti, Zn	20
Solution residues	Electrothermal aerosal generation (graphite rod with trifluoromethane)	ICP–AES	B, Zr, Mo, Cr	21, 22
Solution residues	Electrothermal aerosol generation (graphite filament). (This is a commercial instrument.)	DCP–AES	Al, Bi, Cd, Cr, Fe, Mn, Pb, Ti	23
Biological solution residues	Electrothermal aerosol generation (graphite furnace)	Ar–MIP–AES	Cd, Co, Cr, Cu, Fe, Mg, Mn, Ni, Ti, Zn	24, 25

Solution residues	Electrothermal aerosol generation (graphite furnace)	ICP–AES	Al, Cd, Cr, Cu, Mn	26
Solution residues	Electrothermal aerosol generation (Ta boat)	Energy transfer active N_2 plasma–AES	Ag, Bi, Cd, Cu, Mg, Pb, Ti	27
Desolvated samples of halogenated organics	Electrothermal aerosol generation	He–MIP–AES	Halogens	28
Sulfur compounds in air	Metal-foil collector aerosol	Flame–AES	S	29
Copper metals, pelleted organic powders	Ruby laser	Graphite tube–AAS	Ag, Mn, Pb	30
Solution residues	Graphite rod in torch	ICP–AES	U, Zr, Ti, Mo, B, Cr, Zn, Ag, Cu	31, 32
Coal ash, coal and oil, botanical powder	Graphite cup in plasma	ICP–AES	Ni, Mn, Zn, Cd, Pb	33
Noble metal alloys, slag, coal fly ash, inorganic oxide	Spark source aerosol generator	ICP–AES		34, 35
Air particulate	Nebulize suspended solid	ICP–AES	Fe, Si, Al, Ti, Ca, Zn, Pb, Cu, Mn	36
Residues of petroleum products	Graphite boat in flame	Flame–AAS	Cd	37
Hair, mercurochrome	Separative column to atomizer (column packed with graphite)	Quartz tube–AAS	Cd, Hg	38
Air particulate	In situ collection of air particulate and metal compounds in graphite tube	Furnace–AAS	Pb	39–40

Atomic absorption spectrometry should continue to develop as a technique for solid sample trace element analysis. Reducing the matrix dependency of the signal requires much further research effort. In this regard the introduction of vaporized sample into flame is a useful approach that should be further investigated.

12.2 VOLATILE COMPOUND GENERATION FOR TRACE ELEMENT ANALYSIS

In the previous section some novel methods of solids introduction into atomizers have been discussed. In this section a few others that do not relate either to direct solids introduction or to conventional nebulization will be covered.

The mercury cold vapor and hydride generation methods for producing volatile analyte containing substances have proven to be extremely useful in improving sensitivities with the techniques of atomic spectrometry. It is possible to conceive of ways of producing volatile species of other elements. Examples of these follow.

Moshier and Sievers (41) and Guiochon and Pommier (42) have reviewed the possibilities of producing volatile compounds of the elements using organic ligands. The substances were then determined by gas chromatography. As a recent example of this, Black and Browner (43) reacted iron, zinc, cobalt, manganese, and chromium to form their volatile β-diketonates, which were then introduced into an ICP. They obtained quantitative results for the determination of iron, zinc, and chromium in bovine liver and blood serum. Detection limits for iron, zinc, and chromium in nanograms per liter and are 90 (2), 420 (1), and 950 (10), respectively. (The values in parentheses are those obtained by furnace atomic absorption.) The β-diketones, β-thioketones, and β-ketoamines were used to form volatile metal compounds by Uden and Henderson (44). These were then separated by gas chromatography and determined by AAS and ICP–AES. Ions that were handled successfully were Be(II), Al(III), and Cr(III). Lloyd et al. (45) described a method for the use of Cr(III) trifluoroadylacetonate, CuII *N,N*-ethylenebis(trifluoroacetyl)acetoneimine, and the Ni(II) and Pd(II), analogs to generate metal vapors. These compounds were injected into a DCP after gas chromatographic separation.

Vijan (46) observed that the near atmospheric pressure formation of nickel carbonyl (as used for years in the Mond process) could be employed to generate a $Ni(CO)_4$ vapor suitable for the heated quartz atomizer–atomic absorption determination of Nickel. Lee (47) used this approach to develop a very sensitive method for determining nickel in seawater. In this method soluble nickel is reduced to elemental form using sodium borohydride. The carbonyl is formed in a helium–cobalt stream, and the carbonyl is stripped from solution. Atomization occurs in a flame-heated quartz atomizer. The sensitivity is 0.05 ng Ni.

Brash et al. (48) recorded an atomic absorption–gas chromatographic method for the determination of total zinc, cadmium, copper, nickel, mercury, lead, palladium, and cobalt in biological samples. The chelating agent soldium dethyldithiocarbamate was employed with extraction into carbontetrachloride. The separation was accomplished with 5% OV 101.

Brueggemeyer and Caruso (49) described a new vapor generation method for the atomic absorption determination of lead. In this method Pb(II) was extracted from solution using diethyldithiocarbamate in chloroform. The chloroform was evaporated and the residue treated with methyllithium to give tetramethyllead. The tetramethyllead was trapped on a Poropak Q column and subsequently eluted into a heated quartz tube atomizer.

12.3 ICP–MIP–MASS SPECTROMETRY

In 1975 Grey (50) suggested the interfacing of a quadruple mass spectrometer with an atmospheric pressure dc capillary argon arc plasma. The plasma is used as a source of singly charged ions. Elemental and isotopic analyses are possible with this arrangement, including use of isotope dilution for overcoming matrix problems. Other plasmas have also been employed in this application as well as inductively coupled plasma (51) and microwave plasma (52).

The interface involves coupling of an atmospheric pressure plasma to the vacuum system of the mass spectrometer, into which ions are extracted from the tall flame of the plasma. Two problems are immediately evident. The orifice used in the interface must be a few thousandths of an inch, which can cause clogging problems when solutions of even moderate salt content are analyzed. In addition, the orifice must be wider than it is thick because the orifice has a tendency to melt. Furthermore, Houk et al. (51), using a 50 μm orifice cut into a molybdenum disk, found that the disk became pitted, resulting in an irregular-shaped orifice after about 100 h of operation.

Douglas and French (52) however, described a sampling system using a much larger orifice diameter than that used by Grey (0.030 in. compared with 0.003 in.). This larger size can be used in their instrument because of the higher rate of pumping obtained with their cryogenic pumping system. Because the plasma expands from atmospheric pressure to 10^{-7} torr in about 10 sec, little chance of ion recombination or clustering reactions occurs (52).

Houk et al. (51) and Douglas and French (52) found the dynamic linear working range to be greater than four orders of magnitude. Also, the Douglas group have largely overcome plasma-arcing problems for the orifice system, which greatly improves sampling problems.

In a recent paper, Douglas et al. (53) describe the performance of their instrument for trace element analysis with both an MIP and an ICP source. This is the first paper in which matrix effects (effect of phosphate and aluminum on calcium signal) have been studied in plasma-MS to a significant degree. As would be predicted, the ICP showed greater freedom from matrix effects than the MIP. Isotope dilution can be used relatively easily to overcome such problems.

Results have been presented using the MIP source for the analysis of NBS Orchard Leaves, water, and steel. Direct calibration, standard addition, and isotope dilution techniques were employed. Table 12.2 shows the data obtained for the determination of lead, chromium, copper, and strontium in NBS Orchard Leaves. Reported ICP–MS detection limits for aqueous solutions are given in Table 12.3.

Another feature of this approach compared to ICP optical emission is the relative simplicity of the mass spectrum.

A unique feature of ICP–MIP–mass spectrometry is the relatively rapid rate at which isotopic ratios can be measured. This will be particularly relevant to environmental and clinical work and to high accuracy work by isotope dilution. It is of particular merit in the case of clinical work that it will be possible to use the safer natural nonradioactive isotopes in tracer work.

Table 12.2 Trace Elements in NBS Orchard Leaves[a]

Element	Isotope Dilution	Direct Calibration	Standard Addition	Accepted
Pb		42	65	45
Cr		3.4	3.1	2.6
Cu	13	12	13	12
Sr		160	36	37

[a]In nanograms per liter. From Douglas et al. (53).

Table 12.3 Detection Limits (Aqueous Solutions)[a]

Element	ICP–MS (120)	ICPAES[b]	ICP–MS (54)
Cr	7	30	0.2
Co	18	30	0.8
Cu	50	30	
Mn	6	10	1.3
Al	8	200	0.9
Sr	2		

[a]In nanograms per liter.

[b]Data from my laboratory.

12.4 LASERS

Lasers are finding important applications in trace metal analysis of biological and environmental samples. Two important examples of their use are as a volatilizer–atomizer (plus excitor) in dealing with direct analysis of solids and as a microprobe for the analysis of very small samples and sample areas. The former application has been to some extent covered under the solid sample section.

Measurements of the signal in a laser atomizer can be done in absorption, fluoresence, and emission. Time resolution of signal and background can often be obtained to aid in background correction. In the emission mode a spark discharge through the atom cloud can be used as an excitation booster.

12.4.1 Laser Microprobe

Sometimes it is important to isolate a small area in a sample for analysis. For example, in biological applications it is important to be able to do trace analyses of discrete small areas of a tissue sample. The sample is usually a section of the tissue mounted on a slide and optical microscopy is used to identify the area of interest. Such a device, called a laser microprobe, has been marketed commercially since 1962. Cross-excitation using an auxilliary spark system is also employed. For a comprehensive treatment of this subject the reader is referred to the excellent treatment by Lagua (55) published in 1979.

12.4.2 Microanalysis

In this application the laser can be used to vaporize small discrete samples such as drops of liquid or a small collected sample powder. To illustrate the latter, the air tracer technique employed by Barringer Research Ltd. can be cited (56). Particulate matter from air is sampled on a continuous basis by an aircraft flying at 100 knots. Each 5 sec portion of sample is impacted as a spot on a continuously running strip of adhesive tape. The strip is then rerun in the laboratory through a laser vaporization cell and the vapor from each spot carried into an ICP for multielement analysis. The laser is a pulsed traverse carbon dioxide laser with maximum power of 17 kW.

12.4.3 Other Applications

The laser can be used to deal with samples that are poisonous, radioactive, and so on, which must be kept in a closed container. In this application the laser can be focused on the sample through a window in the box.

Samples that cannot be transferred readily to conventional atomizers can often be analyzed in a remote mode by a laser technique. An example of this application is the analysis of a molten metal in a furnace.

REFERENCES

1. F. J. Langmyhr, *Analyst* **104,** 993 (1976).
2. B. V. L'vov, *Inzh. Fiz. Zhun.* **2,** 44 (1959).
3. B. V. L'vov, *Talanta* **23,** 109 (1976).
4. B. V. L'vov, *Spectrochim. Acta* **33B,** 153 (1978).
5. K. Govindaraju, G. Mevellead, and C. Chouard, *Chem. Geol.* **8,** 131 (1971).
6. H. Ramage, *Nature* **123,** 601 (1929).
7. W. A. Roach, *Nature* **144,** 1047 (1939).
8. F. C. Stewart and J. A. Harrison, *Ann. Bot.* **3,** 427 (1939).
9. E. Preuss, *Angew. Mineral.* **3,** 8 (1940).
10. J. Noar, *Spectrochim. Acta* **9,** 157 (1959).
11. A. Danielsson, F. Lundgren, and G. Sundkvisk, *Spectrochim. Acta* **15,** 122 (1959).
12. T. Kleinmann and V. Svoboda, *Anal. Chem.* **41,** 1029 (1969).
13. E. O. Salin and G. Horlick, *Anal. Chem.* **51,** 2284 (1979).
14. J. W. Robinson, L. Rhodes, and D. K. Walcott, *Anal. Chim. Acta* **78,** 475 (1975).
15. D. J. Koop, M. D. Silvester, and J. C. Van Loon, Pittsburg Conference on Analytical Chemistry and Applied Spectroscopy, Cleveland, Ohio, March 1979.
16. J. Ip. Y. Thomassen, L. R. P. Butler, and J. C. Van Loon, *Anal. Chim. Acta* **10,** 1 (1979).
17. R. K. Skogerboe, D. L. Dick, D. A. Pavlica, and F. E. Licht, *Anal. Chem.* **47,** 568 (1975).
18. F. N. Abercrombie, M. D. Silvester, and G. Stoute, Pittsburgh Conference on Analytical Chemistry and Applied Spectroscopy, February 28–March 4, Cleveland, Ohio, 1977, Abstr. No. 406.
19. D. E. Nixon, V. A. Fassel, and R. N. Knisely, *Anal. Chem.* **46,** 210 (1974).
20. A. M. Gunn, D. L. Millard, and G. F. Kirkbright, *Analyst* **103,** 1066 (1979).
21. G. F. Kirkbright, International Winter Conference 1980, Developments in Atomic Plasma Spectrochemical Analysis, San Juan, Puerto Rico, January 7–9, 1980, Abstr. No. 77.
22. Dr. Hull and G. Horlick, Pittsburgh Conference on Analytical Chemistry and Applied Spectroscopy, Atlantic City, March 10–14, 1980, Abstr. No. 243.
23. T. R. Gilbert and K. J. Hildebrand, *Amer. Lab.* **14**(2), 78 (1982).
24. A. Aziz, J. A. Broekaert, and F. Leis, *Spectrochim. Acta* **37B,** 381 (1982).
25. A. Aziz, J. A. C. Broekaert, and F. Leis, *Spectrochim. Acta* **37B,** 369 (1982).
26. G. Grabi, P. Cavalli, M. Achilli, G. Rossi, and N. Omenetto, *At. Spectrosc.* **3,** 81 (1982).
27. H. C. Na and T. M. Niemczyk, *Anal. Chem.* **54,** 1839 (1982).
28. J. W. Carnahan and J. A. Carnso, *Anal. Chim. Acta* **136,** 261 (1982).
29. R. A. Kagel and S. O. Farwell, Pittsburgh Conference on Analytical Chemistry and Applied Spectroscopy 1983, Abstr. No. 228.
30. K. Wennrich and K. Dittrich, *Spectrochim. Acta* **37B,** 913 (1982).
31. G. F. Kirkbright and Zhong Li-Xing, *Analyst* **107,** 617 (1982).
32. G. F. Kirkbright and S. J. Walton, *Analyst* **107,** 276 (1982).
33. W. E. Pettit and G. Horlick, Pittsburgh Conference on Analytical Chemistry and Applied Spectroscopy, 1983, Abstr. No. 151.

34. R. Belmore, R. Manabe, and J. Beaty, Pittsburgh Conference on Analytical Chemistry and Applied Spectroscopy 1983, Abstr. No. 287.
35. J. Beaty, R. Gordon, D. Belmore, and R. Manobe, Pittsburgh Conference on Analytical Chemistry and Applied Spectroscopy, 1983, Abstr. No. 406.
36. A. Sugimae and T. Mizoguchi, *Anal. Chim. Acta* **14,** 205 (1982).
37. A. N. Rcheulishuili, *Zhur. Analit. Khim.* **36,** 2106 (1981).
38. M. Yanagisawa, K. Kitagawa, and S. Tsuge, *Spectrochim. Acta* **37B,** 493 (1982).
39. G. Torsi, E. Desimoni, F. Palmisana, L. Sabatini, and P. G. Zambonin, Pittsburgh Conference on Analytical Chemistry and Applied Spectroscopy 1983, Abstr. No. 207.
40. G. Torsi, E. Desimon, F. Palmisane, and L. Sabbatini, *Anal. Chem.* **53,** 1035 (1981).
41. R. W. Moshier and R. E. Sievers, *Gas Chromatography of Metal Chelates*. Pergamon Press, New York, 1965.
42. G. Guiochon and C. Pommier, *Gas Chromatography in Inorganics and Organometallics*. Ann Arbor Science, Ann Arbor, MI, 1973.
43. M. S. Black and R. F. Browner, *Anal. Chem.* **53,** 249 (1981).
44. P. C. Uden and D. E. Henderson, *Analyst* **102,** 889 (1977).
45. R. J. Lloyd, R. M. Barnes, and W. G. Elliot, *Anal. Chem.* **50,** 2025 (1978).
46. P. N. Vijan, *At. Spectrosc.* **1,** 143 (1982).
47. D. S. Lee, *Anal. Chem.* **54,** 1182 (1982).
48. G. Brasch, L. V. Meyer, and G. Kauert, *Fresenius Z. Anal. Chem.* **311,** 571 (1982).
49. T. W. Brueggemeyer and J. A. Caruso, *Anal. Chem.* **54,** 872 (1982).
50. A. L. Grey, *Analyst* **100,** 289 (1975).
51. R. S. Houk, V. A. Fassel, G. D. Flesch, H. J. Svec, A. L. Grey, and C. E. Taylor, *Anal. Chem.* **52,** 2283 (1980).
52. D. J. Douglas and J. B. French, *Anal. Chem.* **53,** 41 (1981).
53. D. J. Douglas, E. S. K. Quan, and R. G. Smith, *Spectrochim. Acta* **38B,** 39 (1983).
54. A. L. Grey and A. R. Date, Ninth International Mass Spectrometry Conference, Vienna Austria, August 30–September 3, 1982.
55. K. Lagua, (N. Omenetto, Ed.), *Analytical Laser Spectroscopy*, Wiley, New York, 1979, chapter 2.
56. F. M. Abercrombie, M. D. Silvester, and G. Stoute, Pittsburgh Conference on Analytical Chemistry and Applied Spectroscopy, 1977, Abstr. No. 406.

INDEX